Challenges in Green Analytical Chemistry

RSC Green Chemistry

Series Editors:
James H Clark, *Department of Chemistry, University of York, York, UK*
George A Kraus, *Department of Chemistry, Iowa State University, Iowa, USA*

Titles in the Series:
1: The Future of Glycerol: New Uses of a Versatile Raw Material
2: Alternative Solvents for Green Chemistry
3: Eco-Friendly Synthesis of Fine Chemicals
4: Sustainable Solutions for Modern Economies
5: Chemical Reactions and Processes under Flow Conditions
6: Radical Reactions in Aqueous Media
7: Aqueous Microwave Chemistry
8: The Future of Glycerol: 2nd Edition
9: Transportation Biofuels: Novel Pathways for the Production of Ethanol, Biogas and Biodiesel
10: Alternatives to Conventional Food Processing
11: Green Trends in Insect Control
12: A Handbook of Applied Biopolymer Technology: Synthesis, Degradation and Applications
13: Challenges in Green Analytical Chemistry

How to obtain future titles on publication:
A standing order plan is available for this series. A standing order will bring delivery of each new volume immediately on publication.

For further information please contact:
Book Sales Department, Royal Society of Chemistry, Thomas Graham House, Science Park, Milton Road, Cambridge, CB4 0WF, UK
Telephone: +44 (0)1223 420066, Fax: +44 (0)1223 420247
Email: books@rsc.org
Visit our website at http://www.rsc.org/Shop/Books/

Challenges in Green Analytical Chemistry

Edited by

Miguel de la Guardia and Salvador Garrigues
Departamento de Química Analítica, Universidad de Valencia, 46100 Burjassot, Valencia, Spain

RSC Publishing

RSC Green Chemistry No. 13

ISBN: 978-1-84973-132-4
ISSN: 1757-7039

A catalogue record for this book is available from the British Library

© Royal Society of Chemistry 2011

All rights reserved

Apart from fair dealing for the purposes of research for non-commercial purposes or for private study, criticism or review, as permitted under the Copyright, Designs and Patents Act 1988 and the Copyright and Related Rights Regulations 2003, this publication may not be reproduced, stored or transmitted, in any form or by any means, without the prior permission in writing of The Royal Society of Chemistry or the copyright owner, or in the case of reproduction in accordance with the terms of licences issued by the Copyright Licensing Agency in the UK, or in accordance with the terms of the licences issued by the appropriate Reproduction Rights Organization outside the UK. Enquiries concerning reproduction outside the terms stated here should be sent to The Royal Society of Chemistry at the address printed on this page.

The RSC is not responsible for individual opinions expressed in this work.

Published by The Royal Society of Chemistry,
Thomas Graham House, Science Park, Milton Road,
Cambridge CB4 0WF, UK

Registered Charity Number 207890

For further information see our web site at www.rsc.org

Printed and bound in Great Britain by CPI Antony Rowe, Chippenham and Eastbourne

Preface

The general public worldwide has a poor opinion of chemistry. Almost every day the mass media broadcast bad news about environmental damage caused by uncontrolled industrial practices and accidents. Chemical elements or compounds are identified as being responsible for the pollution of air, water or soil, and also for the deaths of humans, animals and plants.

In such a doom-laden scenario it can be difficult to convince our colleagues and students of the benefits of chemistry. We believe that the chemistry community should adopt a new style of communication in order to promote the idea that chemistry is our best weapon to combat illness, and that chemical methods can solve pollution problems caused by the incorrect use of materials, or by the accumulation and transport of dangerous substances in inappropriate conditions. There is not bad chemistry and good chemistry: there are only bad and good uses of chemistry. The truth is that the advancement of chemistry is a good indicator of the progress of humanity. However, we must look for a new paradigm that can help to build bridges between the differing perspectives of chemists and the general public.

In our opinion 'green chemistry' now represents not only the right framework for developments in chemistry but also the best approach to informing the general public about advances in the subject. The term was first introduced in 1990 by Clive Cathcart (*Chemistry & Industry*, 1990, **21**, 684–687) and the concept was elaborated by Paul Anastas in his 12 principles. Briefly, green chemistry provides a way to predict the possible environmental downsides of chemical processes rather than solving them after the fact. It provides a series of recommendations for avoiding the deleterious side effects of chemical reactions, the use of chemical compounds and their transport, as well as a philosophy for improving the use of raw materials in order to ensure that our chemical development is sustainable. The principles of green chemistry build on the efforts made in the past to improve chemical processes by improving the experimental conditions, but pay greater attention to the use of hazardous materials, the consumption of energy and raw materials, and the generation of residues and emissions. This is consistent with recent regulations that have come into effect in different jurisdictions relating to the registration, evaluation, authorization and restriction of chemical substances, especially the REACH norms established by the European Union.

Within the framework of green chemistry, green analytical chemistry integrates pioneering efforts to develop previously known clean methods of analysis, the search for highly efficient digestion systems for sample preparation, the minimization of analytical determinations, their automation, and the on-line treatment of analytical wastes. These efforts have improved the figures of merit of the methodology previously available, helped to reduce the cost of analysis and improved the speed with which analytical information can be obtained. Along with all these benefits there have been improvements in the safety of methods, both for operators and for the environment. It is therefore not surprising that green analytical chemistry is now a hot topic in the analytical literature.

Two books on green analytical chemistry have appeared in the last year: one by Mihkel Koel and Mihkel Kaljuran, published by the Royal Society of Chemistry, and one by Miguel de la Guardia and Sergio Armenta, published by Elsevier. These books help to clarify the present state of green analytical chemistry and the relationship between the relevant publications in the analytical literature. However, until now there has been no multiauthor book by specialists in the different fields of our discipline describing the various developments made in green analytical chemistry. The present book is an attempt to make such an approach to recent advances in sample preparation, miniaturization, automation and also in various analytical methods, ranging from electroanalysis to chromatography, in order to contribute to the identification of the green tools available in the literature and to disseminate the fundamentals and practices of green analytical chemistry.

We hope that this book will be useful both for readers working in the industrial field, in order to make their analytical procedures greener, and also for those who teach analytical chemistry in universities, to help them see their teaching and research activities in a new light and find ways of making our discipline more attractive to their young students.

This book has been made possible by the enthusiastic collaboration of several colleagues and good friends who have written excellent chapters on their respective fields. The editors would like to express their gratitude for the extra effort involved in this project, generously contributed by people who are continually active in the academic, entrepreneurial and research fields. During the development of this project we lost one of the authors, Professor Lucas Hernández, from the Universidad Autónoma de Madrid, an excellent scientist and a good friend. He became ill while writing his chapter and died before seeing the final version of this book. On the other hand, Professor Lourdes Ramos, from the CSIC, became pregnant and we celebrate the arrival of her baby Lucas. So, in fact this book is also a piece of life, a human project, written by a number of analytical chemists who believe there is a better way to do their work than just thinking about the traditional figures of merit of their methods. We hope that readers will enjoy the results of our labours.

Miguel de la Guardia and Salvador Garrigues
Valencia

Contents

Chapter 1 An Ethical Commitment and an Economic Opportunity 1
M. de la Guardia and S. Garrigues

    1.1 Green Analytical Chemistry in the Framework of the Ecological Paradigm of Chemistry 2
    1.2 Environment and Operator Safety: an Ethical Commitment 4
    1.3 Green Chemistry Principles and Green Analytical Chemistry 7
    1.4 Strategies for a Green Analytical Chemistry 9
    1.5 Cost of Green Analytical Chemistry 10
    Acknowledgements 11
    References 11

Chapter 2 Direct Determination Methods Without Sample Preparation 13
S. Garrigues and M. de la Guardia

    2.1 Remote Sensing and Teledetection Systems 14
    2.2 Non-Invasive Methods of Analysis 19
    2.3 Direct Analysis of Solid and Liquid Samples Without Sample Damage 23
        2.3.1 Elemental Analysis by X-Ray Techniques 23
        2.3.2 Molecular Analysis by NMR 24
        2.3.3 Molecular Analysis by Vibrational Spectroscopy 25
    2.4 Analysis of Solids Without Using Reagents 29
        2.4.1 Electrothermal Atomic Absorption Spectrometry 29
        2.4.2 Arc and Spark Optical Emission Spectrometry 30

		2.4.3	Laser Ablation	31
		2.4.4	Laser-Induced Breakdown Spectroscopy	33
		2.4.5	Glow Discharge	34
		2.4.6	Desorption Electrospray Ionization	37
	2.5	Summary of Present Capabilities of Direct Determinations		38
	Acknowledgements			39
	References			39

Chapter 3 Replacement of Hazardous Solvents and Reagents in Analytical Chemistry 44
Jennifer L. Young and Douglas E. Raynie

	3.1	Green Solvents and Reagents: What This Means		45
	3.2	Greener Solvents		46
		3.2.1	Supercritical Fluids	47
		3.2.2	Ionic Liquids	48
		3.2.3	Water	49
		3.2.4	Green Organic Solvents	51
	3.3	Greener Reagents		56
		3.3.1	Chelating Agents	56
		3.3.2	Derivatization	56
		3.3.3	Preservatives	58
	References			60

Chapter 4 Green Sample Preparation Methods 63
Carlos Bendicho, Isela Lavilla, Francisco Pena and Marta Costas

	4.1	Greening in Sample Preparation		63
	4.2	Microwave-Assisted Sample Preparation: Digestion and Extraction		65
		4.2.1	Microwave-Assisted Digestion	66
		4.2.2	Microwave-Assisted Extraction	69
	4.3	Ultrasound-Assisted Sample Preparation: Digestion and Extraction		70
		4.3.1	Ultrasound-Assisted Digestion	72
		4.3.2	Ultrasound-Assisted Extraction	73
	4.4	Supercritical Fluid Extraction		75
	4.5	Pressurized Liquid Extraction		79
	4.6	Solid-Phase Extraction		81
	4.7	Microextraction Techniques		83
		4.7.1	Solid-Phase Microextraction	83
		4.7.2	Stir Bar Microextraction	86

	4.7.3	Liquid Phase Microextraction	87
4.8	Membrane-Based Extraction		90
4.9	Surfactant-Based Sample Preparation Methods		94
	4.9.1	Surfactant-Based Extraction	94
	4.9.2	Emulsification	98
4.10	Present State of Green Sample Preparation		99
References			99

Chapter 5 Miniaturization of Analytical Methods — 107
Miren Pena-Abaurrea and Lourdes Ramos

5.1	Miniaturization as an Alternative for Green Analytical Chemistry: Strengths and Current Limitations		107
5.2	Miniaturized Analytical Techniques for Treatment of Liquid Samples		110
	5.2.1	Solvent-Based Miniaturized Extraction Techniques	110
	5.2.2	Sorption-Based Miniaturized Extraction Techniques	118
5.3	Miniaturized Analytical Techniques for Treatment of Solid Samples		130
	5.3.1	Matrix Solid-Phase Dispersion	130
	5.3.2	Enhanced Fluid/Solvent Extraction Techniques	133
5.4	Analytical Micro-Systems: From Lab-on-a-Valve to μ-TAS		136
Acknowledgments			138
References			138

Chapter 6 Green Analytical Chemistry Through Flow Analysis — 144
Fábio R.P. Rocha and Boaventura F. Reis

6.1	The Scope of Flow Systems in Chemical Analysis and Green Analytical Chemistry		144
6.2	Brief Description of Flow Systems		145
	6.2.1	Segmented Flow Analysis	145
	6.2.2	Flow Injection Analysis	145
	6.2.3	Sequential Injection Analysis	146
	6.2.4	Monosegmented Flow Analysis	147
	6.2.5	Multicommutation Approach	147
	6.2.6	Multipumping and Multisyringe Flow Systems	149
6.3	Evolution of System Design and Reduction of Waste Generation		149

	6.4	Contributions of Flow-Based Procedures to Green Analytical Chemistry	152
		6.4.1 Replacement of Hazardous Chemicals	152
		6.4.2 Reuse of Chemicals	155
		6.4.3 Minimization of Reagent Consumption and Waste Generation	155
		6.4.4 Waste Treatment	163
	6.5	Future Trends in Automation	164
	References	164

Chapter 7 Green Analytical Separation Methods 168
Mihkel Kaljurand and Mihkel Koel

7.1	Why Green Separation Methods Are Needed in Analytical Chemistry	168
7.2	Green Chromatography	169
	7.2.1 Gas-Phase Separations	169
	7.2.2 Liquid Phase Separations	171
7.3	Miniaturization of Separation Methods	185
	7.3.1 Continuous-Flow Microfluidics	186
	7.3.2 Droplet and Digital Microfluidics	186
	7.3.3 World-to-Chip Interfacing and the Quest for a 'Killer' Application in Microfluidics	189
	7.3.4 Non-Instrumental Microfluidic Devices	191
7.4	Challenges in Miniaturization of Separation Methods	195
References	195

Chapter 8 Green Electroanalysis 199
Lucas Hernández, José M. Pingarrón and Paloma Yáñez-Sedeño

8.1	The Role of Electroanalytical Chemistry in Green Chemistry	199
8.2	Green Stripping Voltammetric Methods for Trace Analysis of Metal and Organic Pollutants	200
	8.2.1 Determination of Trace Metal Ions with Bismuth Film Electrodes	200
	8.2.2 Determination of Organic Compounds with Bismuth Electrodes	201
	8.2.3 Stripping Voltammetry at Other Modified Electrodes	201
8.3	Electrochemical Sensors as Tools for Green Analytical Chemistry	202

	8.3.1 Electrochemical Detection in Flow Injection Analysis and Other Injection Techniques	203
	8.3.2 Microsystems	207
8.4	Alternative Solvents	209
	8.4.1 Ionic Liquids	209
	8.4.2 Supercritical Fluids	211
8.5	New Electrode Materials	212
	8.5.1 Metal Nanoparticles	212
	8.5.2 Hybrid Nanocomposites	213
	8.5.3 Oxide Nanoparticles	213
	8.5.4 Polymers	214
	8.5.5 Solid Amalgams	214
8.6	Electrochemical Biosensors	214
	8.6.1 Environmental Applications	215
	8.6.2 Biosensors Using Ionic Liquids	216
	8.6.3 Natural Biopolymers	218
	8.6.4 Microsystems-Based Biosensors	218
8.7	Future Trends in Green Electroanalysis	220
References		220

Chapter 9 Green Analytical Chemistry in the Determination of Organic Pollutants in the Environment — 224
Sandra Pérez, Marinella Farré, Carlos Gonçalves, Jaume Aceña, M. F. Alpendurada and Damià Barceló

9.1	Green Analytical Methodologies for the Analysis of Organic Pollutants	224
9.2.	Sample Preparation	226
	9.2.1 Solvent-Reduced Techniques	226
9.3	Greening Separation and Detection Techniques	247
	9.3.1 Immunochemical Techniques	247
	9.3.2 Biosensors	251
	9.3.3 Non-Biological Techniques	270
9.4	Future Trends in Organic Pollutants Analysis	273
References		274

Chapter 10 On-line Decontamination of Analytical Wastes — 286
Sergio Armenta and Miguel de la Guardia

10.1	Introduction	286
10.2	Recycling of Analytical Wastes: Solvents and Reagents	287
10.3	Degradation of Wastes	293
	10.3.1 Thermal Degradation	294
	10.3.2 Chemical Oxidation	294

	10.3.3 Photocatalytic Degradation	294
	10.3.4 Biodegradation	295
10.4	Passivation of Toxic Wastes	296
Acknowledgements		298
References		298

Subject Index   302

CHAPTER 1

An Ethical Commitment and an Economic Opportunity

M. DE LA GUARDIA AND S. GARRIGUES

Departamento de Química Analítica, Edificio de Investigación, Universidad de Valencia, C/. Dr. Moliner 50, 46100 Burjassot, Valencia, Spain

The side effects of the use of analytical methodologies may involve serious risks for operators as well as damage to the environment, and for these reasons it is relevant to think about the consequences of our activity as researchers or users of analytical methods.

Both from the point of view of citizens interacting ethically with the environment and as part of a fundamental evaluation of the costs of analytical procedures, we must take into consideration the inherent risks of some types of samples, together with the extensive use of chemical reagents and solvents, the energy consumption associated with modern instrumentation and, of course, the laboratory wastes and emissions resulting from the various steps of analytical procedures. This last aspect involves consumables and also the budget required to avoid or repair environmental damage.

Our view of analytical chemistry therefore involves moral and economic factors. We consider that the greening of analytical methodologies offers excellent business opportunities, as well as being a result of our moral commitment to our society and our future.[1,2]

1.1 Green Analytical Chemistry in the Framework of the Ecological Paradigm of Chemistry

The foundation of chemistry as a scientific discipline can be dated to the publication of the *Traité élémentaire de chimie* by A. L. de Lavoisier in 1789.[3] His work involved organizing chemical knowledge with respect to the experimental evidence, and created the basis of a paradigm focused on the atomic and molecular structure of matter and the relationship between the composition of matter and its behaviour.

As Professor Malissa has clearly explained,[4] the old chemical practices coming under the general heading of 'archeochemistry' were the first paradigm of chemistry providing the basis for the development of metal and alloy technologies, gold analysis, and developments in ceramics. This step was followed by the philosophical and experimental development of alchemy, a type of magic, which was introduced into the early universities through a study of the chemistry of natural products as pharmaceutical tools, thus creating the period of 'iatrochemistry'.[5] In this framework the 'chemiological' era began with scientific evidence of the nature of the chemical composition of matter and the relationship between structure and properties of materials, and, based on the rapid development of synthesis, provided the tools for a 'chemiurgical' period.

For the general public and for our students, most ideas about chemistry are probably based on the capacity of chemical principles and practices to create new materials and to transform our lives. However, it is also clear that as well as its beneficial effects the chemical revolution has caused terrible damage. Today we cannot imagine our life without many of the developments of the chemiurgical period, such as the introduction of petroleum-based fuels, the synthesis of pharmaceuticals and phytosanitary products, and many other industrial products, in spite of the environmental consequences and the risks to our lives caused by the use of chemical compounds.

The bad conscience of chemists and consumers about the side effects of chemicals has created a new view of chemical problems, which Malissa calls the 'ecological paradigm'; this aims to put chemical knowledge within the frame of environmental equilibrium. In the new framework of a sustainable chemistry all problems, from synthesis to individual applications, including analytical methods, must be evaluated in order to avoid collateral damage. This is especially important for the analytical community who, day after day, use large quantities of reagents and solvents to check the chemical composition of samples in every imaginable field, from natural sources to industrial processes and products, from the analysis of soils to that of water and air, not to mention the study of biota and the clinical evaluation of human health.

As Professor George Pimentel said in his *Opportunities in Chemistry* report to the U.S. National Academy of Sciences,[6] there is a need to increase the proportion of research and development devoted to exploratory studies of environmental problems and the detection of potentially undesirable environmental constituents at levels below their expected toxicity, thus increasing the support for analytical chemistry in a prominent way by the Environmental Protection Agency (EPA) and other American institutions.

In the 1990s there was a widespread bad conscience about the deleterious effects of chemistry and the collateral effects of analytical methods, due to the use of toxic reagents and solvents and the generation of dangerous wastes. This was the basis of some of the pioneering effort for greening the methods of analysis through the minimization of risks for operators by using mechanized procedures and closed systems.[7] As a result, initiatives like the development of environmentally friendly analytical methods[8] or clean methods[9] were proposed in 1994. The ethical agreement between chemistry and the environment has emerged from the green chemistry movement under the leadership of Paul Anastas,[10,11] although it was Cathcart[12] who first used the term 'green chemistry'.

In fact, the philosophy of green chemistry can now be considered as the central theory of ecological chemistry. In this framework, analytical chemistry, as a tool to determine the quality of air, water, and soil, can be seen indispensable to demonstrate the side effects of the chemiurgical period. It also provides the data required to establish the development of models for the decomposition of synthetic toxic molecules, in order to reinforce the need for chemical knowledge for the evaluation of environmental risks of the production, transport and use of chemicals. On the other hand, analytical activities can also contribute to damage of ecosystems through the use of toxic reagents and the generation of wastes. The opportunities offered by this discipline must therefore be complemented by a series of commitments to environmental preservation, and by social activities addressed to policy-makers and the general population in order to demonstrate the benefits of chemistry. In short, the use of 'green chemistry' must improve social benefits and avoid collateral damage; this principle should be considered in all fields, including analytical activities. Today, the prestige of our discipline depends heavily on the safety of measurements and the absence of environmental risks.

The increasing social demand for analytical methods and the need for fast, accurate, precise, selective and sensitive methodologies also oblige us to consider the use of reagents that are innocuous, or at least less toxic than those formerly used; to drastically reduce the amounts of samples, reagents and solvents employed; and to minimize, decontaminate and neutralize the wastes generated. For these reasons, a safe and sustainable analytical chemistry must be clearly established from the fundamental, practical and application points of view.

Figure 1.1 shows a schematic evolution of the main objectives of analytical chemistry in the frame of the chemiurgical and ecological paradigms. As this figure shows, the replacement of economic and technological development by the search for an equilibrium between the human race and the biosphere has involved broadening the interest of analysts from the main focus of their methodologies in order to consider the side effects of their practices too. However, in an evolutionary perspective, it is our opinion that good green analytical chemistry must pay attention to the new challenges without renouncing improvements in the basic aspects of analytical methods. We must find an equilibrium between the replacement of toxic reagents by innocuous

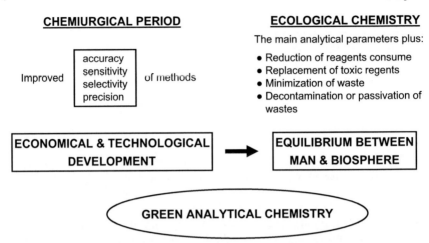

Figure 1.1 Evolution of the objectives of analytical chemistry from the chemiurgical period to the ecological period.

ones, or the reduction of sample, reagent and solvent consumption, and the preservation or enhancement of the accuracy, sensitivity, selectivity and precision of the methods available. Otherwise we could damage the capacity of analytical chemistry to provide valuable data to support our knowledge of the stability, evolution and damage of ecosystems. For this reason, green analytical chemistry must be considered as an balance between the quality of methods and their environmentally friendly character.[13]

1.2 Environment and Operator Safety: an Ethical Commitment

The avoidance of environmental risks, starting by assuming the operator's safety, is a philosophical principle and a social commitment; it is a prevailing concept of green analytical chemistry. Preserving the quality of air, water and land means thinking of future generations. Avoiding the use of dangerous reagents is the best way to guarantee the safety of users. These two aspects are complementary, and sum up the sustainability of green analytical chemistry.

Previously, the reasons for using greener methods were based on the advantages offered by automation and miniaturization in order to reduce the costs of analysis and also increase laboratory productivity. These were the main reasons for downsizing the scale of methods and pushing new ideas such as flow injection analysis,[7] sequential injection analysis[14] or multicommutation,[15] or developing solvent-free sample preparation techniques such as solid phase extraction,[16] solid phase microextraction,[17] single drop microextraction[18] or stir bar sorptive extraction.[19] However, it is clear that these analytical milestones have a new meaning when considered in the framework of the green analytical chemistry philosophy. In fact, the absence of extra costs in green

An Ethical Commitment and an Economic Opportunity

methodologies is one of their most attractive aspects, because it offers a unique opportunity to be socially honest without sacrificing economic benefits.

When we think about the main strategies that green analytical chemistry can use to avoid environmental side effects (see Figure 1.2), it is evident that there is good correlation between environmental and operator benefits due to the reduction of sample and reagent consumption through automation, miniaturization and on-line detoxification of wastes. The best thing is that the costs are reduced to the acquisition of basic equipment, which is easily offset by the reduced consumption of reagents and the enhancement of laboratory productivity. In terms of the analytical figures of merit, only sensitivity can be affected by the change from batch analysis to the use of automation. However, it is clear that when sample volume is reduced, in-batch selectivity can be enhanced by incorporating the physical and chemical kinetic aspects. It is also evident that the mechanization of analytical methods always improves the repeatability and reproducibility of analytical signals, avoiding operator errors.

However, the most important aspect is that green strategies can offer a new perspective of chemistry to the general public, allowing them to appreciate the important role of chemistry in both prevention and remediation of the environmental pollution, and can also counter the common idea that chemistry itself is the main reason for environmental damage. This approach can be highly beneficial in terms of social support for new developments in chemistry. For this reason, in both teaching and publishing, there is a crucial interest in the incorporation of green terminology and environmental considerations in analytical chemistry today. In order to do this the systematic evaluation of green aspects of new and available methodologies is mandatory. Many efforts have

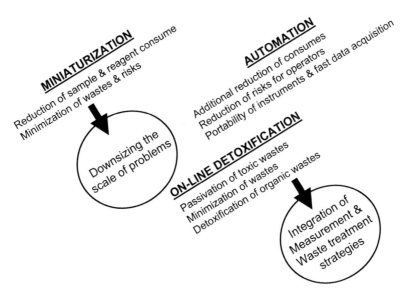

Figure 1.2 Main strategies of green analytical chemistry.

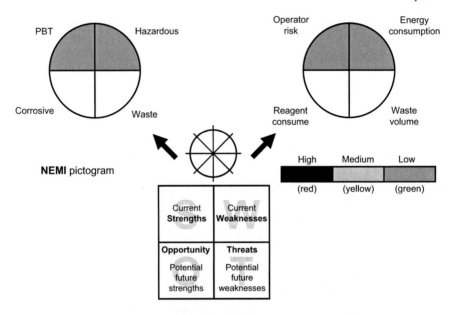

Figure 1.3 Green pictograms and SWOT summary tables employed in the literature to focus on the evaluation of the green parameters of methods.

been made to incorporate SWOT (strengths–weaknesses–opportunities–threats) analysis in the evaluation of green alternatives,[20] and to use green pictograms to identify the environmentally friendly character of available methods.[21] As shown in Figure 1.3, these green symbols can contribute to the visibility of efforts towards improving the safety of available procedures.

In fact an extra effort of communication is needed to transfer the environmentally friendly conscience of the scientific community to method users. This is the intention of recent initiatives which can be seen in editorials of journals specifically devoted to green chemistry, like *Green Chemistry* published by the Royal Society of Chemistry from 1999 or *Green Chemistry Letters and Reviews* published since 2007 by Taylor & Francis. Special issues of analytical journals have been devoted to green methods, such as those published in February 1995 by *The Analyst*, issues of *Spectroscopy Letters* devoted to 'green spectroscopy' in 2009 and *Trends in Analytical Chemistry* concerning green analytical chemistry published in 2010. It is also important to note the publication in 2010 of two books on green analytical chemistry, that of M. Koel and M. Kaljurand published by the RSC[2] and that of M. de la Guardia and S. Armenta published by Elsevier.[22]

Mary Kirchhoff, in an editorial in the *Journal of Chemical Education*,[23] has highlighted the importance of education for a sustainable future, emphasizing the positive contributions of chemistry to human health and environmental preservation as the best way to connect with the way society is moving. Probably one reason for the prevalence of the term 'green analytical chemistry' in preference to other descriptions—such as environmentally friendly,

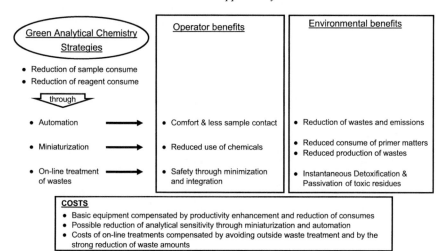

Figure 1.4 Consequences for operators and environment of the main strategies of green analytical chemistry, also introducing the problem of costs.

sustainable, clean, safe or ecological analytical chemistry—is that the word 'green' is commonly used in the mass media and the general public clearly identify its ethical implications with the sustainability of our activities.

To conclude, Figure 1.4 summarizes these discussions about the relationship between green analytical chemistry, operators and the environment, focusing on the benefits created by green strategies in terms of comfort and safety and introducing the problems of costs.

1.3 Green Chemistry Principles and Green Analytical Chemistry

Although many of the basic developments leading to green analytical chemistry took place in the 1970s and 1980s, the 1990 Pollution Prevention Act in the United States provided a political starting point for the green paradigm. As indicated by Linthorst,[24] who focuses on the EPA and the philosophical principles of green chemistry established by Paul Anastas and co-workers,[10,11,25,26] this was the basis of the green revolution which has involved all aspects of today's chemistry, from synthesis and analytical practices to engineering.

Figure 1.5 shows another diagram of green chemistry principles, emphasizing the special concerns of analytical practice. Only the second principle shown on the figure, 'maximize atom economy', has no evident application in the analytical field. Two principles—avoidance of chemical derivatizations and the use of catalysts—can be directly translated into recommendations for method selection. However, on looking for the analytical consequences of green chemistry principles it is clear that two main activities strongly recommended for the greening of analytical methods are absent—the minimization of sample,

GREEN CHEMISTRY PRINCIPLES	ANALYTICAL CONSEQUENCES
1. Prevent waste	Replace toxic reagent by innocuous ones
2. Maximize atom economy	--- --- ---
3. Design less hazardous chemical synthesis	Use less hazardous process
4. Design safer chemicals and products	Use safer reagents
5. Use safer solvents & reaction conditions	Use green solvents
6. Increase energy efficiency	Consume less energy
7. Use renewable feedstock	Use reagent and solvents obtained from renewable sources
8. Avoid chemical derivatives	Avoid chemical derivatization
9. Use of catalyst	Use of catalysts
10. Design for degradation	Use degradable reagents
11. Analysis in real time to prevent pollution	Use remote sensing, in-line or non-invasive methods
12. Minimize the potential accidents	Take care of operator and environment safety

Figure 1.5 Analytical consequences of Paul Anastas's green chemistry principles.

reagent and solvent consumption through automation or miniaturization, and the avoidance (as far as possible) of sample treatment. On the other hand, the use of less hazardous, safe reagents, green solvents, easily degraded reagents, or chemicals obtained from renewable sources could be summarized in just one or two recommendations to avoid redundancy.

So, additional efforts must be made to adapt the green chemistry principles to the analytical field. Namiesnik's attempt to establish the priorities of green analytical chemistry is probably a good starting point. He identified four possible routes:[27]

- Elimination or reduction of reagents and solvents
- Reduction of emissions
- Elimination of toxic reagents
- Reduction of labour and energy.

Our research team has expanded these points into six basic strategies for greening analytical methods:

- Analysis of untreated samples as directly as possible
- Use of alternative (less polluting) sample treatments
- Miniaturization and automation of methods
- On-line decontamination of wastes
- Search for alternative reagents
- Reduction of energy consumption.

The analytical community must establish its own principles. However, it is evident that in many cases green practices are already established in analytical chemistry, preceding the theoretical developments concerning the sustainability of methods. The important thing is to keep looking for the development of new, greener methods for the greening of previously available procedures. We are convinced that this effort could drastically modify the state of the art in our discipline.

1.4 Strategies for a Green Analytical Chemistry

In order to guarantee safety of operators and the environment, some of the main objectives of the green analytical chemistry are simplification, reagent selection, maximization of information obtained from samples, minimization of consumption and detoxification of wastes. These principles, which are clearly compatible with analytical figures of merit, can be directly implemented through the application of a few basic strategies which can now be used to improve the available analytical methodologies or to develop new ones. Figure 1.6 provides a scheme illustrating methodologies that can easily be incorporated into laboratory analysis at both development and application scales.

The objective of simplification is an obvious consequence of the basic green chemistry principles of reduction of steps and avoiding derivatizations. It is exemplified by the use of remote sensing of analytical parameters whenever possible, and of non-invasive, or at least in-line and on-line, determinations. Such methods provide information directly from the system to be evaluated, without any reagent consumption, solvent use or sample treatment, thus completely avoiding the side effects of traditional methods of analysis which always required previous sample dissolution and created problems of sampling, sample transport and sample stock. These procedures also minimize risks to the operator, as well as being non-destructive or causing little damage to samples.

Reagent selection is not specific to analytical chemistry, but the established rule of green chemistry is to avoid the use of toxic or hazardous reagents, for operator safety and environmental reasons, and to choose chemicals obtained from renewable sources. A key factor is to select easily degradable products for use in analytical procedures, in order to facilitate waste decontamination.

The miniaturization of sample, reagents and energy consumption is a desirable aim in improving the figures of merit of green analytical chemistry: it increases safety and reduces costs, as well as providing methods suitable for use with microsamples when required. The use of miniaturized sample preparation and the automation of all analytical steps are complementary strategies in

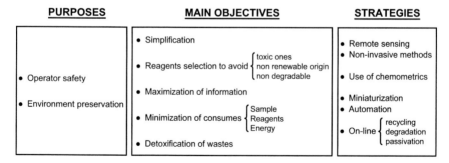

Figure 1.6 Main strategies of green analytical chemistry, derived from the objectives of greening the methods in order to guarantee operator safety and environmental preservation.

modern analytical chemistry. Additionally, new developments in low-energy processes, such as sonication, pressurized techniques and microwave-assisted procedures, are of interest for greening analytical methods in terms of energy consumption, and also provide benefits in terms of speed and laboratory productivity.

In our opinion, automation is the main strategy available for greening analytical methods, partly because of reduced consumption and enhanced sampling frequency but especially because automation is the best way to integrate all the steps of a method, including the on-line treatment of wastes.[13] We are absolutely convinced that the inclusion of a recycling, degradation or, at least, waste passivation treatment at the end of any analytical measurement is the best way to avoid environmental risks and ensure innocuous procedures, in spite of the fact that the sample components to be determined or the required reagents could be hazardous or toxic. The incorporation of recycling strategies, such as precipitation or distillation, or of degradation approaches based on biological, thermal, photochemical or oxidation processes, can assure the complete mineralization of organic reagents without reducing the sampling frequency. On the other hand, for analytical wastes containing metals or other non-degradable residues, the use of on-line passivation strategies based on chemical adsorption or co-precipitation can reduce the scale of analytical wastes from kilograms to grams and minimize the risks of contamination by modifying the chemical nature of the non-degradable pollutants.

Chemometrics,[28,29] which is one of the masterpieces of today's analytical chemistry, is the best way to maximize the information attainable from the samples. The original objective of chemometrics was the improvement of data treatments, but it now offers an excellent tool to reduce the need for external calibrations and specific procedures to determine each of the properties or components of a sample. Chemometrics can therefore be considered as a basic green strategy which can be employed to improve non-invasive or remote sensing methods in order to obtain accurate information from direct signals, avoiding a lot of reagents, energy and labour.

1.5 Cost of Green Analytical Chemistry

Basic components, such as appropriate software for chemometric signal treatment, or basic elements for flow injection analysis (FIA), sequential injection analysis (SIA) or multicommutation in order to automate measurements, together with miniaturized sample treatment set-ups or measurement units, represent the extra costs incurred by an analytical laboratory that would like to green its methods (Figure 1.7).

Some basic elements, such as micro total analytical devices (μ-TAS)[30] or sophisticated miniaturized methods for sample preparation,[17] are relatively expensive. However, these extra costs can be offset by reduced use of consumables. Figure 1.7 shows that the limited costs of greening methods and adapting to the new paradigm have great benefits from a financial point of view and also offer business opportunities. Taking into account energy savings, the

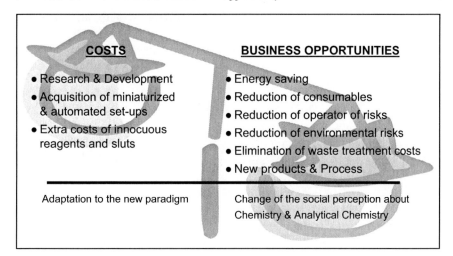

Figure 1.7 Balance between costs and business opportunities offered by greening analytical methods.

substantial reduction of consumables and glassware involved in mechanized procedures, and the elimination of waste treatment costs, it can be seen that green analytical chemistry is also good business. This is one of the reasons why applied laboratories are so interested in it.[31]

Acknowledgements

The authors gratefully acknowledge the financial support of the Generalitat Valenciana Project PROMETEO 2010-055.

References

1. S. Armenta, S. Garrigues and M. de la Guardia, *TrAC, Trends Anal. Chem.*, 2008, **27**, 497.
2. M. Koel and M. Kaljurand, *Green Analytical Chemistry*, Royal Society of Chemistry, Cambridge, 2010.
3. A. L. Lavoisier, *Traité élémentaire de chimie*, Cuchet, Paris, 1789.
4. H. Malissa, in *Euroanalysis VI Reviews on Analytical Chemistry*, ed. E. Roth, Les Ules, France, 1987.
5. M. Meurdrac, *La chimie charitable et facile, en faveur des dames*, Jean d'Honry, Paris, 1666.
6. G. Pimentel, *Opportunities in Chemistry*, National Academic Press, Washington, DC, 1995.
7. J. Ruzicka and E. H. Hansen, *Anal. Chim. Acta*, 1975, **78**, 145.
8. M. de la Guardia and J. Ruzicka, *Analyst*, 1995, **120**, 17N.

9. M. de la Guardia, K. D. Khalaf, B. A. Hasan, A. Morales-Rubio and V. Carbonell, *Analyst*, 1995, **120**, 231.
10. P. T. Anastas and C. A. Farris, *Benign by Design; Alternative Synthetic Design for Pollution Prevention. ACS Symposium Series*, American Chemical Society, Washington DC, 1994.
11. P. T. Anastas and R. Warner, *Green Chemistry, Theory and Practice*, Oxford University Press, New York, 1998.
12. C. Cathcart, *Chem. Ind.*, 1990, **5**, 684.
13. M. de la Guardia, *J. Braz. Chem. Soc.*, 1999, **10**, 429.
14. J. Ruzicka and G. D. Marshall, *Anal. Chim. Acta*, 1990, **237**, 329.
15. B. F. Reis, M. F. Gine, E. A. G. Zagatto, J. L. F. Costa-Lima and R. A. Lapa, *Anal. Chim. Acta*, 1994, **293**, 129.
16. J. T. Stewart, T. S. Reeves and I. L. Honigberg, *Anal. Lett. Pt.B*, 1984, **17**, 1811.
17. C. L. Arthur, K. Pratt, S. Motlagh, J. Pawliszyn and R. P. Belardi, *HRC, J. High Resolut. Chromatogr.*, 1992, **15**, 741.
18. H. H. Liu and P. K. Dasgupta, *Anal. Chem.*, 1996, **68**, 1817.
19. E. Baltussen, P. Sandra, F. David and C. Cramers, *J. Microcolumn Sep.*, 1999, **11**, 737.
20. M. Deetlefs and K. R. Seddon, *Green Chem.*, 2010, **12**, 17.
21. L. H. Keith, L. U. Gron and J. L. Young, *Chem. Rev.*, 2007, **107**, 2695.
22. M. de la Guardia and S. Armenta, *Green Analytical Chemistry: Theory and Practice*, Elsevier, Amsterdam, 2010.
23. M. M. Kirchhoff, *J. Chem. Educ.*, 2010, **87**, 121.
24. J. A. Linthorst, *Found Chem.*, 2009, **12**, 55.
25. P. T. Anastas and T. C. Williamson, *Green Chemistry: Designing Chemistry for the Environment*, ACS Symposium Series, American Chemical Society, Washington DC, 1996.
26. P. T. Anastas and M. M. Kirchhoff, *Acc. Chem. Res.*, 2002, **35**, 686.
27. J. Namiesnik, *J. Sep. Sci.*, 2001, **24**, 151.
28. M. A. Sharaf, D. L. Illman and B. R. Kowalski, *Chemometrics*, John Wiley & Sons, Toronto, 1986.
29. D. L. Massart, B. G. M. Vandeginste, S. M. Deming, Y. Michotte and L. Kaufman, *Chemometrics: A Textbook*, Elsevier Science, Amsterdam, 1988.
30. S. C. Jakeway, A. J. de Mello and E. L. Russell, *Fresenius J. Anal. Chem.*, 2000, **366**, 525.
31. M. Tobiszewski, A. Mechlinska and J. Namiesnik, *Chem. Soc. Rev.*, 2010, **39**, 2869.

CHAPTER 2
Direct Determination Methods Without Sample Preparation

S. GARRIGUES AND M. DE LA GUARDIA

Departamento de Química Analítica, Edificio de Investigación, Universidad de Valencia, C/. Dr. Moliner 50, 46100 Burjassot, Valencia, Spain

The ideal green analysis method would be a method based on direct measurement of samples without sampling, sample transport, addition of reagents, or waste generation. Remote and *in situ* measurements made in real time must be considered as the best choice, because they are based on direct measurements of samples that are untreated or with a minimum treatment.

The U.S. Environmental Protection Agency (EPA) is promoting the 'triad approach', which is an innovative approach to decision-making proactively exploiting new characterization and treatment tools[1,2] based on three primary components: (1) systematic planning, (2) dynamic work strategies, and (3) real-time measurement systems. The last of these is the most important aspect for the development of green analytical chemistry because real-time measurements do not use chemicals for sample preservation or analyte extraction, or use only small amounts of them. So, a triad approach based on real-time measurements reduces the side effects of the analytical steps as well as providing a less expensive analytical methodology. Other alternatives are based on the development of methods to preconcentrate or to extract analytes in the field and store pretreated samples for laboratory analysis. In some cases these new tools permit a direct measurement of the analytes retained without the use of any solvent or reagent or, alternatively, on-line measurement with a reduced volume of reagents, thus providing green alternatives to traditional methods of sample storage and analysis.

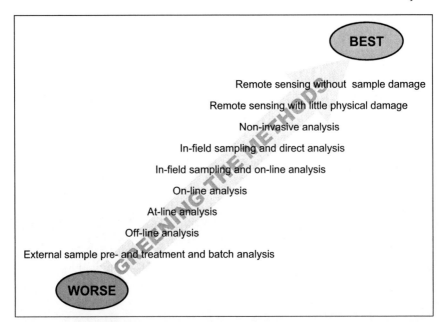

Figure 2.1 Hierarchical organization of the sample analysis approach in order to move from multistep and high pollutant risk methods to reagent-free methodologies.

This chapter focuses on the use of methods based on direct measurement without sample preparation or with minimal sample treatment. Remote and non-invasive measurements, which avoid the contact with and damage to samples, are the most suitable strategies but these options are not possible in all cases and thus direct analysis without chemical treatment and with the minimum possible sample damage are also recommended. Figure 2.1 shows a hierarchical organization of the strategies involved in moving from traditional methodologies, involving a series of steps which can pollute the environment, to really clean analytical methods.

2.1 Remote Sensing and Teledetection Systems

Remote sensing and digital image analysis provide methods for the acquisition of data and the easy interpretation of measurements of an object without any physical contact between the measuring device and the object itself, thus enhancing the information available without any damage to the sample, or using preliminary analytical steps or reagents.

Remote sensing is the science of acquiring, processing, and interpreting images and related data, obtained from aircraft and satellites, that record the interaction between matter and electromagnetic energy.[3] The term takes on a number of different meanings depending on the discipline involved.

Direct Determination Methods Without Sample Preparation 15

Traditionally, remote sensing referred to measurements made, from a distance, of the radiation spectra reflected and emitted from the earth's surface to acquire information without being in physical contact with the object (which in this case is the atmosphere).

In the last decade the use of satellite remote sensing of air quality has evolved dramatically; now, thanks to the increasing spatial resolution afforded by modern instrumentation,[4] global observations are available for a wide range of species including aerosols, tropospheric O_3, tropospheric NO_2, CO, HCHO and SO_2. The role of remote sensing is therefore under scrutiny, given its potential capacity for systematic observations at scales ranging from local to global and the availability of data archives extending back over several decades.

Three major applications of retrieved trace gases and aerosols by satellite remote sensing are available: forecast of events that affect air quality, interference of surface air quality itself (particulate matter, NO_2, O_3 and CO), and estimates of surface emissions (NO_x, VOCs, CO and aerosol sources).

The availability of remote sensing technology also contributed to the decision of the Kyoto Protocol of the United Nations Framework Convention on Climate Change (UNFCC) to limit or reduce greenhouse gas emissions to 1990 levels. Five major areas have been suggested where remote sensing technology could be applied to support the implementation of the Kyoto Protocol:[5]

- Provision of systematic observations of relevant land cover (in accordance with articles 5 and 10).
- Support to the establishment of a 1990 carbon stock baseline (article 3).
- Detection and spatial quantification of change in land cover (regarding articles 3 and 12).
- Quantification of above-ground vegetation biomass stocks and associated changes therein (also articles 3 and 12).
- Mapping and monitoring of certain sources of anthropogenic CH_4 (in accordance with articles 3, 5 and 10).

The first application of satellite remote sensing of aerosols was based on the use of an advanced very high-resolution radiometer (AVHRR) to observe Sahara dust particles over the ocean[6] and later for monitoring volcanic sulfate.[7] The total ozone mapping spectrometer (TOMS) was the first instrument designed for satellite remote sensing of tropospheric trace gases. Initially it was aimed at determining global knowledge of stratospheric O_3, but it also yields information about volcanic SO_2,[8] tropospheric O_3,[9] and ultraviolet-absorbing aerosols.[10] These instruments have been very effective. The last TOMS was deactivated in 2007, but the new generation of ozone monitoring instruments (OMI) has satisfactorily replaced them. Table 2.1 gives some examples of satellite instruments designed for remote sensing of aerosols and chemically reactive trace gases in the lower troposphere.

Satellite remote sensing is not limited to surface air quality. It can be used for ecological applications including land cover classification, integrated ecosystem measurements, and change detection, such as climate change or habitat loss.[11]

Table 2.1 Some examples of satellite remote sensing instruments used for air quality control.

Instrument		Platform	Measurement period from
MOPITT	Measurements of Pollution in the Troposphere	Terra	2000
MISR	Multiangle Imaging Spectroradiometer	Terra	2000
MODIS	Moderate Resolution Imaging Spectroradiometer	Terra Aqua	2000 2002
AIRS	Atmospheric Infrared Sounder	Aqua	2002
SCIAMACHY	Scanning Imaging Absorption Spectrometer for Atmospheric Chartography	ENVISAT	2002
OMI	Ozone Monitoring Instrument	Aura	2004
TES	Tropospheric Emission Spectrometer	Aura	2004
PARASOL	Polarization & Anisotropy of Reflectances for Atmospheric Sciences Coupled with Observations from a Lidar	PARASOL	2004
CALIOP	Cloud-Aerosol Lidar with Ortogonal Polarization	CALIPSO	2006
GOME-2	Global Ozone Monitoring Experiment	MeteOp	2006
IASI	Infrared Atmospheric Sounding Interferometer	MeteOp	2006

This is a very important tool for ecologists and conservation biologists, as it offers new ways to approach their research in order to provide scientific responses to environmental changes. Additionally, remote sensing has been used for environmental and natural resource mapping, and for data acquisition about hydrological sources and soil water and drought monitoring for early warning applications.[12] The use of remote sensing permits the monitoring of soil salinity caused by natural or human-induced processes,[13] the study of groundwater,[14] or the quantitative study of soil properties.[15]

Recent developments in optical remote sensing related to spatial resolution provide powerful tools in precision agriculture, which has the ability to rapidly evaluate the maturation of fruits or cultivars for optimal harvesting, or possible infection by diseases.[16] It also permits assessment of water use by crops and on-farm productivity monitoring—this latter though measuring methane emissions—thus making it possible to increase water use efficiency.[17] Specific properties of the vegetation, *e.g.* healthy or diseased, can be related to the amount and quality of radiation reflected or emitted from the leaves and canopies of plants. Remote sensing can therefore be applied to study plant pathology.[18]

From the instrumentation point of view, remote chemical sensing is a group of techniques. Basically, we can be distinguish between remote electrochemical

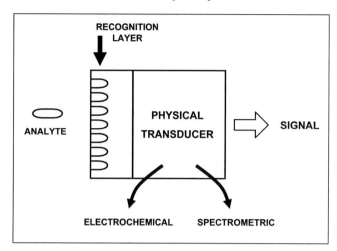

Figure 2.2 Schematic representation of a chemical sensor.

sensors and remote spectroscopy monitoring systems.[19] In addition, geoelectric techniques, such as DC-resistivity sounding, magnetotellurics, ground-penetrating radar, fixed frequency (FEM), and transient electromagnetic (TEM), have been used for remote monitoring of groundwater pollution and for estimation of hydraulic properties of aquifers and sediments.[20]

Chemical and biochemical sensors are based on a combination of a recognition layer and a physical transducer, and their use has been proposed for *in situ* remote monitoring of organic and inorganic pollutants (see Figure 2.2).[21] The introduction of modified electrodes and ultramicroelectrodes, the design of complex biological and chemical recognition layers, molecular devices, and sensor arrays, developments in micro and nanofabrication, as well as in technology of flow detectors and compact, low-powered and user-friendly instruments, have contributed to the development of electrochemical sensor devices for real-time monitoring of a wide range of molecules and contaminants.

On the other hand, two approaches can be distinguished for spectroscopic remote sensing: (1) direct, when both the electromagnetic radiation and the signal measured are used along an open path (*i.e.* atmosphere); (2) indirect; in this case radiation or signal is directed through fibre optics (see the scheme in Figure 2.3). Direct remote systems can be further classified as active, if the instrument contains its own source of radiation, or passive, for external emission sources, such as the sun. Active systems based on infrared or ultraviolet radiation sources are currently used and can utilize Raman scattering, fluorescence, or light absorption as their measuring principle.[22] However, there are other classification criteria for chemical sensor discrimination (see Figure 2.4) and, with reference to the position of the radiation source (transmitter) and the sensor (receiver) optical remote sensing instruments can be classified as: (1) monostatic, when the transmitter and the receiver are located in a single fixed position, and either a topographic target (building wall, ground,

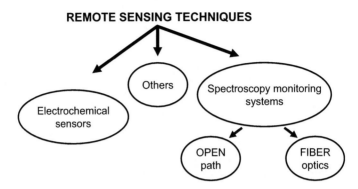

Figure 2.3 Classification of remote sensing techniques.

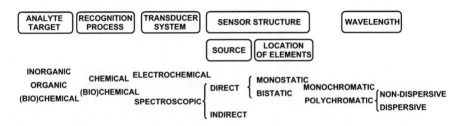

Figure 2.4 Different criteria to classify the chemical sensor.

vegetation), or atmospheric aerosols and molecules, or a retroreflector may be used to reflect the transmitted radiation back to the receiver; or (2) bistatic, in which the radiation source is in one location and the sensor in another, the distance between them being open optical path length.

Remote active sensing systems can be grouped into monochromatic instruments and instruments with a broadband source of radiation, subdivided into non-dispersive and dispersive. Monochromatic instruments are equipped with laser sources that provide spectral lines at microwave, infrared, or ultraviolet frequencies allowing identification and measurement of air pollutants. Non-dispersive analysers have been designed basically for specific constituents of gases. Dispersive instruments make it possible to obtain detailed information about the spectra of molecules and species present in a sample, but their sensitivity can be limited in comparison to that of monochromatic sources.

One technique used for direct optical remote sensing using monochromatic radiation is differential adsorption laser (DAL), in which two laser beams of different wavelengths are passed through the sample, one being coincident with the absorption maximum of the target analyte and the other being a non-absorbing wavelength. The difference between the two beams is proportional to the amount of absorbing molecules. Laser photoacoustic spectrometry (PAS) and light detection and ranging (LIDAR), which use a pulsed laser system, provide systems like radar where the time required to return the reflected radiation is measured and used to determine the distance of the reflecting material. The principles of differential absorption spectroscopy (DOAS) and simultaneous correlation spectroscopy (COSPEC) can be employed for optical absorption measurements of gaseous constituents in the atmosphere in the ultraviolet and near infrared range, using the radiance of the sky as a distributed light source.[23]

For indirect optical remote sensing, the development of fibre optics has been revolutionary. New materials, increased flexibility of fibres, long-range transmission capability, small size, broad bandwidth and imaging capability have made possible a variety of design options. These advances have provided fibre optical devices that can be used over long distances, or as non-invasive techniques for clinical or medical application. Fibre optical sensors in combination with laser-induced plasma spectroscopy can be employed for the determination of elemental sample composition; laser-induced fluorescence spectroscopy provides information about native fluorophores or fluorescent-labelled molecules; Raman spectroscopy is useful to obtain inorganic and organic vibrational structure information. Laser photofragmentation, which measures luminescence from sample fragments; photothermal spectroscopy, which provides inorganic and organic electronic and vibrational structure data; and ultraviolet, visible and infrared absorption spectroscopy, which are suitable for obtaining data about inorganic and organic electronic and vibrational structure, are some of the available tools for remote sensing measurements that have been described in the literature.[24]

2.2 Non-Invasive Methods of Analysis

The use of non-invasive techniques for direct analytical determination of components of packaged products is a very interesting route to the development of clean analytical methodologies, especially for quality control laboratories. It requires non-destructive measurement techniques that can be used through different types of sample containers, such as blisters, bags, vials, or bottles made of various materials. However, it is only possible to make this kind of measurement if the container material is transparent or the package has a suitable window for the source radiation.

NIR and Raman spectroscopy are two techniques that have suitable characteristics for obtaining chemical and physical information from non-destructive and non-invasive measurements of packaged samples.

The use of NIR spectroscopy was proposed to determine residual moisture in lyophilized sucrose through intact glass vials.[25] Common types of glass are virtually transparent to NIR radiation, and powdered samples may be measured in glass vials using the reflectance mode. This approach offers some advantages, for example (1) direct measurement without sample manipulation; (2) conservation of samples inside the vials after analysis which means that they can be employed/consumed or stored; (3) the lack of any deleterious effect on samples, which are not altered by the operator or the laboratory environment.

This last advantage can be very important for labile forensic samples, especially samples with legal relevance, as in the analysis of seized illicit drugs. Moros *et al.* proposed a non-destructive direct determination of heroin in seized illicit street drugs based on diffuse reflectance NIR measurements of samples contained in standard chromatographic glass vials.[26] Since neither chemicals nor time-consuming sample preparation processes are necessary, NIR spectroscopy provides an ideal analytical method for direct and instantaneous measurement of seized drugs. The use of portable and hand-held NIR spectrometers enables rapid checking of this type of samples in routine analysis and police checks.

NIR measurements on solid dosage forms can be performed in diffuse reflectance and this technique has been applied to pharmaceutical analysis, for the determination of active principles or the identification of pharmaceutical excipients inside USP vials[27] or through the blister pack. As can be seen in Figure 2.5, diffuse reflectance NIR spectra of tablets containing acetylsalicylic acid measured directly or measured inside the blister pack using the integrating sphere of the spectrophotometer show the characteristic bands of a standard of acetylsalicylic acid measured within a glass vial. This indicates that the polymeric blister material is adequately transparent to NIR radiation, making it possible to carry out a direct determination of this active principle in pharmaceutical formulations without the need to extract samples from the blister pack.

The above strategy contributes to the implementation of process analytical technology (PAT)[28] that promotes strategies for the control of primary and

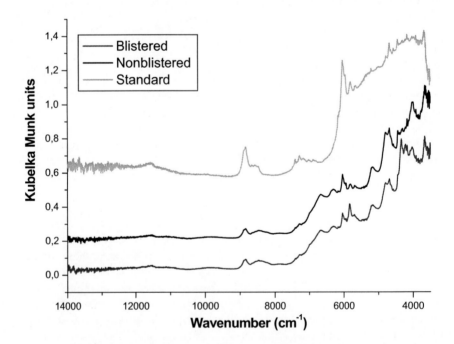

Figure 2.5 Diffuse reflectance NIR spectra of acetylsalicylic acid tablets measured directly and through the blister pack or a glass vial.

secondary manufacturing processes and offers an excellent method for non-destructive and direct analysis of final products stored in blister packs or other radiation-transparent containers.

NIR transmission measurements have been employed for identity confirmation of double-blind clinical trial tablets. The correctness, shipping, packaging, and labelling of the blister packs need to be checked before samples are shipped and a classification tool based on the NIR transmittance spectra has been developed to determine the different strengths of tablets using commercially available NIR instrumentation.[29]

Broad *et al.* proposed the use of NIR spectroscopy for the simultaneous determination of ethanol, propylene glycol and water contents in a pharmaceutical oral liquid formulation by direct transmission measurements through amber polyethylene terephthalate (PETE) bottles.[30] These plastic containers were expected to contribute to the NIR absorption spectrum. However, the background spectrum from an empty bottle was very small and small differences between different bottle spectra were also observed, showing that for calibration samples prepared in their own individual bottles the background absorption and spectral variations will be incorporated in the calibration model and compensated for sample prediction.

NIR transmittance spectroscopy has been proposed as technique for direct determination of the ethanol content of alcoholic beverages, making measurements through the glass bottles.[31] This technique has been applied to commercial instruments such as the Infratec^{TM} 1256 beverage Analyzer (Foss). This instrument, based on NIR dispersive scanning in the range 850–1050 nm or 570–1100 nm, includes a colour module that is suitable for analysis of different types of alcoholic beverages, and permits direct measurement of samples inside their bottles without any sample preparation. The analyser incorporates a selection of ready-to-use calibrations (regression programs) for different types of beverages, based on partial least squares and artificial neural networks. For the analysis of beer samples, alcohol content, original extract and colour can be directly predicted and other calculated parameters are real extract, apparent extract, degree of fermentation, energy, specific gravity, original gravity, present gravity, extract gravity, spirit indication, and refractive index. The complete analysis can be made in less than 45 seconds, making it possible to use of this instrument to control production at-line.[32]

It is difficult to make NIR absorption measurements of aqueous samples in large bottles because of the inadequate energy transmission due to the long optical path length and the high water absorption of infrared radiation. The reproducibility of on-line measurements can be poorer as a consequence of the variation in path length resulting from the lack of reproducibility of bottle positioning. So, well-validated methodologies for manual sampling and NIR transmittance or diffuse reflectance measurements cannot be used for continuous measurements.

Raman spectroscopy is another alternative for the analysis of samples directly through glass or plastic packages. As indicated in Figure 2.6, Raman

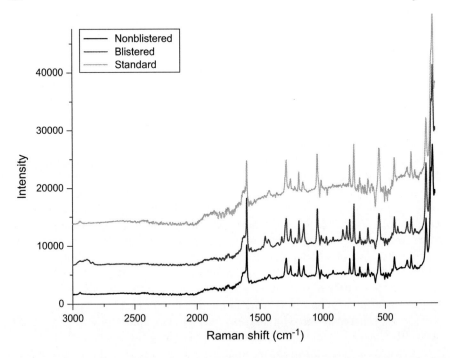

Figure 2.6 Raman spectra of acetylsalicylic acid obtained directly or through the pharmaceutical blister pack (The authors acknowledge the collaboration of Prof. J.M. Madariaga's research group, from the University of the Basque Country, for providing Raman spectra, and specially the help of Dra. S. Fernandez-Ortiz de Vallejuelo.)

spectra of acetylsalicylic acid tablets measured directly or through the blister pack exhibit the characteristic Raman bands of the standard, thus demonstrating that it is possible to measure and quantify active ingredients in blister-packaged pharmaceuticals. Additionally, Raman spectroscopy offers some advantages over NIR spectroscopy because it can provide a simple optical configuration that is easily interfaced for on-line measurements. In addition, as can be seen on comparing the spectra in Figures 2.5 and 2.6, the resolved details from the richer spectral features can be used to relate Raman spectra to molecular structure and composition.

Raman spectroscopy has been employed for the static analysis of ethanol content of spirits (whisky, vodka and sugary alcoholic drinks) in 200 ml (flat) and 700 ml (round) glass bottles, using a 785 nm laser and the Raman ethanol signal at 880 cm^{-1}. The technique only is applicable to the analysis of clear glass bottles because coloured bottles exhibited strong fluorescence.[31]

The quantitative *in situ* analysis of povidone in eyewash solutions contained in low-density polyethylene (LDPE) bottles was also made by Raman spectroscopy.[33] In order to correct the lack of physical and chemical homogeneity in the walls of plastic bottles, and the variation in the sensitivity of the Raman response to the sample position with respect to the focal plane, Raman dispersed radiation was collected using a wide area illumination (WAI) scheme that involves an incident laser with a large surface area ($28.2\,mm^2$) and a long focal length (248 mm). The resulting Raman spectra are much less sensitive to morphological variation of the sample bottles, and the high incident laser spot provides a reproducibility enhancement. Additionally, the use of an isobutyric anhydride external standard in front of the plastic bottles makes it possible to correct Raman intensity and reduce laser fluctuations. This WAI scheme with the use of a synchronous external standard has the potential to allow Raman spectroscopy to be used for quality control (QC) analysis for a wide range of liquid samples contained in glass (clear or amber) or plastic containers.

In a recent study Schmidt *et al.* have developed a prototype hand-held Raman sensor for the *in situ* characterization of meat quality. The Raman sensor head was integrated with a microsystem-based external cavity diode laser module that operates at an excitation wavelength of 671 nm and the Raman signal was guided by an optical fibre to the charge-coupled device (CCD) detection unit. Raman spectra of meat were obtained with 35 mW power within 5 seconds or less, and for measurements of raw and packaged pork meat this Raman sensor head evidenced its capability to detect microbial spoilage on the meat surface, even through the packaging foil.[34]

2.3 Direct Analysis of Solid and Liquid Samples Without Sample Damage

In this section we consider the possibilities offered by several techniques, such as X-rays, nuclear magnetic resonance (NMR) and vibrational spectroscopy to directly analyse solid and liquid samples without any sample damage.

2.3.1 Elemental Analysis by X-Ray Techniques

X-ray fluorescence is based on the secondary X-ray emission of characteristic radiation from the internal electrons of the different atoms present in a sample, as a consequence of the interaction between the primary X-ray and the inner orbital electrons. The secondary X-ray fluorescence spectrum identifies the elements present in the sample by the corresponding transition peaks at characteristic wavelength or energy positions, and their intensity is proportional to the elemental concentration.

Modern X-ray fluorescence analysis systems, based on either wavelength dispersive (WD-XRF) or energy dispersive (ED-XRF) measurements, are well established methodologies offering fast, non-destructive and clean forms of analysis that can routinely provide information about the elemental composition of samples with an adequate accuracy and reproducibility. Conventional XRF systems incorporating vacuum systems can measure elements from Na to U in solids and liquids with a precision better than 0.5% of their relative standard deviation in many cases, the limit of detection being typically in the low parts per million range and as low as 0.1 ppm for some elements. Practically any sample type can be analysed by XRF: pressed powders, glasses, ceramics, metals and alloys, rock, coal, plastics, oil, *etc.* The simplicity of sample preparation, minimum manipulation, and the possibility of analysing some elements, such as sulfur, that are hard to determine by other techniques, have promoted XRF as a useful alternative to conventional molecular and atomic spectroscopy techniques.[35]

In recent years the development of micro-XRF and portable XRF instruments, as a result of advances in miniaturization and semiconductor detector technology, has opened interesting green applications. These instruments make it possible to carry out *in situ* measurements and provide green analytical tools for fast and non-destructive elemental analysis that has been used for geological studies and art analysis.[36]

2.3.2 Molecular Analysis by NMR

The phenomenon of nuclei absorbing resonant radiofrequency in a static magnetic field is called NMR and this process is always accompanied by nuclear relaxation. The resonant frequency of absorption of energy of magnetic nuclei in a magnetic field is proportional to both the strength of the field and the magnetic moment of the nucleus. The resonant NMR frequency is a fingerprint of the local electronic environment of the nucleus, but depends on the external magnetic field. An NMR spectrum is a series of peaks of various widths and shapes that are a reflection of the local molecular environment of the nuclei under observation.

Under certain conditions, the NMR intensity is proportional to the number of resonance nuclei producing the signals, thus providing interesting possibilities for quantitative analysis in addition to the traditional use of NMR signals for the structural analysis of pure compounds.

NMR as an analytical technique has the advantages of non-destructiveness, no need to separate analyte from complex mixtures, and avoidance of the use of toxic reagents when spectra can be obtained directly from solids. The commercialization of high-field NMR instruments and probe improvements have contributed to the development of analytical applications in the field of natural products,[37] pharmaceuticals,[38] agricultural and food and beverage analysis.[39] The main limitation is the high price of NMR instrumentation and its low sensitivity as compared to other spectroscopy techniques. Additionally, in many cases, dissolution or dilution of the sample in a suitable solvent, such as deuterated water or chloroform, is required and direct analysis is limited to solid samples.

The introduction of mobile low-field NMR analysers offers an excellent method for non-destructive and fast measurements, with great potential to be used for on-line QC. Instruments such as these can be used for measurements in the near-surface volume of samples of any size. The mobile probe of the instrument is a pair of anti-parallel polarized permanent magnets joined by an iron yoke, producing a static inhomogeneous magnetic field. A surface coil is placed in the gap between the poles of the permanent magnet, generating the radiofrequency pulses. The measurement volume of the probe is about 5×5 mm in area and 2.5 mm in depth. The probe can be designed such that the first distance beneath the probe surface does not contribute to the NMR signal. As an example, this analyser has been employed for *in vivo* determination of fat content in salmon.[40] Similar instrumentation has been employed for analysis and conservation of art works.[41]

2.3.3 Molecular Analysis by Vibrational Spectroscopy

Vibrational spectroscopy is a well-established set of techniques traditionally used for qualitative molecular information, especially by organic chemists. However, in recent decades developments in instrumentation and the application of chemometrics to data treatment have demonstrated that vibrational spectroscopy provides fast quantitative analytical methods that enable non-destructive analysis and permits, in a green way, the simultaneous determination of multiple components from the same sample in a single instrumental measurement without environmental side effects.[42]

Infrared, both in the middle (MIR) and the near (NIR) region, and Raman spectroscopy are the main vibrational techniques employed for the direct analysis of samples.

Infrared spectroscopy is based on the interaction of electromagnetic radiation with a molecular system, in most cases in the form of absorption of energy from the incident beam. The absorption of infrared radiation induces transitions between the vibrational energy levels of molecular bonds. Different chemical bonds of a molecule absorb at different infrared wavenumber depending on the atoms connected, the surrounding molecules, and the type of vibration of the absorbance give rise to (stretching or bending). Most molecules have infrared bands in the spectral range between 400 and 4000 cm^{-1} (MIR), and most of the intense features of any MIR spectrum can be assigned to fundamental transitions. As a consequence of the anharmonicity of the vibrational energy levels, overtone transitions appear at high wavenumbers that are multiples of fundamental transitions, but they have very weak absorption compared to fundamental bands. When two fundamental vibrational transitions absorb energy simultaneously, a combination band can appear. These overtone and combination bands are more complicated to assign than fundamental bands, and provide weak signals in the NIR region between 12 800 and 4000 cm^{-1}.

On the other hand, Raman spectroscopy is an emission technique in which the sample is radiated with monochromatic visible or NIR laser radiation. This brings the vibration energy levels of the molecule into a short-lived, high-collision state, which returns to a lower energy state by emission of an energy photon. The emitted photon usually has a lower frequency than the laser radiation and in this case corresponds to the Stokes–Raman scattering emission. The difference between the frequency of the excitation laser radiation and that of the scattered photon is called the Raman shift and its units are cm^{-1}. The Raman shift corresponds to the frequency of the fundamental IR absorbance band of the bond involved in the relaxation processes.

IR spectroscopy detects vibrations during electrical dipole moment changes, whereas Raman spectroscopy measurements are based on detection of vibrations during the electrical polarizability changes. This implies that bonds that connect two identical or practically identical parts of a molecule can be more active in Raman than in IR, thus providing complementary spectral information; *i.e.* O–H stretching vibration is very strong in IR, but very weak in Raman. So, for instance, water has high absorption in the MIR region and is practically invisible in Raman spectroscopy.

One of the advantages of IR spectroscopy is its ability to obtain information from samples in many physical states—solid, liquid or gas—and in different measurement modes. Transmittance measurements are possible when samples are highly transparent to the radiation (*i.e.* gas) or when it is possible to obtain a suitable thick film of the sample that can let the IR radiation pass to the detector, this effect being strongly dependent on the IR energy. For example, pharmaceutical tablets can be analysed by transmittance NIR spectroscopy whereas some liquid samples have a higher absorption in the MIR region and cannot be directly measured by transmittance inclusively with a pathlength less than 0.1 mm.

On the other hand, reflection techniques are very suitable for samples that are highly absorbent or non-transparent to the IR radiation. Diffuse reflection is a phenomenon observed when radiation strikes a diffuse surface and is scattered in all directions. This technique is specially interesting for the analysis of powdered samples in the NIR region. Internal reflection or attenuated total reflectance (ATR) occur when radiation moving through a transparent material of high reflective index impinges on the interface with a low-refractive material, like the sample, at an angle greater than the so-called critical angle. Then the radiation penetrates inside the sample, to a small depth of a fraction of the wavelength of the radiation. To obtain a good ATR spectrum the sample must be in optical contact with the internal reflection crystal of the ATR accessory, normally ZnSe, germanium, KRS-5 or diamond. Considering that the effective sample thickness is very small, from a fraction of micrometre to a few micrometres in the MIR region, the fact that internal reflection accessories normally produce multiple reflections means that the surface layer is sampled multiple times, thus increasing the intensity of the resulting spectra. Many basic ATR attachments have been developed for standard laboratory spectrometers and some of them include probes and flow cells to be used for monitoring chemical processes in the laboratory or on-line. Recent designs of ATR for MIR use a composite internal reflection element, combining a diamond sample-contacting surface with ZnSe or KRS-5 parabolic focusing element, this later being equivalent to a beam condenser that increases the sensitivity. The strength, hardness and chemical inertness of diamond make it an exceptionally useful material for internal reflection spectroscopy. The refractive indexes of diamond (2.39), ZnSe (2.43), and KRS-5 (2.38) are very similar, so when diamond is in optical contact with either ZnSe or KRS-5 there is no significant loss of energy at the interface and these materials provide an excellent mechanical support for the diamond. These ATR accessories have been used for the analysis of liquid, solid and paste samples by IR spectroscopy.[43]

For Raman spectroscopy samples must be illuminated with monochromatic radiation as brightly as possible. Then, the scattered radiation must be efficiently collected and the mixture of reflected and elastically scattered radiation must be separated from its much weaker Raman component, and this weak scattered radiation alone must be processed to obtain the Raman spectrum. Traditionally this separation was carried out using multiple monochromators, but current Raman spectrometers incorporate a filter and a single monochromator, spectrograph or interferometer system. The expansion of Raman systems is related to their versatility for sample handling by incorporating microscopes, telescopes or fibre optic bundles to illuminate and view the sample and then redirect the collected radiation to the detector.

Both IR and Raman spectroscopy can be applied for the direct analysis of untreated samples at macroscopic or microscopic scale. Raman microspectroscopy has the potential for improved resolution because of the low wavelength of the radiation used. Additionally, Raman instruments offer confocality and it is possible to focus on different planes below the sample surface. Moreover, the signal-to-noise ratio is much lower in Raman spectroscopy than in IR spectroscopy, and if samples have natural fluorescence obtaining Raman spectra may be impossible. This problem, also present in macro Raman spectroscopy, may be overcome if Raman instrument is equipped with a low-energy NIR laser instead of a laser working in the visible region; this avoids the fluorescence of most molecules, but it decreases the spatial resolution and sensitivity. In addition, the heat generated by the laser may alter the Raman spectra and also destroy the sample during measurements. In order to avoid sample damage a suitable laser power and measurement time may be selected.[44] Modern Raman instruments used for quantitative measurements are equipped with 1064 nm or 785 nm lasers operating at powers of a few milliwatts to avoid sample fluorescence and thermal degradation.

As mentioned previously, one of the advantages of vibrational techniques is their ability to easily obtain spectra of different types of samples in any physical state. From these spectra it is possible to obtain quality global information about the chemical or physical characteristics of samples or of individual components present in them. For this reason IR and Raman spectroscopy have been widely employed in agricultural and food science,[45–47] pharmaceuticals,[48] and petrochemicals.[49] NIR spectroscopy is widely used,[50–52] because of the higher penetration depth of NIR radiation and the feasibility of sample manipulation or the synergistic combination with remote fibre optic probes. However, recent developments in mobile and portable Raman instruments offers the opportunity to make a direct analysis of samples of special interest.[53]

The combined use of chemometrics with vibrational spectroscopy has contributed to the increase in analytical applications based on NIR, MIR and Raman spectroscopy in the past two decades. The main features provided by vibrational spectroscopy combined with chemometrics are:

- The possibility of direct determination of an analyte without the need for a previous separation
- The simultaneous determination of multiple components in the same sample from an unique spectral measurement
- The capacity to directly evaluate several sample properties (*e.g.* physicochemical) or characteristics that do not correspond to a particular analyte or group of analytes but clearly relate to all the components, thus providing information about sample quality, *e.g.* tannins, related to long-term colour stability and astringency of wines[54] or pH of albumen and Haugh units for testing egg quality[55]
- The indirect modelling of spectra for determination of analytes at trace levels. In this sense the concentrations of many compounds present in the same sample can be modelled from the overlapping bands of the spectra; although there is no direct relationship between the presence of trace compounds and specific bands, vibrational modes assigned to different organic molecules, which can be modified by the presence of trace components, offer the possibility of establishing highly predictive models from a series of spectra of well-characterized samples. These indirect determinations are of special interest in the analysis of mineral elements because they offer the possibility of developing quantitative or at least screening methods suitable for obtaining information on trace components from NIR, MIR or Raman spectra of untreated samples, as an alternative to classical methodologies based on wet or dry-ash digestion and dissolution of samples before determination by atomic or ionic techniques. These solvent-free analytical methodologies, based on direct measurements of untreated samples, combined with chemometric techniques, offer a green alternative for studies in environmental and control analysis and for screening purposes in general. Some examples of recently developed applications are the NIR determination of mineral elements in soils,[56] sediments,[57] food,[58] beverages,[59] and forage crops.[60]

To conclude, vibrational spectroscopy techniques appear to be excellent green analytical tools because they are capable of providing high-quality spectral information of samples, in a non-destructive and, in more cases, non-invasive way, avoiding the use of reagents or solvents. Their use can be extended to other exciting fields, such as clinical[61] or microbiological[62] analysis. Additionally, the development of hyperspectral imaging as a technology to integrate conventional imaging and spectroscopy, to obtain spatial and spectral information about samples, contributes to the development of new green analytical applications especially for QC and process analytical technologies.[63,64]

Direct Determination Methods Without Sample Preparation

2.4 Analysis of Solids Without Using Reagents

Sample preparation for solution analysis is usually a labour-intensive process and the most time-consuming step of analysis; it is the major source of accuracy errors such as those concerning analyte loss and/or contamination and lack of repeatability of measurement. Sample digestion often involves health hazards for laboratory technicians and produces a lot of corrosive and toxic emissions and wastes. Moreover, digestion or dissolution inevitably involves dilution of samples, except in the case of dry-ashing sample processes, making the analysis of components at trace levels in digested samples more difficult. Direct solid sample analysis is an alternative that avoids these errors and limitations and could also provide clean analytical methods with greatly reduced use of acids and solvents and a drastic minimization of wastes. Some available techniques for direct analysis of untreated solid samples involving only the alteration or destruction of small amounts of samples are discussed in this section, including electrothermal atomic absorption spectrometry, arc/spark optical emission, laser ablation, glow discharge, laser-induced breakdown spectroscopy (LIBS) and desorption electrospray ionization (DESI) (see Figure 2.7).

2.4.1 Electrothermal Atomic Absorption Spectrometry

Electrothermal atomic absorption spectrometry (ETAAS) and, in particular, the most frequently used graphite furnace (GF-AAS) method, involves the drying and ashing of samples and atomization of the mineral residues in separate and well-controlled steps.[65] It enhances the sensitivity of traditional flame atomic absorption (FAAS) by three orders of magnitude. Additionally, solid samples, instead of dissolved ones, can be introduced into the atomizer, providing a fast alternative which avoids sample dissolution and makes the drying step unnecessary. However, direct solid sampling in GF-AAS has some drawbacks: (1) the difficulty of handling and introducing small sample mass; (2) the difficulty of calibration, which in some cases requires the use of solid standards with characteristics similar to those of the samples to be analysed; (3) the limited linear working range and the

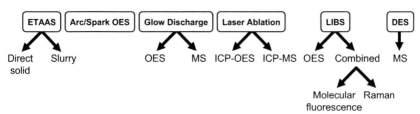

Figure 2.7 Green alternatives to solid sample dissolution methods based on the damage of small amounts of sample for their elemental or molecular analysis.

difficulty of diluting solid samples; and (4) the high imprecision of results due to the heterogeneity of samples, a large number of repeated determinations being required when this is possible.[66]

Some of these disadvantages can be avoided by the use of the slurry technique, but this creates new problems relating to the dilution of samples provided by slurry preparation and stabilization and the need to use some reagents.[67]

Direct solid sampling GF-AAS is preferable to the use of the slurry technique for materials that are very difficult to dissolve, and for the elements most susceptible to contamination. It is not affected by problems such as sedimentation or partial leaching of the analyte, and particle size, which is much less critical in direct solid analysis than in slurry sampling. On the other hand, the maximum sample mass that can be introduced into the graphite furnace for solid sampling is one or two orders of magnitude higher than that for slurry sampling.[68]

With regard to calibration in solid sampling GF-AAS, reasonable results can be obtained with standard calibration with aqueous solutions after a careful optimization of the analytical conditions. It is remarkable that solid sampling GF-AAS offers a fast screening tool, essentially without any sample preparation,[69] and can also be used to determine the homogeneity and microhomogeneity of certified reference materials (CRMs) and other samples.[70] It is of special interest for the determination of trace elements in complex samples that are difficult to dissolve.

Modern solid sampling GF-AAS units may be equipped with a device for automatic weighing and introduction of the solid samples,[71] and use a platform designed for handling high sample amounts and transverse heating of the graphite tube that provides homogeneous temperature distribution. One of the advantages of this approach is related to the potential use of appropriate temperature programming to minimize possible matrix effects. In theory, by controlling the temperature, it is possible to separate in time the atomization of the analyte from the vaporization of the main matrix components. External calibration with aqueous standards should thus be possible after removal of the matrix.[72]

An additional advantage of solid sampling GF-AAS is its great potential for automation; the main problem of this technique, on the other hand, is the lack of capability for multielemental determination.

2.4.2 Arc and Spark Optical Emission Spectrometry

For a long time in the history of spectroscopy, arc and spark excitation were the leading techniques for elemental analysis. Nowadays they have been displaced by the use of inductively coupled plasmas employed as excitation sources for optical emission spectrometry (ICP-OES) and mass spectrometry (ICP-MS), and by microwave-induced plasma for optical emission spectrometry (MIP-OES) in gas phase, with their almost perfect analytical performance. However, it should be pointed out that arc and spark provide a good excitation source for OES and still retain superior solid sampling capability, in comparison with ICP or MIP techniques which require the use of solutions or gaseous phases respectively.[73]

Techniques based on the ablation of the sample by an electrical spark or arc have been used for direct analysis of solid samples, like metal or alloys in metallurgy industry, ores and minerals in geological prospecting, and solid wastes in environmental monitoring. The principle of these techniques is based on the effect of a high potential difference applied between two electrodes in an argon atmosphere (one of them being the sample). A repetitive unidirectional electrical discharge ablates atoms from the surface of the solid sample and produces a high-temperature plasma, in which the atoms extracted from the sample are excited and the corresponding emission can be observed by using an spectrometer.[74]

The use of new excitation sources based on an alternate current (AC) arc with a low acquisition and operational cost and with the capacity to operate at atmospheric conditions without an additional inert gas or the use of a CCD as a detector has create a new generation of low-cost and compact arc OES spectrometers. Some reasonably priced portable analysers can be obtained that may be very useful for *in situ* and fast analysis.

2.4.3 Laser Ablation

Ablation of solids using laser pulses and the subsequent transfer of the released material to the measurement device using a gas flow, usually argon, is an attractive alternative to the nebulization of aqueous solutions for the direct analysis of samples, also providing a depth and spatial distribution of the analytes.

A laser ablation (LA) system combined with inductively coupled plasma optical emission spectrometry (ICP-OES) or mass spectrometry (ICP-MS) provides information about the elemental composition at trace and ultratrace levels. LA-ICP-MS represents the most modern method for direct analysis of the elemental composition of solid samples, with the ability to gain local and depth-resolved information about their distribution, for both conducting and non-conducting samples.

A typical LA set-up consists of a lens (which may be incorporated into an optical microscope so that optical and visual focusing are coincident), an ablation chamber, and an adjustable platform. For LA-ICP the sample is placed in an airtight ablation cell flushed with an inert gas to transport the ablated material to the ICP. The volume and aerodynamics of ablation cells, as well as the length and geometry of the transfer tube, affect dispersion of sample density in the ICP. It has been reported that small cell volumes may reduce the sample-washout time of the cell and a high carrier gas flow through the cell and transfer tubes reduces deposition of ablated material, as well as decreasing memory effects and increasing transport efficiency.[75]

The laser ablation of solids involves processes that include heating, melting and evaporation of sample material at extremely high temperatures and pressures. Because of the complexity of the process, ablated material may be removed from the sample in the form of atoms, molecules, vapour, droplet solid flakes, large particulates, or mixtures of these, and the distribution of the materials will depend on the selection of laser parameters as a function of the sample.

Three main types of lasers—ruby, Nd:YAG, and excimer—have been widely reported and validated for ablation but currently the most commonly used is the Nd:YAG laser because it is relatively cheap, robust, reliable, and easy to operate. Moreover, with this laser it is possible to operate at 1064 nm (fundamental wavelength) or 532, 355 and 266 nm.[76]

Laser pulse frequency and power affect the ablation rate and the particle size of the ablated material, but typical values are of the order of 1–20 µg s^{-1} and less than 100 nm, respectively, so sample consumption is limited to a few micrograms.[77] Taking into consideration sample consumption and laser spot diameter size (from a few microns to 100 µm), LA-based techniques could be considered as almost non-destructive tools from a macro point of view,[78] and thus especially interesting for the analysis of samples that in most cases are unique.

As compared with the use of spark, LA eliminates the restriction associated with the necessity of ensuring electrical conductivity of samples to be analysed but involves other problems and limitations such as (1) the amount of the sample material aerosolised per unit of time is significantly smaller than can be obtained with a spark discharge; (2) fractional evaporation may be observed; (3) the small area of the spot of focused laser radiation is useful, but to determine the general composition of macrosamples the area of the analysed surface has to be increased and a mechanical device for scanning the laser radiation over the sample surface is required.[79]

The major advantage of ICP-MS as a detector for LA analysis is its high sensitivity, wide dynamic range, relative simple spectra, and fast scanning, which provide an excellent method for a direct simultaneous analysis of complex samples without any pretreatment or use of reagents.

Regarding calibration strategies, the best results were obtained using matrix matching for external calibration, because the ablation rate varies with the sample matrix, but this implies the use of CRMs with a similar composition to the samples. However, other alternatives are the use of external calibration to a solid reference standard in conjunction with internal standardization or calibration using solutions when possible.[80]

LA coupled with ICP-MS has been shown to be an excellent analytical tool for the analysis of an extended number of analytes in a wide range of matrices, as can be seen in some examples reported in Table 2.2.

Compared with conventional dissolution techniques, LA avoids the dissolution step and the use of or exposure to potentially hazardous reagents, as well as the risk of introducing contamination or losing volatile components during sample preparation. It also saves analysis time. In addition, LA can be applied to any type of solid sample and there are no sample size requirements, a chemical analysis being possible with only few micrograms of sample. One example of the green analytical characteristics of LA is illustrated by the analysis of radioactive samples: the organic solvents or concentrated acids that are required for conventional radiochemical analysis are not necessary for LA sampling, and less than 1 µg of sample is used, which reduces the risks associated with sample handling and sample contamination; moreover, elemental

Direct Determination Methods Without Sample Preparation 33

Table 2.2 Examples of applications of laser ablation solid sampling combined with ICP mass spectrometry.

Field	Application	Objective
Environmental	Tree ring	Study of changes in atmospheric conditions, soil chemistry and pollution history
	Tree bark	Provide information about the degree of pollution of a certain region
	Seashells	Trace element fluctuations reflect environmental change and major pollution events
	Coral	Concentration of trace elements in coral skeletons provides information related to changes in seawater properties
	Airborne particulates	Information for monitoring air quality and inorganic pollutants
Geological	Geochronology	U-Pb isotopic analysis for dating
	Inclusion analysis	Study of microscopic inclusions in minerals
	Isotopic analysis	Precise measurement of isotopic composition of characteristics elements (Hf, W, Sr, U, Th, Pb, Os) at trace levels
	Bulk analysis	Measurement of rare earth elements or the platinum group elements for geological or economical importance
	In situ analysis	Spatially resolved analysis of elements
Archaeology		Authentication of precious antiques with a minimum damage
Waste-sample		Especially to analyse radioactive samples
Other	Fingerprinting	Forensic chemical analysis of physical evidences
	Film doping and depth profiling	Quantitative analysis of dopant dose implanted in crystalline silicon wafers or the analysis of multilayer coatings systems

and isotopic analysis can be obtained entirely within a hot cell environment, further reducing the risk of environmental contamination.

2.4.4 Laser-Induced Breakdown Spectroscopy

LIBS is an emission spectroscopy technique with the capability to detect, identify and quantify the chemical composition of any material. This technique utilizes a pulsed laser focused on a small area to create a microplasma on the sample surface. The resulting light emission is collected optically and then resolved temporally and spectrally in order to produce an intensity versus wavelength spectrum containing emission lines from the atomic, ionic, and molecular fragments created by the plasma.[81]

In some cases the laser pulses reach the target sample through an optical fibre, and the collection of the plasma light and transport back to the detector system can be done using either the same fibre optic cable or a second one. This method has been shown to work over distances up to 100 m, but requires the optical fibre to be positioned adjacent to the sample. Obviously this use of fibre optics is restricted in its application and even impossible in cases where contaminated or hostile chemical or temperature environments may affect the probe and, evidently, the operators. In these instances an open-path LIBS configuration, in which the laser beam and the returning plasma light are transmitted through the atmosphere, is available. This stand-off LIBS has been suggested for elemental analysis of materials located in environments where physical access is impossible or dangerous, but optical access could be envisaged. Solid samples can be analysed at distances of a few metres by open-path stand-off LIBS using nanosecond laser pulses, while liquid samples can be measured at distances of few metres. The use of femtosecond laser pulses is predicted to extend LIBS capabilities to very long distances because the high-power densities achieved with these lasers can also induce self-guided filaments in the atmosphere which can produce LIBS excitation of a sample at kilometre ranges.[82]

For a long time, the analytical applications of laser-induced plasma spectrometry have been restricted mainly to overall and qualitative determination of elemental composition in bulk solid samples. However, the introduction of new compact and reliable solid state lasers and technological developments in multidimensional intensified detectors have made possible new applications of LIBS for the direct sampling of any material, irrespective of its conductive status, without any sample preparation and with a sensitivity adequate for any elements in different matrices.[83]

LIBS has intrinsic advantages over other analytical techniques for elemental analysis, such as X-ray microprobe or X-ray fluorescence spectrometry; it provides a rapid, spatially resolved or in-field geochemical analysis of elements of low atomic weight, being an excellent tool for mineralogical and petrological analysis, either in the laboratory or in the field using real-time field-portable instruments.[84]

LIBS systems also can be combined with molecular fluorescence or Raman emission for the development of hybrid sensors systems. Figure 2.8 shows a schema of the experimental set-up for a stand-off dual Raman-LIBS mobile sensor prototype developed for the analysis of explosive materials. The system is able to simultaneously measure both Raman spectrum and laser-induced breakdown spectrum, and this approach exploits the energy distribution profile of the laser beam to extract the vibrational fingerprinting from the non-ablated section of the interrogated target within the outer part of the laser beam jointly together with the atomic information gained from the ablated mass by the inner part of the laser beam. By selecting suitable operating conditions (breakdown timing, laser power and acquisition time), molecular and multielemental spectral information, from the same sampling point and at the same laser event, can be obtained without any operator exposure.[85]

2.4.5 Glow Discharge

Glow discharges (GD) have been used as sources for emission and mass spectroscopy. Traditionally GD optical emission spectroscopy (GD-OES) has been a widely used technique for routinely bulk analysis in material sciences, for rapid depth and profile analysis of surfaces, thin films, and coatings. GD

Direct Determination Methods Without Sample Preparation 35

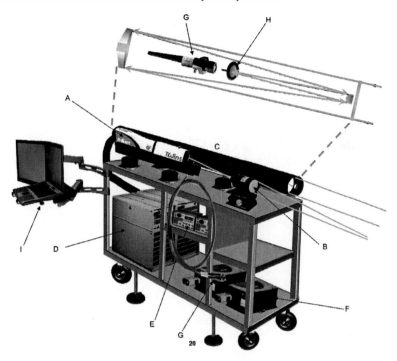

Figure 2.8 Experimental set-up of the stand-off dual Raman-LIBS sensor: (A) Nd:YAG laser (532 nm); (B) beam expander; (C) telescope; (D) laser power sources; (E) pulse and delay generators; (F) spectrographs; (G) bifurcated optical fibre coupled to a collimating lens; (H) holographic SuperNotch filter; (I) personal computer. The inset shows the telescope optical layout. (Reproduced from Ref. 85, with kind permission of the American Chemical Society.)

mass spectrometry (GD-MS) has been much used for direct analysis of conducting solids; it is virtually unrivalled for trace analysis of impurities in high-purity materials, and for monitoring the shallow depth distribution of traces. For bulk analysis of less pure samples, GD-OES competes with spark emission spectroscopy and X-ray techniques, but in many cases it exhibits fewer matrix effects or has lower detection limits than the other techniques.[86] Regarding depth profile and surface analysis, especially of thick coatings, GD-OES does not have much competition because of the low total cost of the instrumentation, the ease of sample handling, and the speed of analysis. For the analysis of refractory materials, glasses, *etc.*, it is exceeded only by X-ray fluorescence in terms of ease of analysis but without providing information about light elements. Auger electron spectroscopy (AES), secondary ion mass spectrometry (SIMS), and secondary neutral mass spectrometry (SNMS) may be preferred because of their possible depth resolution, and the information they provide on lateral distribution and structure, but they are more costly than GD and require longer analysis times.[87]

GD operates in a primary vacuum and uses the bombardment of the sample surface with ions of rather random orientation and high energy for the sputtering or atomization process. The sputtered species, mostly atoms, diffuse into the negative glow area, where they are excited and/or ionized. As compared with other techniques, like SIMS, in GD sputtering and ionization processes are separated in space and time, resulting in only minor variations in sensitivity and little matrix dependence, so quantification is very easy; in some cases it is possible without the absolute need for matrix-matched standards.[88]

The source commonly used in GD is the Grimm-type chamber, which operates in the presence of a noble gas (usually argon) under a reduced pressure. This source consists of an anode tube, usually grounded, and the sample to be analysed (cathode) that is placed perpendicularly in front of this anode tube. A ceramic spacer maintains a distance of less than 0.1 mm between the flat sample surface and the anode tube, and an O-ring allows this mount to be sufficiently vacuum tight. The electrical energy to ignite and maintain the plasma is fed into the plasma chamber directly through the sample. To avoid excessive heating, samples are cooled. Sputtering is caused by bombardment of the sample surface, and several particles are responsible for this process.

GD has been considered as a fairly rough tool only able to analyse bulk materials and rather thick coatings in the micrometre range, making it possible to characterize the elemental composition of a sequence of layers of varying thickness, ranging from nanometres to several micrometres, in a single analysis step without any sample preparation. Using a Grimm-type configuration, however, GD is capable of performing surface and interface analysis with a depth resolution in the nanometre range, possibly even at the atomic layer level.

In Grimm-type sources a typical anode tube has an internal diameter of 4 mm, but working with smaller anode tubes (*e.g.* 1 mm) the lateral resolution can be improved but the sensitivity is reduced. This limits the capability of GD for doing microspot analysis as compared with other techniques such us LA-ICP-MS or SEM-X-ray.

GD-OES has been extensively applied for quantitative depth profile analysis of hard coatings, and it is possible to analyse quantitatively a wide variety of commercial and experimental hard coatings with a single calibration if the coatings are conductive.[89]

For electrically conducting materials, the performance of direct current (DC) and radiofrequency (RF) powered GD sources is very similar in both depth resolution and sensitivity. However, non-conductive coatings or samples can be analysed only with RF-powered discharges. Most electrically isolating materials are also poor thermal conductors. The energy deposited at the sample surface by the sputtering process therefore cannot easily be dispersed and causes an considerable increase in temperature of the analysed material, which can lead to sample destruction. Pulsing the RF power supply can help, by reducing the average power dumped at the sample surface without reducing the instantaneous energy available for the sputtering and excitation processes.[90] Using pulsed RF-GD makes possible the analysis of thermally sensitive materials such as thin coatings on glass samples.[91] Combining the capacity of

RF-GD to sputter non-conductive materials with the quasi-simultaneous coverage of a large mass range, time of flight (TOF) mass spectrometers offer a new range of applications, not only to determine the elemental information of a layered sample but also to obtain molecular information.[92] An example is the analysis of multilayer structures composed of different polymers, including polystyrene (a polymer that has not yet been successfully sputtered by cluster ion beams typically used for TOF-SIMS analysis of polymers), the polymers being distinguished by their characteristic molecular fragments.[93]

Trends in GD spectroscopy are focused on the development of new sources and interfaces.[90] New ionization sources have been designed for direct mass spectrometric analysis of solid materials at atmospheric pressures. These techniques allow the fast, versatile, and extremely sensitive analysis of real samples with minimal or no sample preparation. In recent years an atmospheric-pressure glow discharge (AP-GD) source has been developed for the generation of reagent ions that, coupled to a TOF-MS, allows the detection of a wide variety of compounds, both polar and non-polar, on a broad range of solid substrates from glass, to plastics, textiles, wood, *etc.*, with a high analytical sensitivity. The analysis can be made in less than 1 minute and might become an important tool for screening unknown samples.[94] Another important advance in atmospheric pressure discharges is the reduction in size and power consumption of the discharge. These microplasmas can be used for portable, battery-operated instruments which are smaller than laboratory-scale instruments. These novel sources can be used as small and inexpensive detectors for chromatography and electrophoresis separations in combination with lab-on-a-chip systems. The advantages are that microplasmas uses smaller volumes of reagents and are therefore cheaper, quicker, and less hazardous to use, and more environmentally friendly than conventional systems.

Regarding developments in reduced pressure sources for GD based on Grimm-type sources, the design of new fast flow sources has been directed to high gas flow rate, improving the efficiency of ion transport by gas convection and thus reducing diffusion losses.

2.4.6 Desorption Electrospray Ionization

DESI coupled to mass spectrometry (DESI-MS) provides a new tool for direct analysis of solid surfaces without sample pretreatment and can give molecular information about different analytes in various solid surfaces. It is of great interest for the simultaneous determination of polar and non-polar compounds in liquid and solid samples.[95,96]

A DESI source, with a spray solvent and a gas nebulizer, generates a gas jet that impinges on the sample surface and creates positive and negative ions which can be sampled into a mass spectrometer. The method is fast, highly selective and sensitive; it provides absolute detection limits in the subnanogram to subpicogram range and constitutes a powerful tool for *in situ* molecular analysis and also for elemental speciation.[97] Recent developments based on the use of an appropriate internal standard have improved the quantitative capabilities of this technique (G. Hieftje, personal communication).

2.5 Summary of Present Capabilities of Direct Determinations

Direct analysis of samples without any previous treatment is clearly the greenest alternative for any kind of determination and, if we take into account the different strategies presented through this chapter, it is nowadays relatively easy to find the appropriate tool to do any kind of determination. However, it is clear that not all possible problems can solved by remote sensing without any direct contact of the operator with the sample and without sample damage. As shown in Figure 2.9, a hierarchical classification of methods can be established in order to obtain as green an analysis as possible.

Fortunately many suitable are available techniques to do a direct analysis of untreated samples without damaging them such as vibrational methods and NMR, followed by X-ray. It is possible to use MS for the direct determination of gaseous molecules or gaseous molecules in equilibrium with liquids or solids. However, in most applications a physical treatment of samples is mandatory to obtain a small quantity of ionized molecules which could be accelerated and determined in the mass spectrometer.

An interesting aspect of modern instrumentation is that it has been developed in order to reduce the size of components and their power requirements, thus providing portable instruments with reduced cost and tremendous possibilities for *in situ* analysis.

On comparing ablation strategies with earlier methods involving sample dissolution it is clear that we can drastically reduce the energy consumption and

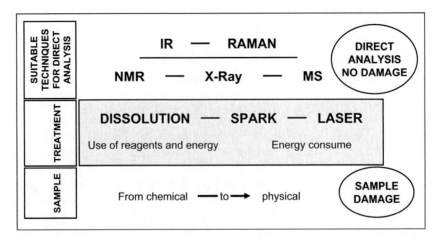

Figure 2.9 Analytical techniques suitable for direct analysis without and with sample damage.

the amount of sample required and, once again, the deleterious side effects on the environment can be reduced.

Dreaming about the future, we can imagine that both remote sensing and direct analysis with little or no sample damage can offer simultaneous information about many analytes and properties of samples by incorporating multivariate data treatments. The extension of chemometrics in direct methods is an exciting possibility which can improve the level of information obtained without the use of any solvent or reagent and also avoiding the generation of wastes.

It is also clear that direct analysis strategies offer fast alternatives to traditional methods based on sample pretreatments. This capability of direct methods to reduce analysis time and thus improve decision-making must also be taken into consideration in selecting the greenest alternative.

Finally, we can show that green technology is not at all detrimental to traditional analytical figures of merit. On the contrary, saving analytical steps improves analytical precision and sampling throughput, avoids causes of both systematic and random errors, and also provides economic opportunities, as we argued in Chapter 1.

Acknowledgements

The authors gratefully acknowledge the financial support of the Generalitat Valenciana Project PROMETEO 2010–055.

References

1. L. H. Keith, L. U. Gron and J. L. Young, *Chem. Rev.*, 2007, **107**, 2695.
2. U.S. EPA, *Triad Resource Center*, http://www.triadcentral.org.
3. F. F. Sabins, *Remote Sensing. Principles and Interpretation*, 3rd edn. W. H. Freeman, New York, 1997.
4. R. V. Martin, *Atmos. Environ.*, 2008, **42**, 7823.
5. A. Rosenqvist, A. Milne, R. Lucas, M. Imhoff and C. Dobson, *Environ. Sci. Policy*, 2003, **6**, 441.
6. R. S. Fraser, *Appl. Opt.*, 1976, **15**, 2471.
7. L. L. Stowe, A. M. Ignator and R. R. Singh, *Geophys. Res. Lett.*, 1992, **19**, 159.
8. A. J. Krueger, *Science*, 1983, **220**, 1377.
9. J. Fishman, C. E. Watson, J. C. Larsen and J. A. Logan, *J. Geophys. Res.*, 1990, **95**, 3599.
10. J. R. Herman, P. K. Bhartia, O. Torres, C. Hsu, C. Seftor and C. Celarier, *J. Geophys. Res.*, 1997, **102**, 16911.
11. J. T. Kerr and M. Ostrovsky, *Trends Ecol. Evol.*, 2003, **18**, 299.
12. A. M. Melesse, Q. Weng, P. S. Thenkabail and G. B. Senay, *Sensors*, 2007, **7**, 3209.
13. G. I. Metternicht and J. A. Zinck, *Remote Sens. Environ.*, 2003, **85**, 1.
14. M. W. Becker, *Ground Water*, 2006, **44**, 306.

15. E. Ben-Dor, *Advan. Agron.*, 2002, **75**, 173.
16. A. Hall, D. W. Lamb, B. Holzapfel and J. Louis, *Aus. J. Grape Wine Res.*, 2002, **8**, 36.
17. T. G. van Niel and T. R. McVicar, *Aus. J. Agr. Res.*, 2003, **85**, 1.
18. H. E. Nilsson, *Annu. Rev. Phytopathol.*, 1995, **15**, 489.
19. J. Namiesnik, *Crit. Rev. Anal. Chem.*, 2000, **30**, 221.
20. T. Mills, P. Moekstra, M. Blohm and L. Evans, *Ground Water*, 1988, **26**, 771.
21. J. Wang, *TrAC-Trends Anal. Chem.*, 1997, **16**, 84.
22. A. Szczurek, *Environ. Prot. Eng.*, 1996, **22**, 29.
23. P. Weibring, H. Edner, S. Svanberg, G. Cecchi, L. Pantani, R. Ferrara and T. Caltabiano, *Appl. Phys.*, 1998, **B67**, 1.
24. V. Panne, *TrAC, Trends Anal. Chem.*, 1998, **17**, 491.
25. M. S. Karmat, R. A. Lodder and P. P. Deluca, *Pharm. Res.*, 1989, **6**, 961.
26. J. Moros, N. Galipienso, R. Vilches, S. Garrigues and M. de la Guardia, *Anal. Chem.*, 2008, **80**, 7257–7265.
27. H. R. H. Ali, H. G. M. Edwards and I. J. Scowen, *Spectrochim. Acta Pt. A Mol. Biol.*, 2009, **72**, 890.
28. J. Workman, M. Koch and D. J. Veltkamp, *Anal. Chem.*, 2003, **75**, 2859.
29. R. De Maesschalck and T. Van den Kerkhof, *J. Pharmaceut. Biomed. Anal.*, 2005, **37**, 109.
30. N. W. Broad, R. D. Jee, A. C. Moffat, M. J. Eaves, W. C. Mann and W. Dziki, *Analyst*, 2000, **125**, 2054.
31. A. Nordon, A. Mills, R. T. Burn, F. M. Cusik and D. Littlejohn, *Anal. Chim. Acta*, 2005, **548**, 148.
32. S. Garrigues and M. de la Guardia, in *Beer in Health and Disease Prevention*, ed. V. R. Preedy, Elsevier/Academic Press, Oxford, 2009, pp. 943–961.
33. M. Kim, H. Chung, Y. Woo and M. Kemper, *Anal. Chim. Acta*, 2007, **587**, 200.
34. H. Schmidt, K. Sowoidnich and H. D. Kronfeldt, *Appl. Spectrosc.*, 2010, **64**, 888.
35. E. Marguí, M. Hidalgo and I. Queralt, *Spectrochim. Acta, Part B*, 2005, **60**, 1363.
36. P. J. J. Potts and M. West (ed), *Portable X-ray Fluorescence Spectrometers: Capabilities for In Situ Analysis*, Royal Society of Chemistry, Cambridge, 2008.
37. I. W. Burton, M. A. Quilliam and J. A. Walter, *Anal. Chem.*, 2005, **77**, 3123.
38. U. Holzgrabe, R. Deubner, C. Schollmayer and B. Waibel, *J. Pharmaceut. Biomed. Anal.*, 2005, **38**, 806.
39. L. I. Nord, P. Vaag and J. O. Duus, *Anal. Chem.*, 2004, **76**, 4790.
40. E. Veliyulin, C. van der Zwaag, W. Burk and U. Erikson, *J. Sci. Food Agric.*, 2005, **85**, 1299.
41. E. Del Federico, S. A. Centeno, C. Kehlet, P. Currier, D. Stockman and A. Jerschow, *Anal. Bioanal. Chem.*, 2010, **396**, 213.
42. J. Moros, S. Garrigues and M. de la Guardia, *TrAC, Trends Anal. Chem.*, 2010, **29**, 578.

43. J. Fitzpatrick and J. A. Reffner, in *Handbook of Vibrational Spectroscopy*, (ed) J. M. Chalmers and P. R. Griffiths, John Wiley & Sons, Chichester, 2002.
44. L. G. Thygesen, M. M. Lokke, E. Micklander and S. B. Engelsen, *Trends Food Sci. Technol.*, 2003, **14**, 50.
45. E. C. Y. Li-Chan, P. R. Griffith and J. M. Chalmers (eds), *Applications of Vibrational Spectroscopy in Food Science*, John Wiley & Sons, Chichester, 2010.
46. M. M. Mossoba, V. Milosevic, M. Milosevic, J. K. Kramer and H. Azizian, *Anal. Bioanal. Chem.*, 2007, **389**, 87.
47. N. Viereck, T. Salomonsen, F. Van den Berg and S. B. Engelsen, in: *Raman Spectroscopy for Soft Matter Applications*, M.S. Amer (ed), John Wiley & Sons, Chichester, 2009.
48. T. Van Keirsbilck, A. Vercauteren, W. Baeyens, G. Van der Weken, F. Verpoot, G. Vergote and J. P. Remon, *TrAC, Trends Anal. Chem.*, 2002, **21**, 869.
49. M. T. Bona and J. M. Andrés, *Anal. Chim. Acta*, 2008, **624**, 68.
50. T. Woodcock, G. Downey and C. P. O'Donnell, *J. Near Infrared Spectrosc.*, 2008, **16**, 1.
51. Y. Roggo, P. Chalus, L. Maurer, C. Lema-Martínez, A. Edmond and N. Jent, *J. Pharm. Biomed. Anal.*, 2007, **44**, 683.
52. H. Chung, *Appl. Spectrosc. Rev.*, 2007, **42**, 251.
53. P. Vandenabeele, K. Castro, H. Hargreaves, L. Moens, J. M. Madariaga and H. G. M. Edward, *Anal. Chim. Acta*, 2007, **588**, 108.
54. K. Fernández and E. Agosín, *J. Agric. Food Chem.*, 2007, **55**, 7294.
55. N. Abded-Nour, M. Ngadi, S. Prasher and Y. Karimi, *Int. J. Poultry Sci.*, 2009, **8**, 170.
56. G. Siebelec, G. W. McCarty, T. I. Stuczynski and J. B. Reeves III, *J. Environ. Qual.*, 2004, **33**, 2056.
57. J. Moros, M. C. Barciela-Alonso, P. Pazos-Capeans, P. Bermejo-Barrera, E. Peña-Vázquez, S. Garrigues and M. de la Guardia, *Anal. Chim. Acta*, 2008, **624**, 113.
58. D. Cozzolino and A. Morón, *Anim. Feed Sci. Tech.*, 2004, **111**, 161.
59. D. Cozzolino, M. J. Kwiatkwski, R. G. Daunbergs, W. U. Cynkar, L. J. Janik, G. Skouroumounis and M. Gishen, *Talanta*, 2008, **74**, 711.
60. J. L. Halgerso, C. C. Sheaffer, N. P. Martin, P. R. Peterson and S. J. Weston, *Agron. J.*, 2004, **96**, 344.
61. G. Hosafci, O. Klein and G. Oremek, *Anal. Bioanal. Chem.*, 2007, **387**, 1815–1822.
62. M. Harz, P. Rösch and J. Popp, *Cytometry Part A*, 2009, **75**, 104.
63. C. P. O'Donnell, P. J. Cullen, G. Downell and J. M. Frias, *Trends Food Sci. Technol.*, 2007, **18**, 590.
64. D. W. Sun (ed), *Hyperspectral Imaging for Food Quality and Control*, Academic Press, New York, 2010.
65. M. Resano, F. Vanhaecke and M. T. C. de Loos-Vollebregt, *J. Anal. At. Spectrom.*, 2008, **23**, 1450.

66. M. G. R. Vale, N. Oleszczuk and W. N. L. dos Santos, *Appl. Spectrosc. Rev.*, 2006, **41**, 377.
67. M. J. Cal-Prieto, M. Felipe-Sotelo, A. Carlosena, J. M. Andrade, P. Lopez-Mahia, S. Muniategui and D. Prada, *Talanta*, 2002, **56**, 1.
68. M. Hornung and V. Krivan, *Spectrochim. Acta, Part B*, 1999, **54**, 1177.
69. M. A. Belarra, M. Resano and J. R. Castillo, *J. Anal. At. Spectrom.*, 1999, **14**, 547.
70. C. S. Nomura, C. S. Silva, A. R. A. Nogueira and P. V. Oliveira, *Spectrochim. Acta, Part B*, 2005, **60**, 673.
71. K. C. Friese and V. Krivan, *Spectrochim. Acta, Part B*, 1998, **53**, 1069.
72. M. A. Belarra, M. Resano, F. Vanhaecke and L. Monees, *TrAC, Trends Anal. Chem.*, 2002, **21**, 828.
73. Z. Zhou, K. Zhou and X. Hou, *Appl. Spectrosc. Rev.*, 2005, **40**, 165.
74. V. B. E. Thomsen, *Modern Spectrochemical Analysis of Metals: An Introduction for Users of Arc-Spark Instrumentation*, ASTM International, Materials Park, OH, 1996.
75. D. Günther, I. Horn and B. Hattendorf, *Fres. J. Anal. Chem.*, 2000, **368**, 4.
76. N. S. Mokgalaka and J. L. Gardea-Torresdey, *Appl. Spectrosc. Rev.*, 2006, **41**, 131.
77. A. Kehden, J. Flock, W. Vogel and J. A. C. Broekaert, *Appl. Spectrosc.*, 2001, **55**, 1291.
78. F. C. Alvira, F. R. Rozzi and G. M. Bilmes, *Appl. Spectrosc.*, 2010, **64**, 313.
79. N. N. Gavrilyukov, V. N. Samoplyas and V. V. Mandrygin, *Inorg. Mater.*, 2008, **44**, 1547.
80. R. E. Russo, X. Mao, H. L. Liu, J. Gonzalez and S. S. Mao, *Talanta*, 2002, **57**, 425.
81. R. S. Harman, F. C. DeLucia, C. E. McManus, N. J. McMillan, T. F. Jenkins, M. E. Walsh and A. Miziolek, *Appl. Geochem.*, 2006, **21**, 730.
82. B. Salle, P. Mauchien and S. Maurice, *Spectrochim. Acta, Part B*, 2007, **62**, 739.
83. J. M. Vadillo and J. J. Laserna, *Spectrochim. Acta, Part B*, 2004, **59**, 147.
84. C. Fabre, M. C. Boiron, J. Dubessy, A. Chabiron, B. Charov and T. M. Crespo, *Geochim. Cosmochim. Acta*, 2002, **66**, 1401.
85. J. Moros, J. A. Lorenzo, P. Lucena, L. Miguel Torabia and J. J. Laserna, *Anal. Chem.*, 2010, **82**, 1389.
86. T. Nelis and J. Pallosi, *Appl. Spectrosc. Rev.*, 2006, **41**, 227.
87. N. Jakubowski, R. Dorka, E. Steers and A. Tempez, *J. Anal. At. Spectrom.*, 2007, **22**, 722.
88. V. Hoffmann, M. Kasik, P. K. Robinson and C. Venzago, *Anal. Bioanal. Chem.*, 2005, **381**, 173.
89. R. Payling, M. Aeberhard and D. Delfosse, *J. Anal. At. Spectrom.*, 2001, **16**, 50.
90. Ph. Belenguer, M. Ganciu, Ph. Guillot and Th. Nelis, *Spectrochim. Acta, Part B*, 2009, **64**, 623.

91. A. C. Muñiz, J. Pisonero, L. Lobo, C. Gonzalez, N. Bordel, R. Pereiro, A. Tempez, P. Chapon, N. Tuccitto, A. Licciardello and A. Sanz-Medel, *J. Anal. At. Spectrom.*, 2008, **23**, 1239.
92. M. Hohl, A. Kanzari, J. Michler, T. Nelis, K. Fuhrer and M. Gonin, *Surf. Interface Anal.*, 2006, **38**, 292.
93. N. Tuccitto, L. Lobo, A. Tempez, I. Delfanti, S. Canolescu, N. Bordel, P. Chapon, J. Michler and A. Licciardello, *Rap. Commun. Mass Spectrom.*, 2009, **23**, 549.
94. F. J. Andrade, J. T. Shelley, W. C. Wetzel, M. R. Webb, G. Gamez, S. J. Ray and G. M. Hieftje, *Anal. Chem.*, 2008, **80**, 2654.
95. Z. Takats, J. M. Wiseman, B. Gologan and R. G. Cooks, *Science*, 2004, **306**, 471.
96. J. F. García-Reyes, A. U. Jackson, A. Molina-Díaz and R. G. Cooks, *Anal. Chem.*, 2009, **81**, 820.
97. Z. Lin, M. Zhao, S. Zhang, C. Yang and X. Zhang, *Analyst*, 2010, **135**, 1268.

CHAPTER 3
Replacement of Hazardous Solvents and Reagents in Analytical Chemistry

JENNIFER L. YOUNG[1] AND DOUGLAS E. RAYNIE[2]

[1] ACS Green Chemistry Institute®, American Chemical Society, 1155 Sixteenth Street, NW, Washington, DC 20036, USA; [2] Department of Chemistry and Biochemistry, South Dakota State University, Brookings, SD 57007, USA

This chapter touches on two components of analytical chemistry that can have a significant impact on the greenness of analysis, but that are often overlooked: solvents and reagents. Both are integral parts of sample preparation and analysis. In the 12 principles of green chemistry outlined by Anastas and Warner,[1] prevention of waste, atom economy, safer solvents and reagents, energy efficiency, renewability, reducing derivatives, real-time analysis, and inherently safer chemistry for accident prevention can be linked to the solvents and reagents chosen for the analytical technique. Green analytical chemistry is influenced by these principles,[2,3] though all of them may not apply in every situation. The goal is to strive for improved greenness and continual improvement.

Several approaches have been attempted to bring a level of greenness to the field of analytical chemistry. One review of environmentally friendly extractions described reduced-solvent approaches like supercritical fluid extraction, pressurized liquid extraction, microwave-assisted extraction, solid-phase microextraction, and more.[4] Solid-phase microextraction and similar approaches do not use a traditional extraction solvent. Rather, a stationary phase is used to adsorb the analyte from the sample solution. These sorbent-based methods are not discussed in this chapter and life-cycle analysis or other green assessments have not been conducted for these methods. A trend toward solvent minimization, beyond the micro-scale approach, has involved the modification of existing techniques or the creation of new analytical approaches. For example, in 2009, the journal *Spectroscopy Letters* published a series of articles enumerating this approach, such as techniques like flow-through solid-phase spectroscopy;[5] direct analysis;[6] tungsten-coil atomic spectroscopy, long pathlength spectrophotometry, flow-based methodology, and surfactant-mediated extractions;[7] and miniaturization, reagent replacement, on-line analysis, and spectroscopy.[8] Of course, complete elimination of any solvent would be the ideal situation and as a result the use of reflectance spectroscopy is becoming increasingly popular. However, many analytical methods still rely on the use of solvents and reagents, and reducing their impact on human health and the environment rather than the specific analytical techniques are the focus of this chapter.

3.1 Green Solvents and Reagents: What This Means

A few research studies have qualitatively and quantitatively measured the greenness of analytical techniques. In one approach, the greenness of analytical methods was considered with respect to four criteria: the use of persistent, bioaccumulative, and toxic (PBT) chemicals, use of hazardous chemicals, corrosiveness based on pH during the analysis, and amount of waste generated from the analysis.[3] The study compared the greenness of over 500 environmental testing methods in the National Environmental Methods Index (NEMI), an on-line database, based on a scale of pass/fail for each specifically defined criterion. More specifically, a method is considered 'less green' according to the following criteria:[3]

- *PBT:* a chemical used in the method is listed as a PBT as defined by the Toxic Release Inventory (TRI) of the U.S. Environmental Protection Agency (EPA)
- *Hazardous:* a chemical used in the method is listed on the TRI or one of the U.S. Resource Conservation and Recovery Act (RCRA)'s D, F, P, or U hazardous waste lists
- *Corrosive:* the pH during the analysis is <2 or >12
- *Waste:* the amount of waste generated is >50 g.

Among the findings,[3] two-thirds of the methods in NEMI failed the waste criterion, which means the methods generated more than 50 g of waste. In many of those instances, the large quantities of waste were a result of solvent extractions to isolate the analyte prior to analysis. It should be noted that the study did not take into account the amount of solvent used as a mobile phase, such as in HPLC (high-performance liquid chromatography), which would further add to the amount of waste generated in analyses that use carrier fluids or mobile phases. Also adding to waste generation are the strong mineral acids used for preservation or digestion of samples.

Half of the methods in NEMI failed the hazardous criterion, primarily because of the solvents and reagents used in the method.[3] Again, those solvents were mainly used for extractions and the reagents for derivatization, digestion, or preservation. A small number of methods in NEMI (5%) failed with respect to the use of PBTs in the analysis, primarily from the use of mercury or lead in some part of the sample preparation. Overall, the greener methods in NEMI use smaller quantities of less hazardous solvents and reagents.

Although NEMI is a limited set of methods for analysing environmental samples, this general methodology and the related knowledge can be applied to all types of analytical methods as a way to compare the relative greenness of methods. For example, the same criteria were applied to all of the laboratory experiments in a popular college-level analytical chemistry textbook.[9]

In another assessment of the greenness of analytical methods, the categories were expanded to include five categories: health, safety, environmental, energy, and waste.[10] The criteria are based on toxicity, bioaccumulation, reactivity, waste generation, corrosivity, safety, energy consumption, and related factors. Furthermore, each category receives a score on a scale of 1–3 using readily available chemical data. By expanding the categories and scoring scale, the comparison of two analytical methods can provide more information and further distinction between methods.

These are just two examples of evaluating metrics on analytical techniques that provide some measure of greenness. Clearly, it is difficult to quantitatively define how green an analytical method is, let alone the greenness of the solvents and reagents. In the remainder of this chapter, specific types of solvents and reagents will be examined further.

3.2 Greener Solvents

Solvents are vital to the analytical process. In nearly all analytical methods, analytes must be taken into solution for separation, volumetrically diluted to the appropriate analytical concentration, and analysed. When organic solvents are used, issues surrounding toxicity and flammability are often of concern; for aqueous systems, wastewater generation becomes a potential problem. In all cases, the energy of evaporation of nonvolatile solvents adds to the green concerns. The largest use of solvents in analytical chemistry is in separation techniques, including extraction and liquid chromatography mobile phases. This solvent use results in the vast majority of the waste generated in an analytical procedure and this waste may be toxic, flammable, or possess other deleterious properties.

Solvents are probably the most active area of green chemistry research[11] and several references address the topic,[12–14] usually from a focus on synthetic chemistry. Directions in green solvents have been identified as (1) substitution of hazardous solvents with those that show better environmental, health, and safety properties; (2) bio-derived solvents; (3) supercritical fluids; and (4) ionic liquids (ILs).[15] On the other hand, Jessop informally surveyed key researchers in the field, asking 'If the adoption of greener solvents over the next 20–30 years will reduce environmental damage from human activities, then the adoption of what class of solvents will be responsible for the greatest reduction in environmental damage?'[16] These research leaders answered: supercritical carbon dioxide (30.2%), water (22.9%), organic solvents (18.8%), ILs (12.5%), switchable solvents (6.3%), glycerol (4.2%), solventless (3.1%), and bio-derived (2.3%) solvents. This contrasts with actual publication numbers. Jessop reported that during the first nine months of 2010, research articles in *Green Chemistry* focused on ILs (41.0%), water (28.4%), solventless methods (11.5%), carbon dioxide (7.1%), glycerol and ethers (3.3%), and all others (including alcohols, switchable solvents, liquid polymers, fluorous solvents, and methyl tetrahydrofuran) (8.7%). However, these solvents were primarily used as reaction media in syntheses, catalyst recovery following synthesis, and biomass processing, rather than for analytical chemistry.[16] Jessop set out the challenges that researchers should ensure that green solvents are available as replacements for any type of non-green solvent, requiring detailed characterization such as using the Kamlet–Taft solvatochromic parameters; recognizing green solvents through energetic, environmental impact, or life-cycle assessment approaches; developing an easy to remove polar aprotic solvent; and eliminating the need for distillation.[16] From an analytical perspective, the 'easy to distil' criteria can be changed to 'easy to concentrate for analysis', generally by evaporation or solvent exchange.

When selecting a solvent for an analytical procedure, criteria such as solute solubility, viscosity, and compatibility with the analytical method are still the major factors. However, cost and 'greenness' must also be of concern. As previously mentioned, we will reach a fully developed green chemistry awareness when chemists have more instinctive knowledge of green concerns, similar to their instinctive knowledge of solute solubility. This recognition of what is a green solvent is somewhat developed for traditional molecular solvents, but lacking for alternative solvents like supercritical fluids and ILs. Additionally, the search for green solvent alternatives in a given analytical procedure can take the approach of finding a green solvent replacement or finding an approach which will make the overall analytical method greener. In the context of green solvent research, bio-derived solvents have made minor impact in analytical chemistry, as have switchable solvents. Each of these solvents shows intriguing potential that may come into play in the future. In this section, we specifically explore supercritical fluids, ILs, water, and green organic solvents.

3.2.1 Supercritical Fluids

It is becoming widely known that substances at temperatures and pressures near or above the critical point—supercritical fluids—possess solvent properties favourable for analytical purposes. As a general rule, these fluids have liquid-like solvating power and gas-like diffusivity. Furthermore, these properties may be varied as a function of temperature and pressure. Perhaps the most important property of supercritical fluids for separation processes is diffusion; after all, regardless of green properties, the solvent must perform. The well-developed rate theory of chromatography relates chromatographic efficiency, and the hot-ball model of extraction[17] shows that solubility and diffusion are significant in obtaining quantitative extraction yield.

The most widely used fluid is carbon dioxide, with critical parameters of 31.1 °C and 73 atm (7.39 MPa). This fluid has been used broadly in chromatography (SFC) and extraction (SFE) especially since the 1980s. Carbon dioxide is nontoxic, nonflammable, readily available, and inexpensive. As a supercritical fluid, it behaves as a nonpolar, or polarizable, solvent, and low molar mass alcohols (co-solvents) are often added in small amounts to alter the solvent polarity. Because carbon dioxide can be depressurized to the gaseous state, the solvent is easily removed and supercritical fluid-based separation methods are easily coupled with subsequent analysis. Supercritical carbon dioxide is routinely used for analyses of lipids, essential oils, and flavour and fragrance compounds in foods and natural products; polycyclic aromatic hydrocarbons, polychlorinated biphenyls, pesticides, and related compounds from environmental samples; polymer additives; and less commonly, ions and metals. A recent revival of SFC activity is spurred by the pharmaceutical industry, looking at the technique for chiral separations, as a replacement for normal-phase liquid chromatography, and in process scale-up. Key applications and the current state of the art can be found in recent publications.[18,19]

Other fluids than carbon dioxide have been reported in the literature, but these solvents, usually organic, do not possess any distinct advantages, may be flammable, and often have high critical temperatures. Inorganics like ammonia have neurotoxicity and nitrous oxide is a strong oxidizing agent.[20] Water has a very high critical temperature, but has unique properties in the subcritical region and will be summarized separately.

3.2.2 Ionic Liquids

ILs are showing increased use in analytical chemistry in the past decade. These liquids are generally composed of large cations and smaller anions, such that the coordination (electrostatic attraction) between the ions is somewhat weak. Hence, they are commonly defined as liquids at less than 100 °C. ILs have low volatility, somewhat tunable viscosity and miscibility, and electrolyte conductivity. Their interest as green solvents primarily stems from the low volatility, as volatile organic compounds (VOCs) are generally not emitted in ionic liquid processes. However, green is a relative term and several ILs are also toxic. Several reviews describe the use of ionic liquids in analytical chemistry.[21–27] Another review cites the most important recent advances as unique multi-functional ionic liquids and calls for increased understanding of their chemical and physical properties as ILs.[28] One example of the development of task-specific hydrophobic ILs is for the isolation of Cd(II) in water and foods with the work-up coupled directly with flame atomic absorption.[29]

One limitation to the use of ILs is their viscosity. Thus, a suitable application of ILs in separations is as the acceptor phase in supported-liquid membrane extraction.[30] Similarly, these liquids can be used as stationary phases in chromatography or sorptive extractions. For instance, phosphonium-based liquids were immobilized on solid supports for preparation of a novel solid-phase system.[31] A non-separation analytical technique taking advantage of the viscosity of ILs is matrix-assisted laser-desorption ionization (MALDI) mass spectrometry, where ionic liquids can replace glycerol or other sample matrices.[32]

While investigating the role of ILs in analyte solvation, the lyotropic theory was employed to consider all ions in solution and describe ion-pairing, ion-exchange, and hydrophobic interactions.[33] Similarly, pH and the presence of salts impact solvating ability and extraction with imidazolium-based ionic liquids.[34]

3.2.3 Water

Liquid water has several desirable (and green) solvent properties. It is nontoxic, safe, inexpensive, pure, and readily available. Water can dissolve a host of polar and ionic materials and is noted as the universal solvent. On the other hand, liquid water has two decidedly non-green concerns—clean-up of the generated wastewater and the energy necessary for solvent removal. Supercritical water is an oxidizing agent that is rarely used analytically because of its extreme critical parameters—374 °C and 218 atm (22 MPa).

A unique, green use of water is in hot-water extraction, also called near-critical extraction, subcritical water extraction, or pressurized hot-water extraction. Water under these conditions is also being used as a chromatographic mobile phase. Actually, there are two modes of water-based separation processes. Under mild heating (perhaps with the application of pressure), solvating ability increases, kinetics are more favourable, diffusion is faster, and viscosity is lessened. Thus in a system such as accelerated solvent extraction (also known as pressurized solvent extraction), hot water extracts the same types of analyte as traditional aqueous extractions, but much faster and with up to 95% less solvent use. However, when heated to above its atmospheric boiling point, but under enough pressure to keep it as a liquid, water begins to lose its effective polarity. For example, at room temperature, water has a dielectric constant of 80; at 250 °C and 50 bar (5 MPa), the dielectric constant drops to 27. This relationship between temperature and the dielectric constant of water is shown in Figure 3.1.[35] Under practical conditions, temperatures employed are generally in the 100–200 °C range, with higher temperatures reserved for the most nonpolar analytes. The hot-water extraction approach can be used for aliphatic and aromatic hydrocarbons and other solutes of environmental interest. As temperature increases, and effective polarity decreases, previously

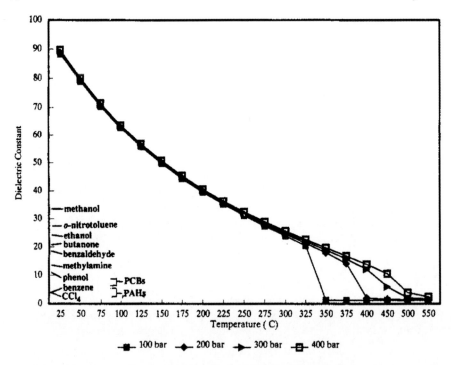

Figure 3.1 Effect of temperature on the dielectric constant of water as a function of temperature, with comparison to selected organic compounds. (Reproduced from Ref. 35, with kind permission of the American Chemical Society)

water-insoluble compound classes can be extracted with water. This approach was first used in the mid 1990s[35] but has been slow to catch on because of lack of commercial instrumentation and the high temperatures used.

Researchers have shown that temperature is the most important consideration in these extractions and that time and flow rate are most important when the extraction is solubility limited.[36] This verifies the thermodynamic and kinetic models of hot-water extraction previously developed.[37] As a consequence, and verified by on-line monitoring of the extraction process, a static extraction step prior to dynamic extraction is important to shorten overall extraction time.[38] However, one concern with performing extractions at these high temperatures is analyte decomposition. This has been studied during the isolation of anthocyanins from red onion at 110 °C.[39] Both extraction and degradation rates were determined and used, for example, to explain the overall extraction rate curve shown in Figure 3.2. Here, the extraction follows the previously described thermodynamic hot-ball model (upper curves) while analyte degradation occurs simultaneously (lower curve). The middle curve represents the actual, observed extraction rate. However, since these extractions occur in a closed (air-free) system, if the water or other solvent is degassed, analyte decomposition is less of a concern.

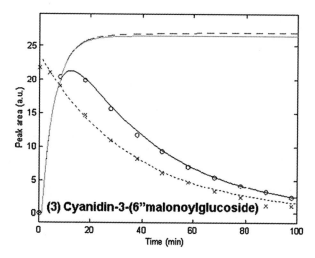

Figure 3.2 Extraction rates for the isolation of cyanidin-3-(6′-malonoylglucoside) from red onion using water at 110 °C. The upper trace represents the theoretical rate of extraction, while the lower curve represents the measured analyte degradation rate. The resulting curve (middle trace) is the actual observed extraction rate. (Reproduced from Ref. 39, with kind permission of Elsevier)

3.2.4 Green Organic Solvents

For consideration of green organic solvents, we can break them down into two classes: 'traditional' organic solvents and non-traditional, 'newer' organic solvents. To a greater extent, the green attributes of traditional solvents have been compared and contrasted in research studies. The newer solvents have been more recently employed to address some specific aspect of green chemistry, such as renewability and inherent safety. Though many references to greener traditional and newer solvents have been made in other contexts, particularly in synthesis, the same conclusions can be carried over to the use of these solvents in analytical chemistry.

One comprehensive investigation of 26 solvents used an environmental, health and safety (EHS) approach combined with life-cycle assessment.[15] The EHS approach examined nine categories: release potential, fire/explosion, reaction/decomposition, acute toxicity, irritation, chronic toxicity, persistence, and air and water hazards. The results from the EHS approach are summarized in Figure 3.3. The combined EHS and life-cycle results were evaluated by Pareto analysis, with the EHS-preferred conventional organic solvents being methanol, ethanol, and methyl acetate, whereas the life-cycle-preferred solvents were hexane, heptane, and diethyl ether. However, it should be noted that hexane presents a significantly greater toxicity concern than other aliphatic hydrocarbons. Solvent mixtures have not been extensively examined, and mixtures, such as hexane–acetone, can have an even greater toxicity than the individual solvent components. Non-recommended solvents (from an environmental perspective) were dioxane, acetonitrile, acids, tetrahydrofuran, and

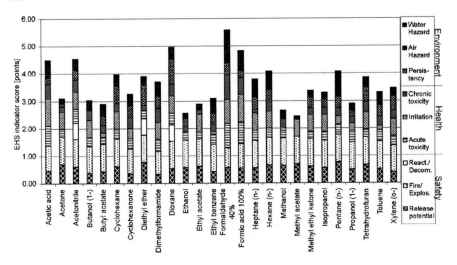

Figure 3.3 Environmental, health and safety assessment of 26 common organic solvents. The lower the indicator score, the more favourable the green assessment of the solvent. (Reproduced from Ref. 15, with kind permission of The Royal Society of Chemistry)

formaldehyde; other solvents with high environmental impact included pentane, cyclohexane, dimethylformamide, and cyclohexanone. The researchers also pointed out that ILs and supercritical carbon dioxide were not included in the study because of lack of data.

A similar study shows the cumulative energy demand and environmental effects.[40] This is shown in Table 3.1. The study found hexane and cyclohexane preferred from an energy perspective and ethyl acetate preferred from an environmental perspective.

Another study applied life-cycle assessment to solvent selection on 47 solvents.[41] The research built on a previously reported solvent selection guide (relative ranking based on environmental, health, and safety issues) to include cradle-to-grave life-cycle impacts. Among the higher scoring (greener) solvents for life cycle were ethylene glycol, ethanol, t-butanol, methanol, dimethyl carbonate, formamide, acetic acid, methyl tert-butyl ether, and diisopropyl ether. However, it should be noted that with the exception of ethylene glycol, all other of these solvents had at least one poor score in another category (environmental waste, environmental impact, health, or safety). For solvents with a low (poor) life cycle score, the authors indicate that solvent recycling or recovery and minimization of use become even more important if the solvent is to be utilized.

A fourth example of a solvent selection guide provides several simplified tables of preferred and replacement solvents.[42] Although the lists of solvents appear simple, they were generated from thorough and systematic evaluation of the solvents regarding worker safety, process safety, and environmental and regulatory considerations. The solvent selection guide (Table 3.2) lists solvents

Table 3.1 Cumulative energy demand and environmental effects of selected solvents.

Solvent	Water	Acetonitrile	Dichloro-methane	Ethyl Acetate	Benzene	Toluene	o-Xylene	n-Hexane	Cyclo-hexane
Cumulative Energy Demand (CED)									
CED for solvent supply (MJ/L)	0.01	50.6	40.3	43.6	53.0	61.0	60.7	43.2	47.6
CED for heating (MJ/L)[1]	9.0	8.7	unstable	7.9	8.5	5.4	5.7	unstable	8.5
CED for work-up (MJ/L)	14.0	3.9	2.7	2.5	2.5	2.5	2.6	1.7	2.2
Thermal disposal credit (MJ/L)	0	−18.0	−7.0	−17.1	−27.9	−27.7	−33.2	−23.7	−27.3
Environmental Effects									
Acute toxicity for humans	low	medium	low-medium	low-medium	medium	low-medium	medium	medium	medium
Chronic toxicity for humans	low	low	medium	low	high	low-medium	low	medium	low
Acute toxicity for aquatic organisms	low	medium	low	low	high	medium	medium	medium	medium
Persistency in the environment	low	low-medium	low-medium	low	low-medium	low-medium	low-medium	low-medium	medium
Bioaccumulation	low	low	low-medium	low	low-medium	low-medium	medium	medium	medium

[1] T = 80 deg. C, t = 5.75 h.
(Adapted from Ref. 40, with kind permission of John Wiley & Sons, Inc.)

Table 3.2 Pfizer solvent selection guide for medicinal chemistry.

Preferred	Usable	Undesirable
Water	Cyclohexane	Pentane
Acetone	Heptane	Hexane(s)
Ethanol	Toluene	Diisopropyl ether
2-Propanol	Methylcyclohexane	Diethyl ether
1-Propanol	Methyl t-butyl ether	Dichloromethane
Ethyl acetate	Isooctane	Dichloroethane
Isopropyl acetate	Acetonitrile	Chloroform
Methanol	2-methyltetrahydrofuran	Dimethyl formamide
Methyl ethyl ketone	Tetrahydrofuran	N-Methylpyrrolidinone
1-Butanol	Xylenes	Pyridine
t-Butanol	Dimethyl sulfoxide	Dimethyl acetate
	Acetic acid	Dioxane
	Ethylene glycol	Dimethoxyethane
		Benzene
		Carbon tetrachloride

Reproduced from Ref. 42, with kind permission of The Royal Society of Chemistry.

Table 3.3 Solvent replacement table.

Undesirable solvents	Alternative
Pentane	Heptane
Hexane(s)	Heptane
Diisopropyl ether or diethyl ether	2-MeTHF or tert-butyl methyl ether
Dioxane or dimethoxyethane	2-MeTHF or tert-butyl methyl ether
Chloroform, dichloroethane or carbon tetrachloride	Dichloromethane
Dimethyl formamide, dimethyl acetamide or N-methylpyrrolidinone	Acetonitrile
Pyridine	Et$_3$N (if pyridine used as base)
Dichloromethane (extractions)	EtOAc, MTBE, toluene, 2-MeTHF
Dichloromethane (chromatography)	EtOAc/heptane
Benzene	Toluene

Reproduced from Ref. 42, with kind permission of The Royal Society of Chemistry.

as preferred, usable, and undesirable. This guide was developed for chemists in the pharmaceutical industry but can easily be applied to analytical methodologies since many of the same solvents are utilized.

Taking the guide another step forward, the authors offer a table of replacement solvents for the undesirable solvents (Table 3.3). The table recommends using heptane instead of pentane or hexane; 2-methyl tetrahydrofuran instead of diisopropyl ether and diethyl ether; and ethyl acetate instead of dichloromethane, among other replacements.[42] The authors point out that the recommendation of dichloromethane over other chlorinated solvents is only if a chlorinated solvent must be used.

The purpose of showing these various solvent selection guides is not meant to be confusing, but to illustrate the complexity of the issue as well as the common themes between the guides. In three of the studies presented above,[15,41,42] ethanol is among the preferred solvents. Ethanol/water mixtures are becoming more common as replacements for other solvents in analytical chemistry, particularly in HPLC. Ethanol is considered a greener alternative to methanol and acetonitrile, which are common solvents for HPLC analyses. Some examples are the separation of sunscreens and pesticides.[43] A systematic study of two test mixtures, a series of alkylbenzenes and a mixture of compounds of different functional group classes including caffeine and *p*-hydroxybenzoic acid, demonstrated the ability of ethanol to replace acetonitrile and methanol in HPLC despite the higher viscosity of ethanol.[44] Method condition modifications included the solvent gradient, UV detection wavelength, flow rate, and temperature. Another HPLC example utilized butyl alcohol and demonstrated comparable results to classical methods using methanol and acetonitrile.[45] In this study, several vitamins (A, E, D_3, and K_1) in food and pharmaceutical supplements were separated using a C-18 column modified with sodium dodecyl sulphate (SDS) surfactant and eluted with a solution of the surfactant in butyl alcohol.[45]

Non-traditional, 'newer' organic solvents are being studied, in part, because of the greenness that they can offer. While methanol and ethanol can be derived from renewable sources and already find uses in analytical chemistry, other solvents from renewable sources may have applications as alternative solvents for separations and analyses. Other organic solvents may be favoured because they have low volatility and/or are less hazardous.

Ethyl lactate is a renewable solvent that has low toxicity, is biodegradable and is finding industrial uses in the cleaning, pharmaceutical, and paint industries.[46] Glycerol, a waste product of biodiesel production, is nonvolatile, nontoxic, nonflammable and biodegradable and has been studied as a solvent[47] along with its derivatives.[48] A natural product, D-limonene, has been studied as a hexane replacement for the extraction of lipids (fats and oils) from food using Soxhlet extraction and Clevenger distillation.[49] Another renewable solvent, 2-methyl tetrahydrofuran (2-methyl THF) has been demonstrated as a replacement for THF and other solvents,[50] though peroxide formation in non-stabilized 2-methyl THF has been demonstrated.[51]

Some other types of solvents are not renewable but have other advantages. Cyclopentyl methyl ether (CPME) has been studied as a replacement for other ethereal solvents and has lower peroxide formation than 2-methyl THF and THF.[52] Polyethylene glycols with a range of molecular weights and properties have low toxicity, low volatility, and are biodegradable.[53]

Another new class of solvents worth mentioning is switchable solvents.[54,55] For green analytical chemistry, the advantage of switchable solvents is in separations. These solvents are designed to react to a trigger, *e.g.* exposure to carbon dioxide or nitrogen, by switching a property such as polarity or hydrophilicity, thus causing a separation. Furthermore, the switchable solvent can be restored to its original composition and recycled in the process. Some other classes of solvents that have been applied in green chemistry, not included in this chapter, are fluorous solvents and deep eutectic solvents.

3.3 Greener Reagents

Reagents are used to a lesser extent in analytical chemistry. Three instances where reagents are used are for chelation, derivatization, and preservation. Chelating agents enable the analysis of ions in solution. They are high molecular weight auxiliaries that bind with the metal ions and ultimately become part of the waste stream. The most common, ethylenediamine tetraacetic acid (EDTA), is hazardous, cumulative, and persistent. Reagents are also used to make derivatives of the analyte to enable characterization and these auxiliaries also ultimately become part of the waste stream. With the advances in analytical techniques, derivatization is not as common as it once was, but nonetheless is still a reagent usage in analytical chemistry to consider in relation to the greenness of the method. Lastly, preservatives are another use of reagents in analytical techniques, particularly for maintaining the composition of aqueous and biological samples prior to analysis. The green chemistry considerations for preservatives include the hazardousness and energy consumption. Greener alternatives will be discussed for chelation, derivatization, and preservation.

3.3.1 Chelating Agents

Despite the prevalence of chelating agents in sequestering metals during analytical procedures, little research has been done on the development of greener chelating agents. Most commonly, EDTA is used both industrially and in analytical laboratories. However, since the rate at which EDTA is discharged into the environment far exceeds the rate at which is degrades, EDTA is considered bioaccumulative. One isomer of EDTA is ethylenediamine disuccinate (EDDS) and this, especially the S,S-isomer, is readily biodegradable. EDDS has found use industrially, for example in laundry detergents. One recent study showed that EDDS can successfully replace EDTA in the titration of aqueous divalent cations.[56] For example, using EDDS as a replacement for EDTA in U.S. EPA Method 130.2 to determine water hardness via titration using Eriochrome Black T indicator, two analysts reported accuracies of 102% relative to the standard method. Separately, it was determined that the hydrophobicity of ILs allows their use as chelating agents,[57] but more work is needed in this area.

3.3.2 Derivatization

In the context of this chapter, derivatization refers to chemically reacting the molecule of interest, usually by adding functionality, in order to change the properties of the molecule such that the molecule can be more easily analysed and characterized.

One of the principles of green chemistry is avoiding the use of chemical derivatives. The reason for this is the contribution of derivatives toward poor atom economy. Atom economy is defined as:[58]

$$\text{atom economy (\%)} = (\text{MW desired product}/\text{MW all reactants}) \times 100 \quad (3.1)$$

Derivatization adds to the molecular weight of the reactants but is not part of the desired product, thus decreasing the atom economy. Ultimately, derivatization increases the amount of waste generated.

In the early days of chemistry, until analytical instrumentation advanced in the ability to identify chemical structures, derivatization was commonly employed for chemical analysis. In particular, if a chemical did not crystallize, a derivative was formed that would crystallize, for characterization of the chemical by melting point determination. Even today, beginning chemistry students are commonly taught this technique in secondary school and university laboratory experiments.

Derivatization today has become more sophisticated; derivatives of compounds are generated not only for crystallinity, but also for changing the solubility properties of the compound, improving separation of similar compounds (such as isomers), and enabling detection.

The use of derivatives to change the solubility or retention in chromatography is used for separation of isomers and especially for chiral separations for analysis and characterization, particularly for pharmaceutical compounds and other fine chemicals. Chromatography separations include gas chromatography, planar chromatography, and capillary electrophoresis. Derivatization of compounds can also enable spectroscopic detection, such as chemiluminescence, bioiluminescence, phosphorescence, infrared spectroscopy, and ultraviolet/light-absorbing spectroscopy.

One example where derivatization may be greener than the alternative technique is a new protein analyser technique. CEM Corporation was awarded a Presidential Green Chemistry Challenge Award in 2009 for the Sprint™ Rapid Protein Analyzer which measures protein content, such as in food (baby formula, pet food, *etc.*) and distinguishes between melamine and protein, unlike the common Kjeldahl method.[59] The method works by tagging histidine, arginine, and lysine, common amino acids found in proteins, with a tagging solution containing an acidic group to attach to the basic amino acids, and also contains an extensive aromatic group for colorimetric measurement. The bound protein is removed by filtration and the remaining tagging solution is measured by colorimetry. The solution is nontoxic, nonreactive, and water soluble and most samples run in 2–3 minutes in comparison to 4 hours for Kjeldahl, with dramatic waste savings in comparison to the Kjeldahl method, eliminating 5.5 million pounds (2.6×10^6 kg) of hazardous waste in the United States each year.[59] In this example, the use of derivatives for analysis save tremendous amounts of hazardous waste and time in comparison to the alternative method of analysis.

Next time a procedure calls for derivatization, consider avoiding derivatives if possible. Some questions to ask are:

- Is there another way to separate or analyse without involving derivatives?
- Is there a direct analysis technique available with different instrumentation, without derivatization?

The small quantities associated with analytical samples add up, and derivatization only increases the amount of waste. If derivatization must be used, consider the health and environmental effects of the chemicals used for the derivatization. Consider if the benefits of derivatization outweigh the consequences.

3.3.3 Preservatives

Preservation of samples before analysis is another component to include when considering the greenness of an analytical method. The human health effects of preservatives may include toxicity and irritation due to corrosivity and environmental effects may include bioaccumulation and energy usage, depending on the method of preservation.

There are a number of important reasons why certain analytical samples are preserved. Preservatives are used to prevent changes within the sample, including physical change (volatilization), chemical change or reaction (oxidation, photochemical reaction, precipitation, heat-induced reaction), and biological change (degradation, enzymatic reaction, microbial growth). Certain types of samples are more prone to these changes and require some type of preservation, especially aqueous samples, soil, food, and biological samples. There are also instances where preservative may be used in another part of the methodology, such as in the standard or in the extractant.

Some of the approaches taken to preserve a sample include chemical approaches such as addition of acid or base and non-chemical energy-intensive methods such as freezing, refrigeration, freeze drying, irradiation, or microwave. Preservation can also be achieved by the type of container used and limiting the hold time prior to analysis. A summary of approaches is shown in Table 3.4.[60]

Table 3.4 Summary of preservation methods.

Sample	Preservation method	Container type	Holding time
pH	—	—	Immediately on site
Temperature	—	—	Immediately on site
Inorganic ions			
Bromide, chloride, fluoride	None	Plastic or glass	28 days
Chlorine	None	Plastic or glass	Analyse immediately
Iodide	Cool to 4 °C	Plastic or glass	24 h
Nitrate, nitrite	Cool to 4 °C	Plastic or glass	48 h
Sulfide	Cool to 4 °C, add zinc acetate and NaOH to pH 9	Plastic or glass	7 days
Metals			
Dissolved	Filter on site, acidify to pH 2 with HNO_2	Plastic	6 months
Total	Acidify to pH 2 with HNO_2	Plastic	6 months

Table 3.4 (*Continued*)

Sample	Preservation method	Container type	Holding time
Cr(VI)	Cool to 4 °C	Plastic	24 h
Hg	Acidify to pH 2 with HNO_2	Plastic	28 days
Organics			
Organic carbon	Cool to 4 °C, add H_2SO_4 to pH 2	Plastic or brown glass	28 days
Purgeable hydrocarbons	Cool to 4 °C, add 0.008% $Na_2S_2O_3$	Glass with Teflon septum cap	14 days
Purgeable aromatics	Cool to 4 °C, add 0.008% $Na_2S_2O_3$ and HCl to pH 2	Glass with Teflon septum cap	14 days
PCBs	Cool to 4 °C	Glass or Teflon	7 days to extraction, 40 days after
Organics in soil	Cool to 4 °C	Glass or Teflon	As soon as possible
Fish tissues	Freeze	Aluminium foil	As soon as possible
Biochemical oxygen demand	Cool to 4 °C	Plastic or glass	48 h
Chemical oxygen demand	Cool to 4 °C	Plastic or glass	28 days
DNA	Store in TE (pH 8) under ethanol at −20 °C; freeze at −20 or −80 °C		Years
RNA	Deionized formamide at −80 °C		Years
Solids unstable in air for surface and spectroscopic characterization	Store in argon-filled box; mix with hydrocarbon oil		

Reproduced from Ref. 60, with kind permission of John Wiley & Sons, Inc.

Historically, chemical preservatives have included some hazardous chemicals, such as mercuric chloride, tributyl tin, formaldehyde, chloroform, and dichloromethane.[61] Now that scientists are more aware of the hazards (bioaccumulation, toxicity), many of these chemical preservatives are no longer used. For improving the greenness of preservation, some questions to consider are:

- If a hazardous chemical is used as the preservative, is there a less hazardous alternative?
- Can chemical preservation be avoided entirely through a shorter hold time and/or refrigeration?
- If freezing is recommended, is refrigeration an option? If freezing is necessary, can the hold time be minimized?
- If freeze drying is recommended, is microwave a viable alternative?

Seeking an alternative preservation method can improve the overall environmental profile of the analytical method.

References

1. P. T. Anastas and J. C. Warner, *Green Chemistry: Theory and Practice*, Oxford University Press, New York, 1998.
2. M. Koel and M. Kaljurand, *Pure Appl. Chem.*, 2006, **78**, 1993.
3. L. H. Keith, L. U. Gron and J. L. Young, *Chem. Rev.*, 2007, **107**, 2695.
4. M. Tobiszewski, A. Mechlinska, B. Zygmunt and J. Namiesnik, *TrAC, Trends Anal. Chem.*, 2009, **28**, 943.
5. J. F. Garcia-Reyes, B. Gilbert-Lopez and A. Molina-Diaz, *Spectrosc. Lett.*, 2009, **42**, 383.
6. S. Armenta and M. de la Guardia, *Spectrosc. Lett.*, 2009, **42**, 277.
7. F. R. P. Rocha, L. S. G. Teixeira and J. A. Nobrega, *Spectrosc. Lett.*, 2009, **42**, 418.
8. M. L. Cervera, M. de la Guardia, S. Dutta and A. K. Das, *Spectrosc. Lett.*, 2009, **42**, 284.
9. D. C. Harris, *Exploring Chemical Analysis: Lab Experiments*, 4th edn, W. H. Freeman, San Francisco, CA, 2008.
10. D. Raynie and J. L. Driver, *13th Green Chemistry and Engineering Conference*, Washington, DC, 2009.
11. P. Anastas and N. Eghbali, *Chem. Soc. Rev.*, 2010, **39**, 301.
12. W. Leitner, P. G. Jessop, C. J. Li, P. Wassersheid and A. Stark (eds), *Handbook of Green Chemistry—Green Solvents*, Wiley-VCH, Hoboken, NJ, 2010.
13. W. M. Nelson, *Green Solvents for Chemistry: Perspectives and Practice*, Oxford University Press, New York, 2003.
14. F. M. Kerton, *Alternative Solvents for Green Chemistry*, Royal Society of Chemistry, Cambridge, 2009.
15. C. Capello, U. Fischer and K. Hungerbuhler, *Green Chem.*, 2007, **9**, 927.
16. P. G. Jessop, *Green Chemistry*, 2011, Advance Article.
17. K. D. Bartle, A. A. Clifford, S. B. Hawthorne, J. J. Langenfeld, D. J. Miller and R. Robinson, *J. Supercrit. Fluids*, 1990, **3**, 143.
18. M. Herrero, J. A. Mendiola, A. Cifuentes and E. Ibanez, *J. Chromatogr. A.*, 2010, **1217**, 2495.
19. M. Zougagh, M. Valcarcel and A. Rios, *TrAC, Trends Anal. Chem.*, 2004, **23**, 399.
20. D. E. Raynie, *Anal. Chem.*, 1993, **65**, 3127.
21. P. Sun and D. W. Armstrong, *Anal. Chim. Acta*, 2010, **661**, 1.
22. Z. Li, J. Chang, H. Shan and J. Pan, *Rev. Anal. Chem.*, 2007, **26**, 109.
23. X. Han and D. W. Armstrong, *Acc. Chem. Res.*, 2007, **40**, 1079.
24. J. L. Anderson, D. W. Armstrong and G. Wei, *Anal. Chem.*, 2006, **78**, 2893.
25. S. Pandey, *Anal. Chim. Acta.*, 2006, **556**, 38.
26. M. Koel, *Crit. Rev. Anal. Chem.*, 2005, **35**, 177.

27. J. F. Lui, G.-B. Jiang and J. A. Jonsson, *TrAC, Trends Anal. Chem.*, 2005, **24**, 20.
28. R. J. Soukup-Hein, M. M. Warnke and D. W. Armstrong, *Ann. Rev. Anal. Chem.*, 2009, **2**, 145.
29. N. Li, G. Fang, B. Liu, J. Zhang, L. Zhao and S. Wang, *Talanta*, 2010, **26**, 455.
30. M. Matsumoto, T. Ohtani and K. Kondo, *J. Membrane Sci.*, 2007, **289**, 92.
31. G. V. Myasoedova, N. P. Molochnikova, O. B. Mokhodoeva and B. F. Myasoedov, *Anal. Sci.*, 2008, **24**, 1351.
32. D. W. Armstrong, L. K. He and M. L. Gross, *Anal. Chem.*, 2001, **73**, 3679.
33. A. Berthod, M. J. Ruiz-Angel and S. Carda-Broch, *J. Chromatogr. A.*, 2008, **1184**, 6.
34. J. Fan, Y. Fan, Y. Pei, K. Wu, J. Wang and M. Fan, *Separ. Purif. Technol.*, 2008, **61**, 324.
35. S. B. Hawthorne, Y. Yang and D. J. Miller, *Anal. Chem.*, 1994, **66**, 2912.
36. J. Kronjolm, K. Hartonen and M. L. Riekkola, *TrAC, Trends Anal. Chem.*, 2007, **26**, 396.
37. A. Kubatova, B. Jansen, J. F. Vaudoisot and S. B. Hawthorne, *J. Chromatogr. A.*, 2002, **975**, 175.
38. S. Morales-Munoz, J. L. Luque-Garcia and M. D. Luque de Castro, *Anal. Chem.*, 2002, **74**, 4213.
39. E. V. Petersson, J. Liu, P. J. R. Sjoberg, R. Danielsson and C. Turner, *Anal. Chim. Acta.*, 2010, **663**, 27.
40. D. Kralisch, A. Stark, S. Korsten, G. Kreisel and B. Ondruschka, *Green Chem.*, 2005, **7**, 301.
41. C. Jiménez-González, A. D. Curzons, D. J. C. Constable and V. L. Cunningham, *Clean Techn. Environ. Policy*, 2005, **7**, 42.
42. K. Alfonsi, J. Colberg, P. J. Dunn, T. Fevig, S. Jennings, T. A. Johnson, H. P. Kleine, C. Knight, M. A. Nagy, D. A. Perry and M. Stefaniak, *Green Chem.*, 2008, **10**, 31.
43. E. Destandau and E. Lesellier, *Chromatographia*, 2008, **68**, 985.
44. C. J. Welch, T. Brkovic, W. Schafer and X. Y. Gong, *Green Chem.*, 2009, **11**, 1232.
45. V. Kienen, W. F. Costa, J. V. Visentainer, N. E. Souza and C. C. Oliveira, *Talanta*, 2008, **75**, 141.
46. S. Aparicio and R. Alcalde, *Green Chem.*, 2009, **11**, 65.
47. Y. Gu and F. Jérôme, *Green Chem.*, 2010, **12**, 1127.
48. J. I. García, H. García-Marín, J. A. Mayoral and P. Pérez, *Green Chem.*, 2010, **12**, 426.
49. M. Virot, V. Tomao, C. Ginies and F. Chemat, *Chromatographia*, 2008, **68**, 311.
50. D. F. Aycock, *Org. Process Res. Dev.*, 2007, **11**, 156.
51. V. Fábos, G. Koczó, H. Mehdi, L. Boda and I. T. Horváth, *Energy Environ. Sci.*, 2009, **2**, 767.
52. K. Watanabe, N. Yamagiwa and Y. Torisawa, *Org. Process Res. Dev.*, 2007, **11**, 251.

53. J. Chen, S. K. Spear, J. G. Huddleston and R. D. Rogers, *Green Chem.*, 2005, **7**, 64.
54. P. G. Jessop, L. Phan, A. Carrier, S. Robinson, C. J. Dürr and J. R. Harjani, *Green Chem.*, 2010, **12**, 809.
55. D. Vinci, M. Donaldson, J. P. Hallett, E. A. John, P. Pollet, C. A. Thomas, J. D. Grilly, P. G. Jessop, C. L. Liotta and C. A. Eckert, *Chem. Commun.*, 2007, **14**, 1427.
56. D. E. Raynie, G. Degam, B. A. Anderson and K. J. Odegaard. *14th Green Chemistry and Engineering Conference*, Washington, DC, 2010.
57. K. Kidani, N. Hirayama and H. Imura, *Anal. Sci.*, 2008, **24**, 1251.
58. B. M. Trost, *Angew. Chem. Int. Ed. Engl.*, 1995, **34**, 259.
59. *The Presidential Green Chemistry Challenge Award Recipients 1996*, United States Environmental Protection Agency, 744K10003, 2010.
60. S. Mitra (ed.), *Sample Preparation Techniques in Analytical Chemistry*, John Wiley & Sons, Hoboken, NJ, 2003.
61. L. L. M. Nollet (ed.), *Handbook of Water Analysis*, 2nd edn, CRC Press, Boca Raton, FL, 2007.

CHAPTER 4
Green Sample Preparation Methods

CARLOS BENDICHO, ISELA LAVILLA, FRANCISCO PENA AND MARTA COSTAS

Analytical and Food Chemistry Department; Faculty of Chemistry; University of Vigo, Campus As Lagoas-Marcosende s/n, 36310 Vigo, Spain

4.1 Greening in Sample Preparation

Since the introduction of the 12 principles of green chemistry more than 10 years ago,[1] awareness has arisen among scientists and engineers worldwide concerning the ways to reduce/eliminate the use or generation of feedstock, products, by-products, solvents, *etc.* Although activities in the analytical laboratory intrinsically involve small-scale chemical processes, their continuous repetition on a routine basis represents a real risk to humans and the environment. From the point of view of the green chemistry goals, analytical chemistry can be considered in two different ways. On the one hand, analytical processes serve the purpose of controlling chemical processes, being a tool for giving information about chemical substances, occurrence in the environment, foods, *etc.*, which helps decision-making about human and environmental health. On the other hand, performing analytical methods constitutes itself a potential risk by repetition of operations that need solvents, chemicals, and energy and yield wastes. Then, as pointed out by Anastas and Warner,[1] the tools needed to guarantee the accomplishment of green chemistry principles can turn into the source of the problem themselves. In this sense, analytical chemistry should become the object of application of the 12 principles of green chemistry, in a similar manner to other areas of chemistry and chemical technology. Principles directly related to analytical chemistry are: (1) prevention (principle 1); (2) safer solvents and auxiliaries (principle 5); (3) design of energy efficiency (principle 6); (4) decreased use of derivatives (principle 8); (5) real-time analysis for pollution prevention (principle 11); and (6) inherently safer chemistry for accident prevention (principle 12). Progress in analytical methodology has been driven by the need to improve analytical characteristics such as accuracy, precision, detection limits, sample throughput, *etc.*, through the enhancement of features such as automation, miniaturization and acceleration of analytical processes. It means that analytical chemists have long been aware of the benefits of those trends, which in many cases include a decrease in hazardous wastes, minimum sample consumption, less energy requirements and use of safer chemicals and procedures. Modern analytical methods usually address features related to 'clean' or 'green' characteristics, and these words appear increasingly in the analytical literature. It is clear that a revaluation of many old analytical procedures used on a routine basis is required to meet the green chemistry requirements. The term 'green analytical chemistry' (GAC) was first coined by Namiesnik,[2,3] and includes both design of new analytical methods under the auspices of green chemistry and modifications of existing methods so that they approach this concept more closely. GAC principles can be implemented at each stage of the analytical process. Some review articles have been published on this topic highlighting the role of GAC.[4,5]

Some initiatives have aimed at the comparison of greenness profiles of analytical methods. The main difficulties have stemmed from the lack of clear discriminatory criteria for comparison, making it clear that the first criteria for the choice of an analytical method should concern its analytical characteristics (sensitivity, detection limit, acceptable bias, precision, *etc.*). Recently, the Green Chemistry Institute of the American Chemical Society (ACS) has proposed 'greenness' criteria for more than 1000 environmental analytical methods included in the National Environment Methods Index (NEMI) database so that greenness profiles can be established.[6] The choice of a particular method can be facilitated on the basis of these profiles. 'Greenness' profiles are characterized by four key criteria: (1) persistent, bioaccumulative and toxic (PBT); (2) hazardous; (3) corrosive; (4) waste. In this chapter, we focus on sample preparation, which is typically considered the 'bottleneck' of analysis. Sample preparation operations are characterized by their diversity, complexity, slowness and difficulty of automation. It should be kept in mind that the choice a particular sample preparation procedure considering greenness issues is not always easy to make. Often, sample preparation is conditioned by the detection technique to be used. As an example, for trace element analysis in biological solid samples, much more intensive organic matter decomposition is needed when electroanalytical techniques are employed than for atomic spectrometric techniques. In the latter case, sample decomposition can be overcome using approaches for direct analysis of the solid material, without resorting to concentrated mineral acids and drastic reaction conditions (*i.e.* high temperature and pressure).

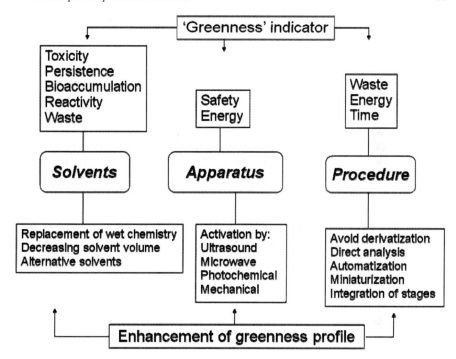

Figure 4.1 Schematic diagram showing the relevant greenness indicators of sample preparation procedures as well as possible paths for their enhancement.

Figure 4.1 shows the relevant greenness indicators for solvents, apparatus, and procedures, along with possible ways of enhancing the greenness profile of sample preparation methods. Examples of green sample preparation methods based on those premises will be discussed in the following sections.

4.2 Microwave-Assisted Sample Preparation: Digestion and Extraction

Nowadays, microwave energy plays a vital role in sample preparation. Microwave heating is a more efficient and rapid method, without temperature gradients, than conventional systems, saving energy consumption and therefore cost.

Microwaves (MWs) are electromagnetic waves with a frequency in the range 300–300 000 MHz. Microwave energy is a non-ionizing radiation consisting of an electric and a magnetic field, though only the electric field transfers energy to the medium. The theory of microwave irradiation was predicted in 1864 and physically demonstrated to exist in 1888.[7] Microwave energy interacts with matter, causing migration of ions and rotation of dipoles, but it does not cause changes in the molecular structure. Therefore, there are two fundamental mechanisms for transferring energy from the electric field of the MWs to the substance to be heated: dipole rotation and ionic conduction.

Dipole rotation is an interaction in which polar molecules try to align themselves with the rapidly changing electric field of MWs. The rotational motion of the molecules when they try to orient with the field results in an energy transfer. As the electric field decreases, thermal disorder is restored with release of thermal energy. At 2450 MHz (a common frequency used in commercial systems), the alignment of the molecules followed by their return to disorder occurs 4.9×10^9 times per second, which results in rapid heating. Ionic conduction occurs when there are free ions or ionic species present in the substance being heated.[8,9]

The dissipation factor of the sample, tan δ, is a measure of how efficiently microwave energy is converted into thermal energy and can be expressed as tan $\delta = \varepsilon''/\varepsilon'$, where ε'' is the electric loss factor, which measures the ability of a sample to convert electromagnetic energy into heat, and ε' is the dielectric constant or relative permittivity, which is related to the ability of a sample to obstruct the passage of microwave energy through it.[9]

Microwave-assisted sample preparation has undergone an impressive growth in analytical chemistry procedures since Abu-Samra *et al.* published the first application of microwave technology for biological sample mineralization in 1975.[10] Since then, microwave energy has shown a potential as an efficient tool for sample preparation. Principles, important developments, microwave equipment and applications of microwave-assisted techniques have been described in many comprehensive reviews[11–16] and book chapters.[9,17,18] Although microwave energy can activate many processes, the discussion here focuses on the most extensive uses in solid sample treatments such as digestion and extraction.

4.2.1 Microwave-Assisted Digestion

The term 'acid digestion' or 'wet digestion' refers to an attack on the matrix by mineral acids at high concentration, which entails the release of the analytes bound to it. This definition is widely accepted[17] and is recommended in the International Union of Pure and Applied Chemistry (IUPAC) guidelines,[19,20] although many others terms can be seen in the literature describing the same concept, *e.g.* wet-ashing, mineralization, decomposition, *etc.* Wet digestion with oxidizing acids is in widespread use for sample decomposition prior to trace element analysis. With traditional heating systems (*e.g.* sand bath, hot plate, Bunsen burner, *etc.*), convective heating occurs. The use of microwave heating allows the speeding up acid digestion procedures and increases the efficiency of digestion. Old methods for acid digestion were performed in open vessels, which have several disadvantages such as risk of analyte volatilization (*e.g.* Hg, Cd, Se, *etc.*), use of large amounts of acids as a result of continuous replacement, and limited oxidizing power of acids used at the maximum temperature allowable at atmospheric pressure (*i.e.* boiling point). This last issue makes it necessary to use $HClO_4$ for applications with poorly oxidizable matter.

The risks inherent to the use of $HClO_4$, due to its explosive reaction with easily oxidizable organic matter, need to be considered. Addition of H_2SO_4 is required in order to increase the boiling point of the sample solution, thereby achieving higher digestion temperature at atmospheric pressure. Another issue when dealing with trace element analysis is contamination by reagents, which should be kept as low as possible.

One way to reduce the amount of acids and increase their oxidizing power is to operate under pressure in closed reactors. Although acid attack in high-pressure closed vessels heated by thermal convection (digestion bombs) leads to very complete digestion products, the main problem is the long time required to accomplish the dissolution process (e.g. >3 h).[21] MWs have largely improved these methods, since direct heating of the sample occurs without previous heat transfer through the vessel walls, and fast and efficient digestions can be performed. MWs only heat the liquid phase; the gas phase does not significantly absorb this energy. As a result, very high temperatures can be reached at relatively low pressures, thus enhancing the acid reactivity.

The digestion efficiency depends on the nature and mass of the sample, the acid(s) used, pressure setting, temperature, time, and output power. Among these parameters, the right choice of the acid mixture is essential. The oxidizing power of a digestion reagent shows a marked dependence on temperature. Both low-pressure and high-pressure digestion can be employed in microwave-assisted digestion (MAD).[17]

The best reagent for MAD is HNO_3 but combinations of oxidizing acids (H_2SO_4, $HClO_4$), non-oxidizing acids (HCl, HF, H_3PO_4) and H_2O_2 are also used, depending on the MW system (closed-vessels heated in MW ovens, focused-MW systems, high-pressure MW systems or flow systems).

The current composition of liners and cups employed for the construction of closed-vessels heated in MW ovens is a fluorinated polymer, such as polytetrafluoroethylene (PTFE) or perfluoroalkoxy (PFA). With these systems, HNO_3 alone or mixed with HCl or H_2O_2 is typically employed for digestion of biological matrices. Addition of HF is needed for digestion of silicate-bearing materials. PTFE closed-vessels with volumes of 50–120 ml are typically employed. Apart from relevant analytical advantages such as low blanks, several greenness-related issues, such as time required for sample decomposition, energy required, volume of acids and safety considerations are remarkably improved. However, the temperature reached in these closed vessels is not always high enough to efficiently decompose very thermoresistent materials. For instance, digestion of samples with high fat content requires temperatures higher than those possible in PTFE closed vessels. High-pressure MW digestion (focused MWs) systems facilitate digestion of these materials, still avoiding in many cases the use of extremely oxidizing acids such as $HClO_4$, but safety requirements due to the high pressure and temperature are increased with these systems.[22,23] In focused-MW heated systems, MW irradiation occurs over a restricted area where the sample is placed, so that it is subjected to a much stronger electric field.[15] Non-pressurized MW-assisted digestion in open vessels using focused MWs is limited to efficient digestion of simple matrices, and additionally, very oxidizing acid mixtures have to be employed, as in conventional acid digestions. On-line digestion in flow systems is attractive since some of the limitations inherent to MAD using closed vessels are eliminated. From the GAC point of view, an advantage of on-line MW-assisted digestion is the drastic decrease in the time required for sample preparation and the possibility of handling with reactions that would be too dangerous in closed vessels.[17] However, on-line digestion is appropriate only in cases where mild conditions can be applied for digestion (*e.g.* liquids and slurries).[24]

As mentioned previously, MW digestion vessels usually have a capacity of 50–120 ml. A further strategy for decreasing the acid volume in MAD is the use of small-volume vessels (7 ml) placed inside conventional PTFE closed-vessels. This strategy has been used for organic and biological samples with only 1 ml of HNO_3[25] and even as little as 0.3 ml.[26] Small-volume polystyrene and glass liners have been also proposed for low-volume MAD.[27]

In comparison with conventional MAD, low-volume MAD has clear advantages from a green perspective: (1) it uses small volumes of acids; (2) three samples can be simultaneously treated in each conventional MW digestion vessel when small-volume vessels are placed inside; (3) short heating programs with low MW power can be used.

In short, the evolution of MW laboratory systems and their automation have provided new and better tools for acid digestion in a greener, safe, and reliable way. Needless to say, inorganic trace analysis does not always require total matrix decomposition, and this will ultimately depend on the analytical technique used for detection. Very matrix-sensitive techniques such as anodic stripping voltammetry will require the most complete sample decomposition so that residual carbon is negligible. Likewise, vapour generation techniques in atomic spectrometry are quite matrix sensitive, and many samples contain the analyte in a form that is not amenable to the reduction reaction (*e.g.* certain organometals).

With techniques such as flame atomic absorption spectrometry (FAAS) or inductively coupled plasma optical emission spectrometry (ICP-OES), a simple MAD procedure may be enough, since complete decomposition will occur in the sample cell. At the opposite end, some techniques will require little or no sample decomposition, and even solid samples can be directly analysed (sampling of direct solid or slurry) by techniques such as electrothermal atomic absorption spectrometry (ETAAS) or totally reflecting X-ray fluorescence spectrometry (TXRF). Therefore, the greenness of the sample digestion procedure will depend to some extent on the final stage of detection. The use of mineral acids at extreme conditions in terms of temperature and pressure, inherent to acid digestion, makes it difficult to meet the GAC principles. That said, digestion strategies that avoid the most oxidizing acids (*e.g.* $HClO_4$) as well as extremely high pressure and temperature, and minimize the consumption of acids in both batch and flow systems, should fulfil those principles in a better way.

Microwave heating has been also used for accelerating Kjeldahl digestion. Conventional Kjeldahl digestion involves boiling the sample in sulfuric acid in the presence of a catalyst (which can be Hg, Se or Cu). Kjeldahl nitrogen is converted to ammonium (sulfate) that is subsequently distilled as ammonia.

Conventional Kjeldahl digestion is time-consuming (2–4 h) and as a result it is costly, especially when the number of samples to be analysed is large, and entails an enormous drain on resources and manpower.[18] MW energy has been used to develop rapid, compact and safe Kjeldahl digestions.[28] Feinberg et al. introduced sophisticated open-vessel focused MW digestion systems for Kjeldahl digestion of food.[29] They found that using a mixture of H_2SO_4 and H_2O_2 for decomposition, the catalyst (generally toxic) was unnecessary. Mason et al. combined a MW Kjeldahl system with a flow injection manifold, hence increasing the sample throughput.[30] Kjeldahl MAD is considered as a more environmentally friendly alternative to conventional Kjeldahl digestion since it offers some greener advantages: (1) the use of a catalyst and the subsequent hazardous waste can be avoided; (2) reduced digestion time (up to 20 times faster); (3) reduced energy consumption; (4) increased automation capability.[31]

Another important process involving digestion is the acid hydrolysis of proteins or peptides. Classical protein hydrolysis has been the rate-determining step in amino acid analysis.[32] Usually, classical protein hydrolysis is performed by heating the sample in 6 M HCl under vacuum at 110 °C for 24 h.[33] Microwave-assisted hydrolysis (MAH) allows fast heating of samples, thus obtaining a total protein hydrolysis in less than 30 min, i.e. more than two orders of magnitude faster.[34] Moreover, the concentration of HCl used as hydrolysing agent can be decreased from 6 M to 3 M when using MAH.[35] MAH also reduces the racemization that usually occurs with traditional protein hydrolysis.[35,36]

4.2.2 Microwave-Assisted Extraction

Microwave-assisted extraction (MAE) is considered one of the milestones in the development of sample preparation strategies from a greener point of view.[37] Although the use of MW energy in sample treatment was initially applied to the mineralization of samples, as previously noted, there is growing interest in the use of MWs to assist the extraction of organic and organometallic compounds from different matrices.[8] In MAE, an extractant (usually a liquid organic solvent) in contact with solid or liquid samples is heated by MW energy. The heating is localized and therefore the temperature increases rapidly near or above the boiling point of the solution, leading to very short extraction times. This process is quite different from classical solvent extraction since the solvent diffuses into the matrix and analytes are extracted by solubilization.

The main parameters influencing MAE are the nature of the solutes (polar or non-polar), extraction parameters (temperature and time) and the nature of the matrix (particle size and water content). For extraction of most organic compounds, organic solvents are used. However, most MAE applications involve mixtures of non-polar solvent and water. It is common to perform MAE with the same solvent as is prescribed for traditional extraction, but ionic liquids (ILs), considered as green solvents, have also been successfully applied in MAE.[38]

Compared with Soxhlet extraction (*i.e.* the method typically used as reference), MAE has the following advantages:[39–41]

- It requires shorter extraction times (3–30 min); in Soxhlet extraction, typically operation times are 3–48 h
- The amount of solvent required is decreased to 10–40 ml, in contrast to 100–500 ml used in Soxhlet
- Sample throughput is high, so multiple samples can be extracted simultaneously
- Waste generation is reduced
- Automation is possible.

MAE can be developed in two modalities: pressurized MW-assisted extraction (PMAE) and focused MW-assisted extraction (FMAE). PMAE is performed under controlled pressure and temperature and employs a MW-transparent closed extraction vessel and a solvent of high dielectric constant. The solvents can be heated above their boiling point under standard pressure conditions since boiling does not occur because the vessel is pressurized. FMAE is performed at atmospheric pressure and employs open vessels and solvents with low dielectric constants. As solvents are essentially MW-transparent, they absorb very little energy, and extraction can therefore be performed in open vessels. The temperature of the sample increases during extraction, because it usually contains water and other components with high dielectric constants. Extraction procedures can be developed either on-line or off-line. This mode of operation can be used to extract thermolabile analytes and it is usually preferred for extraction of organometallic compounds since the precise control of the energy delivered to the sample prevents destruction of the carbon–metal bonds.[42]

Atmospheric and pressure systems have been compared for the extraction of organic compounds. No significant differences were found using both extraction systems but PMAE had advantages over FMAE in terms of efficiency, solvent consumption, extraction time and labour.[43,44]

Apart from extraction of organic compounds, MAE has been widely used in speciation of organometals. The preservation of the organometallic moiety is a necessary requirement in the extraction of these species from the solid material. For this purpose, focused-MW systems have proved very efficient.[42] Thus, fast extractions of compounds such as organotin,[45] organomercury,[46] *etc.* have been reported in the literature. Extraction times for the above compounds were shortened to less than 3 min.

4.3 Ultrasound-Assisted Sample Preparation: Digestion and Extraction

Ultrasound has been used in sample preparation for different processes such as extraction, dissolution, and digestion. The use of this energy allows designing greener analytical methodologies. In general, acceleration of the above processes along with the use of less concentrated reagents can be achieved.

Sound waves are made up of high and low pressure pulses that propagate through a solid, liquid or gas (*i.e.* mechanical waves).[47] Sound waves with a frequency greater than 20 kHz constitute the region of ultrasound. In turn, the ultrasound region can be divided into two parts; one where the cavitation phenomenon takes place (20–100 kHz), called power ultrasound, and the other where no cavitation occurs (5–10 MHz), used for diagnostics. Generally, ultrasound for sonochemical applications extends up to 2 MHz.

Sonochemical effects depend on the cavitation phenomenon. According to the 'hot spot' theory, each cavitation bubble behaves like a microreactor, which, in aqueous systems, generates instantaneous temperatures of several thousand degrees and pressures in excess of 1000 atmospheres.[48] When water is subjected to ultrasound irradiation, a series of radicals, such as H$^{\cdot}$ and OH$^{\cdot}$, are formed at the gas-phase interface of the cavitation bubbles, and to a lesser extent, in the bulk solution, which are responsible for the enhanced reactivity. The shock wave arising at the collapse of the cavitation bubble can, in turn, enhance some processes when ultrasound is transmitted in heterogeneous media (*e.g.* particulate solids suspended in liquid media), which benefit extraction and dissolution processes. A variety of ultrasonic processors can be employed for implementation of ultrasound in laboratory applications, *e.g.* ultrasonic bath, ultrasonic probe and cup-horn sonoreactor, each with a different performance.[49] Ultrasound can promote extraction, dissolution and digestion when applied to solid materials as a result of the following conditions: (1) high local temperatures occurring in the microbubbles on cavitation in the liquid medium, which improve analyte solubility and diffusivity of solvent inside the solid particles; (2) high pressure occurring during microbubble implosion, which improves solvent penetrability and transport; (3) particle fragmentation, which causes surface renewal so that more analyte comes in contact with the solvent; (4) for biological samples, cell disruption, fragmentation of cell membranes and subsequent release of encapsulated analytes present inside the cells occur; (5) formation of oxidizing radicals (*e.g.* OH$^{\cdot}$) and H_2O_2, which helps organic matrix oxidation. Conditions 2–4 are mainly effective when cavitation occurs near the particle surface or in the surface itself. Ultrasound-assisted sample pretreatment for both organic and inorganic trace analysis constitutes a greener approach than more classical pretreatments that use large amounts of solvents and long operation times. In short, ultrasound-assisted sample preparation provides the following advantages:

- A dramatic acceleration of the extraction processes, with consequent saving in energy
- Use of smaller amounts of solvents and/or at lower concentration
- Safer procedures, since operations are performed at almost room temperature and atmospheric pressure
- Fewer opportunities for contamination and/or analyte losses during treatment

- Achievement of ecofriendly and low-cost methods with increased productivity.

Ultrasound-assisted treatment of solid samples allows shortening operating procedures for sample pretreatment for organic and inorganic trace analysis and also for speciation analysis. Apart from the increased sample throughput achieved, this means a saving in energy from the perspective of green chemistry. In many cases, a significant reduction of solvents or their use at lower concentration is also feasible. Depending on the goal pursued, different methodologies can benefit from the application of ultrasound to solid samples prior to analysis. Several reviews,[50–52] books,[53,54] and book chapters[55,56] have been published on the use of ultrasound for sample pretreatment.

4.3.1 Ultrasound-Assisted Digestion

In spite of being very efficient, extended methods for sample decomposition prior to element analysis such as MAD in closed vessels demand high inertness of the vessel walls because impurities can be leached from the walls at high temperatures. Vessel materials transparent to MWs such as PTFE cannot be used with H_2SO_4 (b.p. 330 °C) at temperatures above 200 °C and others such as quartz are not compatible with HF.

New possibilities for acid digestion can arise when ultrasound is applied, and several papers have been published on this subject.[57–60] Ultrasound irradiation aimed at matrix decomposition allows mild conditions (room or nearly room temperature, atmospheric pressure) to be used. Moreover, acid digestion is carried out in open vessels in a safer way as compared to acid attacks under pressure. However, concentrated acids or their mixtures are still needed. Since probe sonication is not recommended with HF, samples requiring this acid to destroy the silicate matrix have usually been digested in ultrasonic baths, which provide much lower sonication power than ultrasonic probes. For many matrices, a pseudodigestion (*i.e.* partial matrix decomposition) is achieved.[61] Analytical techniques include mainly atomic absorption spectrometry (AAS) and ICP-OES. In general, when using ultrasonic baths, longer digestion times (around 1 h) are reported than for MW-assisted digestion in closed vessels.

Other interesting applications regarding the use of ultrasound to assist the sample digestion have also been reported. A greener determination of Kjeldahl nitrogen in milk without inorganic salts can be carried out in 30 min.[62] Conventional chemical oxygen demand (COD) methods involve long treatments (*i.e.* 2 h) in the presence of toxic oxidants (*e.g.* Cr(VI)) in strong acid media (H_2SO_4) and toxic catalysts (*e.g.* Hg(II), Ag(I)). Thus, the digestion time can be decreased from 2 h to 2 min by the use of ultrasound delivered through a glass sonotrode immersed in an open vessel.[63,64] The replacement of traditional chemical oxidants and acids employed in the COD determination by a green oxidizing agent such as Mn(III) has been recently published.[65] In this method, the greenness of the COD determination is dramatically enhanced since apart from an environmentally friendly oxidizing agent, a shortening of the total digestion time is achieved. Enzymes have been involved in sample pretreatment for determination of total element contents and also for speciation analysis.[66] In the context of GAC, digestion using hydrolytic enzymes is attractive, since mild conditions can be employed such as pH close to 7 in most cases, room temperature, and atmospheric pressure. However, long treatments (*i.e.* several hours) are generally needed to achieve an efficient digestion and subsequent element release from solid samples. In recent years, a few papers have addressed the increase in enzymatic kinetics by high intensity ultrasound (*i.e.* ultrasonic probe systems).[67]

4.3.2 Ultrasound-Assisted Extraction

The use of ultrasound for speeding up solid–liquid extraction from different matrices has been exploited in different ways: (1) for determination of total element contents using diluted acids; (2) for determination of organic compounds; (3) for speciation analysis using suitable extractants; (4) for dissolution of different solid phases in environmental samples, such as organic matter, oxides, carbonates, *etc.*, with the aim of obtaining information about the potential element mobility and bioavailability.

A variety of acid media have been employed for ultrasound-assisted metal extractions depending on the matrix. Studies have pointed out that highly efficient sonication such as that provided by probe systems make it possible to use very diluted acids, hence allowing low cost and environmentally friendly methods. For instance, HNO_3 at a concentration of only 3% v/v is enough for quantitative extraction from many biological samples.[68] Ultrasonic baths used for this purpose generally lack sufficient sonication power to achieve fast extractions, and consequently, higher acid concentrations, close to those employed for digestion, are typically recommended.[69] Ultrasonic baths are suitable for those situations where extractions can be accomplished with soft extractants in a short time. This is particularly true for many environmental and industrial hygiene samples (*e.g.* airborne particles, dust, filters, *etc.*) where pollutants can bind onto the surface.[70] Therefore, from the standpoint of GAC, powerful probe and cup-horn sonication systems more closely meet the conditions required for green applications. In this context, ultrasound-assisted extraction (UAE) is advantageous over other techniques discussed in this chapter such as MW- and ultrasound-assisted digestion. UAE is carried out at atmospheric pressure and room temperature, with small amount of acids and at low concentration. Analyte separation from the matrix may occur, which would facilitate the removal of interfering effects. A distinction between digestion and extraction processes assisted by ultrasound is sometimes difficult to establish and it depends on the type of extractant and its concentration, the equipment employed for ultrasound irradiation, and the nature of the substrate. Thus, with acids such as HCl at low concentration, the possibilities of matrix decomposition are minimal, and then extraction predominates.

However, the use of HF even at low concentration in the extractant solution can easily attack the crystal lattice of silicates, causing matrix decomposition to some extent, so this is more likely to be a digestion process. A significant number of applications have arisen during the last decade concerning metal extraction by ultrasound using acidic extractants.

Several applications have been also reported using continuous ultrasound-assisted extraction for food,[71] clinical samples,[72] *etc.* with good performance and low acid concentration. Despite the lower ultrasonic energy reaching the sample inserted in a minicolumn as compared to direct insertion of an ultrasonic probe into a suspension of the powdered material, the repeated passage of the acidic extractant through the column containing the solid material allows efficient extractions with minimum acid consumption and high sample throughput in an automated way. Apart from metals, specific applications of ultrasound to the extraction of anions have also been reported.[73]

For solid–liquid extraction of organic analytes, Soxhlet extraction is the reference method for comparison.[74] As in other modern extraction techniques addressed in this chapter, such as pressurized liquid extraction (PLE) and supercritical fluid extraction (SFE), the amounts of solvents required to accomplish the process are much smaller, but extraction efficiency is sometimes less than that obtained with the other techniques and a clean-up step is generally required. Extraction occurs generally at room temperature and no pressure is involved, which makes this technique very favourable in the context of GAC. When UAE is compared to Soxhlet extraction, the operation time is much shorter. With probe sonication, the extraction time can be as short as 3 min. A possible risk in the application of ultrasound waves for assisting the extraction of organics is the potential degradation that can occur on sonication.[75]

Application of soft extractants under mild conditions is generally a necessary requirement in speciation analysis. The main goal in sample preparation in this field is to preserve the species distribution while achieving an acceptable recovery.[76] Ultrasound has been tried for speeding up extraction of metal species from solid samples.[77–79] Greener extraction methods for speciation analysis have developed in parallel with a substantial improvement of other features such as shortening and/or elimination of some pretreatment steps, simplification, and automation. As hyphenated techniques between chromatography or electrophoresis and specific detection (atomic or mass spectrometries) are typically applied, clean-up of extracts is generally needed.[80] Then, sample preparation is similar to that for typical organic trace analysis applications using this methodology.

Sequential extraction schemes (SESs) are typically applied for establishing the metal content associated with relevant geochemical phases in a variety of environmental samples (*e.g.* sediments, soil, fly ash, sewage sludge, *etc.*). An important issue limiting the application of SESs is the time required to accomplish extraction procedures. SESs such as those provided by Tessier[81] and by the Community Bureau of Reference (BCR) (now the Standards, Measurement and Testing Programme, SM&T),[82] extensively used in many studies, need an overall operation time of 18 and 51 h, respectively. Application of ultrasound by means of a probe makes it possible to decrease these times to only 20–30 min.[83]

The slurry technique allows the analysis of finely powdered samples for elemental determinations without the need for an intensive decomposition method such as acid digestion.[84] Slurries are prepared by suspending an amount of the solid material, previously ground and sieved, in an appropriate diluent. Although the benefits of this technique from the point of view of GAC are evident, other features of the analytical procedure are enhanced, such as decreased risk of sample contamination and analyte losses and increased sample throughput. Among the different approaches available for the mixing and stabilization of slurries, ultrasonic slurry sampling (USS) has proved very efficient in combination with analytical techniques such as ETAAS, since apart from a fast slurry homogenization, solid–liquid extraction of analytes occur to some extent.[85]

A comparison of several sample pretreatment methods addressed to trace element analysis is shown in Table 4.1; both analytical and greenness-related considerations are taken into account. Ultrasound-assisted sample preparation shows improved greenness profiles in comparison with other well-established methods based on the use of concentrated mineral acids at extreme conditions of pressure and temperature, since operations are performed using mostly dilute acids at atmospheric pressure and room temperature.

4.4 Supercritical Fluid Extraction

The use of safer solvents for sample pretreatment is one of the main challenges in developing greener analytical processes, in accordance with the GAC principles. In this sense, the use of supercritical fluids has become very popular as an environmentally friendly alternative to hazardous organic solvents.

A supercritical fluid is any substance that is in its supercritical state (*i.e.* both temperature and pressure are above their critical values). Supercritical fluids are characterized by their unique properties, intermediate between those of gas and liquids, depending on the pressure, temperature, and composition of the fluid.[86] Thus, supercritical fluids display density values similar to those of liquids, which enhances their solubilization ability, while showing viscosity and diffusivity values closer to those of gases as well as negligible surface tension, which enhance mass transfer processes. These favourable properties make them a powerful alternative to conventional extraction methods such as Soxhlet for solid–liquid extraction.

Several books,[87–89] book chapters,[90] and review articles[91] give an in-depth description of supercritical fluids and SFE systems, being highly recommended for interesting readers.

In general, the choice of a supercritical fluid is based on practical issues, such as its availability at high purity and low cost.[92] Water should be therefore the greenest candidate, since it fulfils these requirements and it is considered the most benign solvent. However, its supercritical conditions ($T > 374\,°C$, $P > 22\,MPa$) makes it unsuitable because of its extreme reactivity and

Table 4.1 Comparison of selected sample pre-treatment methods for element trace analysis according to both analytical and greenness-related criteria.

Method	Reagents and time required	Stages	Operating conditions	Greenness-related issues	Analytical considerations
Dry-ashing	$Mg(NO_3)_2$ as ashing-aid; 6–8 h	Ashing; cooling; dissolution of ashes in diluted acid	High temperature combustion with air/O_2 at c. 450 °C	High energy requirements; use of small volume of diluted acids	Loss and contamination risks; possibility of treating large amounts of sample; very efficient for complete removal of organic matter
Wet digestion (open vessel)	HNO_3, $HClO_4$, HCl; overnight treatment; use of catalysts (e.g. SeO_2, $HgSO_4$, $CuSO_4$) and concentrated acids; several hours	Predigestion; heating to boiling (use of hot plate, Kjeldahl flask, etc.); evaporation	Acid digestion with concentrated acids at atmospheric pressure	Risk of explosion; large amounts of corrosive wastes; high energy requirements	Loss and contamination risks
Dry-ashing (combustion flask)	Diluted HCl typically used as absorbent; a few min required for complete combustion	Combustion in a closed vessel; absorption of gases	High-temperature ashing in closed system (atmospheric pressure)	Low energy requirements; use of acid at low concentration	Sample mass limited to the oxygen in the flask; labour-intensive; complex apparatus; risk of incomplete oxidation
Dry-ashing (O_2 plasma)	O_2 stream; ashing time > 1 h, in general.	Attack of the sample by highly reactive 'excited O_2'	Use of a flowing stream of pure O_2 at low pressure, which is converted into 'excited O_2' after passing through a high-frequency electric field	Low-temperature ashing with O_2 as the only reagent. High energy requirements	Low volatilization risks; attack of sample limited to the surface; in general, it is a slow process.

Green Sample Preparation Methods 77

MW-assisted digestion in closed vessels	Concentrated mineral acids; <1 h	Typically several stages with different pressure and temperature	Attack of sample by acids heated above their b.p.	Less corrosive wastes and energy requirements than conventional acid digestion in open vessels	Low loss and contamination risks; drastic reduction of digestion time; very efficient sample decomposition
Ultrasound-assisted extraction	Diluted acids (depending on the ultrasonic processor); 1 min–1 h depending on matrix and ultrasonic processor	Ultrasound irradiation of a finely powdered sample into an acid diluent	Application of ultrasound at the appropriate power	Small amounts of wastes generated; low energy requirements; very safe procedure (operation at room temperature and atmospheric pressure)	Low risk of losses and contamination; low blanks; elimination of interferences due to the solid-liquid extraction; careful optimization of extraction in order to achieve quantitative results
Slurry sample introduction	Diluted acids; slurry homogenization carried out in <1 min with some approaches	Suspension of a finely powdered sample into an acid diluent	Use of mixing or stabilizing agents so that to maintain an homogeneous slurry during sampling	Small amounts of wastes generated; low energy requirements; very safe procedure (operation at room temperature and atmospheric pressure)	Low risk of losses and contamination; low blanks; possible interferences since neither matrix decomposition nor separation is performed; possible influence of inhomogeneity for small sample masses

corrosivity.[93] The low critical temperature (31.1 °C) and moderate critical pressure (7.38 MPa, 72.8 bar) of carbon dioxide makes it the substance of choice for most analytical applications of SFE. Moreover, carbon dioxide is chemically inert, non-toxic, non-flammable, and non-corrosive. However, the non-polar nature of carbon dioxide limits its application to the extraction of non-polar or relatively low polar substances. Two strategies are generally used to facilitate the extraction of polar and ionic compounds: either to increase the polarity of the supercritical fluid or to reduce the polarity of analytes.[94] In both cases, an enhanced analyte solubility is pursued. In the former case, the addition of a low volume of polar organic solvent, known as a modifier, to the supercritical fluid increases the polarity of the mixture, thereby allowing a wider range of applications. Methanol is commonly used as modifier in the range 1–10% v/v, although other solvents may be more efficient for some applications. In the latter case, different strategies can be carried out to decrease the polarity of the analyte, such as *in situ* derivatization, complex formation, or ion-pair formation.[95]

The combination of supercritical carbon dioxide with ILs represents a green technology with tremendous potential for chemical reaction and downstream separation in one system and effective separation of products from ILs. The separation process is based on the solubility of supercritical carbon dioxide in the IL (controlled by pressure) and the insolubility of the IL in supercritical carbon dioxide.[95–98] The use of polar co-solvents, such as ethanol, acetone, or n-hexane, has been reported to increase the solubility of ILs in supercritical carbon dioxide.[99,100]

The SFE process consists of two steps, *i.e.* extraction of target analytes by using a supercritical fluid followed by their collection or trapping. The latter step is as important as the former, being responsible for low extraction recoveries when not controlled carefully.[101] The collection of analytes can be carried out on-line by coupling SFE to chromatographic techniques, or off-line by depressurizing the supercritical fluid into a collection device. Choosing the most appropriate collection mode depends on properties of samples and analytes, extraction variables and the analytical technique used for the detection, solvent collection or solid phase trapping of the analytes being the most used off-line collection modes.[101] Solid-phase trapping offers higher trapping efficiency for volatile analytes than solvent collection, since the trap temperature can be reduced to –30 °C. Moreover, the use of smaller volumes of organic solvents (*ca.* 2 ml) allows the direct analysis of the extract and makes the solid phase trapping mode greener with respect to solvent collection.[101]

Apart from being simple and selective, the most significant advantages of the SFE technique include the preconcentration effect achieved, safety, speed and its quantitative possibilities.[102] Furthermore, automated SFE systems are commercially available for unattended performance, and the coupling of the extraction step with gas, liquid, or supercritical-fluid chromatography is feasible. However, SFE-based methodologies are not free from drawbacks, mainly relating to the very high investment cost needed, the poor robustness of early

SFE commercial systems, lack of an SFE method that works for every analyte in every kind of sample, and the inefficient sample clean-up.[103] SFE processes are energy efficient when compared with conventional Soxhlet extraction, although it is worth highlighting that the energy cost needed to obtain and keep a substance in its supercritical state is pretty high.

Thus, SFE-based methodologies are not widespread and their use is currently decreasing, in spite of the great analytical potential shown by supercritical fluids as green extractants of target analytes from both solid and liquid matrices.[103]

4.5 Pressurized Liquid Extraction

The use of pressurized hot solvents for extraction purposes is another alternative to Soxhlet extraction for solid sample preparation. This technique is commonly called PLE and was introduced in the mid 1990s. It can also be found in the literature under other names such as accelerated solvent extraction (ASE), pressurized hot solvent extraction (PHSE), and high pressure solvent extraction (HPSE) among others. From the GAC perspective, the use of water as extractant phase is particularly attractive, this technique then being called subcritical water extraction (SWE), pressurized hot water extraction (PHWE), *etc*. In this chapter, we will refer to ASE when organic solvents are involved as extractants.

ASE involves the use of organic solvents at a high pressure (1500–2000 psi, 10–14 MPa) and temperature (50–200 °C), without reaching their critical point, in order to extract the desired analytes from solid samples. The use of liquid solvents leads to enhanced performance as compared to classical approaches carried out at room temperature and atmospheric pressure as a result of solubility, mass transfer effects, and disruption of surface equilibria.[104]

The increased temperature used in ASE gives rise to enhanced solubility of target analytes in the selected solvent. Faster diffusion rates are observed in these conditions when compared with low-temperature extractions. Furthermore, generation of a concentration gradient by introducing fresh solvent during a static extraction step boosts the mass transfer rate according to Fick's first law of diffusion.

The use of high temperatures can also disrupt the strong non-covalent interactions (van der Waals, hydrogen bonding, and dipole interactions) produced between analytes and the active sites of the matrix, and leads to decreased viscosity and surface tension of solvents, analytes, and matrix, which, on the whole, allows better contact of the analytes with the solvent. Finally, the use of high pressures facilitates the extraction process since the solvent is forced to be in contact with matrix pores and poorly accessible areas of the matrix where the analyte could be trapped.

A number of excellent review articles covering ASE have been published.[86,105–108] Green aspects of this technique, when compared to classical approaches such as Soxhlet extraction, sonication with non-polar solvents, or blending, are its lower consumption of organic solvents (10–40 ml) and the speed of determinations for solid matrices.[86] Moreover, the reduced time needed to perform the extraction (a complete extraction procedure needs 20 min) has an impact on energy saving as compared to time-consuming conventional procedures. In addition, commercially available ASE systems provide a high degree of automation, thus allowing the sequential extraction of several samples. PLE can also be adapted to other stages of the analytical process, enabling its partial or total automation.[109] Nonetheless, clean-up procedures are usually needed after ASE because of the limited selectivity provided by organic solvents, whose consumption is not as low as would be desirable from the point of view of green chemistry. Furthermore, energy requirements to maintain the pressure and temperature conditions are quite high.

The SWE approach is identical to ASE, but water replaces organic solvents as extractant. The use of water is a valuable alternative to organic solvents because it is non-flammable, non-toxic, readily available, and environmentally benign. Hawthorne et al. demonstrated the huge potential of superheated water at a pressure high enough to maintain the water at liquid state for sequentially extracting polar, moderately polar, and non-polar organic pollutants.[93] SWE takes advantage of the decreased dielectric constant of water when the temperature is increased at moderate pressures due to reduced hydrogen bonding.[4] Thus, a sharp decrease of the dielectric constant from about 90 to 20 occurs when the temperature is raised ambient temperature to 300 °C, hence allowing a change in polarity.

SWE is therefore a modern sample pretreatment technique suitable for complex matrices that can easily be coupled on-line to chromatographic techniques. However, the use of subcritical water gives rise to dilution of the analytes in the extract (then requiring preconcentration steps prior to the analytical measurement) and lower selectivity in comparison with SFE. Like SFE and ASE, degradation of thermally labile analytes could occur during the hot-water extraction process.[86] In order to overcome the latter drawback, the use of binary mixtures of water with miscible organic solvents has been proposed for extraction, giving rise to acceptable extraction efficiencies and reduced risks of analyte decomposition. The addition of modifiers and additives to water modifies its relevant physicochemical properties and the critical temperature and pressure. Quantitative extraction of analytes showing a wide range of polarities is feasible by temperature and pressure tuning, although SWE may not be recommended for the extraction of both non-polar and thermolabile analytes.[110]

Although SWE is a solvent-free sample preparation technique, the dilute extract obtained and its incompatibility with several analytical techniques, as a result of its aqueous nature, makes a concentration/extraction step necessary before the analytical measurement. Different possibilities have been described for this purpose, such as:[111]

- Extraction of the analytes present in the aqueous extract with a small volume of solvent

- Solid-phase extraction (SPE), solid-phase microextraction (SPME) or stir bar sorptive extraction (SBSE) of the aqueous extract
- *In situ* trapping of the analytes by adding a trapping agent to the extraction vessel
- Direct coupling of the extraction system to the assay procedure.

However, when this concentration/extraction step is compulsory, green aspects of SWE, such as the solvent-free nature of the technique, energy, and time saving, as well as simplicity and convenience, are diminished. From the different possibilities listed above, the use of microextraction techniques such as SPME or SBSE after SWE could be considered the greenest alternative owing to their solvent-free nature. Thus, a completely solvent-free sample pretreatment may be performed in solid matrices with this combination.

The state of the art of SWE is presented in book chapters and reviews, which also discuss a wide variety of applications.[106,110,112]

4.6 Solid-Phase Extraction

SPE is a mature sample preparation technique used for the isolation of target analytes from a fluid phase. The principal goals of SPE are preconcentration of target analytes, sample clean-up, and medium exchange (transfer from the sample matrix to a different solvent or to the gas phase).[113] Even though experimental trials concerning sample preparation using a solid sorption material (carbon filters) were published almost 60 years ago,[114] the expansion and introduction of SPE was produced in the early 1980s as a result of the need to decrease the organic solvent usage in laboratories.[115] SPE is currently the most popular sample preparation technique for liquid samples and organic analytes, being used in 22 official methods.[116] There is an excellent retrospective review dealing with the evolution of SPE.[117]

When compared with conventional liquid–liquid extraction (LLE), SPE reduces the consumption of organic solvents, which has a dramatic effect on the reduction of wastes. In addition, high enrichment factors and elevated recoveries are feasible with SPE using short times.

SPE is a widely used sorbent technique that involves a partitioning of solutes between two immiscible phases, *i.e.* a liquid (sample solution) and a solid (sorbent) phase.[118] The efficiency of the extraction process is therefore dependent on the type of sorbent material employed. SPE media can be classified as inorganic oxides, low-specificity sorbents, and compound or class-specific sorbents.[113] Silica-based, chemically bonded sorbents are by far the most popular for SPE. On the other hand, inorganic oxide adsorbents have limitations derived from their irreversible adsorption and catalytic breakdown of sensitive analytes that could give rise to low recovery values for some analytes. With the main goal of enhancing the selectivity in SPE applications, different sorbents based on ion exchange, bioaffinity, molecular recognition, and restricted access materials have also been proposed in the literature, constituting promising SPE sorbents.

Several SPE formats (syringe barrels, cartridges, and discs) are commonly used. Although they all use the same sorbent technology, their different designs make them useful for different applications. Cartridges are generally the preferred option, since they can be easily prepared in the laboratory, making them a cheaper alternative to discs. Nevertheless, discs are recommended for some applications, especially for large sample volumes containing suspended particles. Discs allow higher sample flow rates, hence providing shorter sample processing times than cartridges or syringe barrels.

In general, four consecutive steps are needed to accomplish the SPE procedure, regardless of the SPE format:[92,118]

1. *Wetting and conditioning the sorbent.* The sorbent must be wetted by an appropriate solvent, followed by its conditioning with a solvent or buffer similar to the sample solution. This ensures an appropriate contact between the analytes and the sorbent material.
2. *Loading the sample.* The sample is then forced through the sorbent material by gravity, pumping or aspiration by vacuum. The analytes of interest will be exhaustively retained by the sorbent. Some compounds present in the sample matrix will pass through the solid sorbent without being retained. Thus, preconcentration of analytes and purification of the sample are achieved in this step.
3. *Rinsing the sorbent to elute interferences.* Washing of the sorbent material with a suitable solvent allows eliminating matrix compounds retained in the solid sorbent without displacing the analytes.
4. *Elution of target analytes.* Target analytes are finally eluted from the solid sorbent by using the minimum solvent volume. It is advisable to carry out the elution of the analytes by using two different aliquots of the corresponding solvent. In addition, the solid sorbent should be soaked with the solvent before the elution.

As can be easily noted, different proportions of organic solvents are needed to perform the different SPE steps. Although the solvent consumption is on the whole substantially decreased as compared to classical extraction approaches, the use of such solvent volumes (5–15 ml) make SPE less green than miniaturized extraction techniques (see section 4.7) in relation to solvent consumption, waste generation, and occupational exposure to hazardous chemicals. In this respect, reduced solvent volumes are needed for both conditioning and elution steps when discs are used instead of other SPE formats.[119]

Automation of SPE systems provides several advantages over manual operation. Reduction of processing time, as well as improvement of accuracy and precision, are the most notable benefits of automation. In addition, minimum operator intervention is produced with automated SPE, which gives rise to negligible solvent exposure. Improved sample throughput is obtained with automated parallel-sample processing systems, which are capable of processing up to 400 samples per hour. Among the different automated SPE systems available, commercial 96-well workstations have become very popular, especially in clinical and pharmaceutical laboratories.[119,120] Automation of SPE is, however, not free from drawbacks, mainly due to analyte carryover and sample stability.

SPE is used as a sample pretreatment technique in many areas of chemistry, including environmental, pharmaceutical, clinical, food, and industrial chemistry. The availability of different sorbent materials makes sorptive techniques advantageous over other extraction techniques. However, the final extract solvent is not always compatible with the corresponding analytical technique, thus involving additional steps.

4.7 Microextraction Techniques

Downsizing the scale of the analytical methods through miniaturization is one of the major challenges of GAC. Chapter 5 is devoted to this aspect. However, microextraction techniques as one of the powerful advantages in green sample preparation are considered here also.

Removal or minimization of solvent volumes is a priority towards the development of solvent-free sample preparation approaches, in accordance with the GAC principles.[121] In this context, miniaturization and improvement of classical sample preparation techniques such as SPE and liquid–liquid extraction (LLE) have given rise to the development of different microextraction techniques. Inherent benefits derived from the solvent-free character of these techniques include negligible (or non-existent) waste generation, reduced exposure of the analyst to hazardous chemicals, and simplification and promptness of sample preparation. Furthermore, microextraction techniques integrate sampling, extraction, preconcentration, and even derivatization in a single step.

A schematic representation of the different microextraction techniques discussed in this section is shown in Figure 4.2.

4.7.1 Solid-Phase Microextraction

Solid-phase microextraction (SPME) is a sample handling technique characterized as being a real non-solvent technique. It was introduced in 1990 by Arthur and Pawliszyn.[122] There is an abundant literature about SPME theory and applications.[123–128]

In SPME, a fibre coated with a small amount of polymeric phase (typically less than 1 μl) is exposed to the sample (directly immersed in the bulk sample solution or exposed to the headspace above the sample) in such a way that the analytes present in the sample solution are partitioned between the sample and the coating until equilibrium is reached or the extraction process is stopped. Analytes adsorbed on the coating are then thermally desorbed for analysis, mainly in the injection port of a gas chromatograph. When SPME is combined with liquid chromatography, analytes are eluted from the coating by using a minimum volume of an appropriate organic solvent.

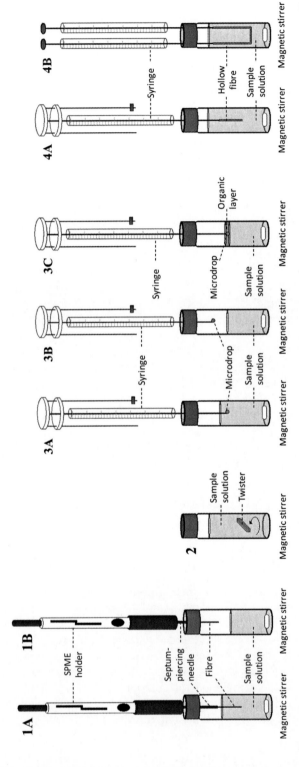

Figure 4.2 Schematic representation of different green sample preparation techniques: (1A) direct-SPME; (1B) HS-SPME; (2) SBSE (Twister™); (3A) direct-SDME; (3B) HS-SDME; (3C) LLLME; (4) HF-LPME configurations, with one syringe (4A); and two syringe needles (4B).

Three SPME modes can be described, depending on the properties of the sample and/or the analytes to be extracted:

1. Direct-SPME (Figure 4.2(1A)): The most appropriate SPME coating is directly exposed, by immersion, to the stirred sample solution placed in a vial. This is the mode recommended for extraction of target analytes, regardless of their volatility, present in clean matrices.
2. Headspace-SPME (HS-SPME) (Figure 4.2(1B)): The fibre is exposed to the headspace of the sample solution for extraction and preconcentration of volatile or semivolatile analytes. Derivatization reactions could also be carried out for the generation of volatile derivatives of the analytes of interest. HS-SPME allows the use of extreme conditions (such as pH of the sample solution) without damaging the SPME fibre.
3. Membrane protected-SPME: Membrane protection should only be used when neither of the first two modes is applicable. This is the case for non-volatile analytes present in very complex and dirty samples.

The choice of extracting phase is essential in SPME to achieve selective and efficient extraction of the analytes to be determined. Extracting phases covering a wide range of polarities are commercially available, polydimethylsiloxane (PDMS) being the most commonly used coating. High molecular weight polymeric liquids or solid adsorptive coatings can also be used as SPME extracting phases. Typically, the chemical nature of a target analyte determines the type of coating to be used, in accordance with the rule 'like attracts like'.[128]

SPME can be considered as a very green sample preparation technique, because of its solvent-free characteristics. SPME allows preconcentrating the analytes in an easy way without the need for solvent evaporation after the extraction process, as typically occurs with conventional extraction techniques. The total removal of organic solvents for sample preparation avoids or minimizes the waste generation derived from the sample preparation step. Moreover, it can be used for the extraction and preconcentration of target analytes present in very small samples because of the convenient dimensions of the SPME system. The amount of extracted analyte depends not on the sample volume but on the concentration of the analyte in the matrix, so direct sampling of the ambient air, water, production stream, *etc.* is possible. A high degree of automation of SPME systems is nowadays possible with the use of Combi-PAL™ and TriPlus autosamplers, introduced by CTC Analytics and Thermo Fisher Scientific, respectively. SPME is, however, not free from drawbacks, such as the relatively high cost and fragility of the fibres, as well as the possibility of carryover between analyses. Moreover, significant lot-to-lot variations are typically observed. The low volume of polymeric phase leads to non-exhaustive extractions when the partition coefficient of the analytes is not very large and the coating thickness is too thin. In addition, a limited number of SPME fibres are commercially available, although a high number of home-made coatings have been proposed in the literature and may be used for specific applications.[129]

4.7.2 Stir Bar Microextraction

SBSE was first reported by Baltussen *et al.* in 1999.[130] SBSE is a solvent-free sample preparation technique based on the same principles as those of SPME, *i.e.* sorptive extraction of analytes on to a polymeric material.[131,132]

SBSE has been developed as an attempt to circumvent the limitation of SPME to achieve complete extraction efficiencies. The fibre used in SPME has a bound stationary phase volume (0.5 µl of PDMS) which leads to incomplete extraction of analytes due to the phase ratio between the aqueous and PDMS phase, even when the distribution constant is favourable. Unlike SPME, in SBSE a polymeric material is placed not on a fixed fibre but as a coating on a magnetic stir bar of 1–4 cm length. The use of stir bars allows the use of larger amounts of sorbent material (50–300 µl of PDMS) as compared to SPME. As a consequence of the significantly increased coated phase volume, SBSE shows higher recoveries and sensitivity than SPME.

PDMS coated stir bars are commercially available (Twister^{TM}, Gerstel GmbH) with two different dimensions, 10×3.2 mm and 40×3.2 mm, for the analysis of 1–50 ml and 100–250 ml sample volumes, respectively.

As in SPME, the extraction of the analytes is controlled in SBSE by the distribution constant of the analytes between the polymer coating and the sample matrix and by the phase ratio between the polymer coating and the sample volume.[132]

Applications of SBSE are commonly carried out by direct immersion of the coated stir bar into the sample solution, although a headspace mode of SBSE (HS-SBSE) may be used for extraction of volatile and semivolatile compounds from liquid and solid samples. Thus, direct-SBSE is performed by simply adding the PDMS-coated stir bar to the sample solution, which is then stirred (1000–1500 rpm) for the prescribed time, commonly more than 30 min. On the other hand, HS-SBSE is carried out by exposing the coated stir bar to the headspace of a stirred sample. Once the extraction step is finished, the stir bar is removed with tweezers and dried with a soft tissue. The analytes are subsequently desorbed from the coated stir bar either thermally, when gas chromatography (GC) is used, or with a low volume of organic solvent mainly, for high performance liquid chromatography (HPLC) analysis. Figure 4.2(2) shows a basic configuration for Direct-SBSE.

Thermal desorption is the preferred, greener option when volatile analytes are determined after GC separation. The desorption process is, however, relatively slow (5–15 min) owing to the large amount of sorbent used in SBSE. In addition, a special interface is mandatory for thermal desorption of target analytes due to the dimensions of the stir bar used in SBSE. Apart from thermal desorption, liquid desorption with an appropriate organic solvent may also be performed when liquid chromatography is used. This alternative broadens the applicability of SBSE but at the expense of losing the solvent-free nature of this sample preparation technique. Both thermal and liquid desorption can be performed using automated systems nowadays marketed by Gerstel.[133]

SBSE is mainly useful for extraction and preconcentration of organic compounds in liquid samples. Miniaturization and solvent-free operation are the outstanding characteristics of this sample preparation technique. The main advantage of SBSE is directly derived from the design of the coated stir bars and the volume of sorbent employed, which allows quantitative extractions and therefore provides lower limits of detection when compared with SPME. However, the extraction time required for the complete extraction of analytes is commonly excessive for routine analysis, so non-equilibrium extractions could be performed to achieve a more convenient sample throughput. Some drawbacks of SBSE systems are due to the incompatibility of the coated stir bar design with the injection port of the GC, which makes necessary the use of a thermal desorption unit. Moreover, the limited range of commercial coating polymers, PDMS being the only polymeric material used, limits the applicability of SBSE. In addition, automation of the extraction step is not an easy task, being generally performed by manual handling.

4.7.3 Liquid Phase Microextraction

LLE is a sample preparation technique that has been widely used in analytical labs. It involves the use of organic solvents for the extraction and preconcentration of the analytes present in a given sample on the basis of their relative solubility in the two immiscible phases. LLE is not free from drawbacks, especially relating to the use of large volumes of hazardous organic solvents that results in an excessive waste generation and high exposure risks. Furthermore, LLE is a time-consuming technique that yields small sample-to-extractant phase volume ratios which, in turn, can give rise to insufficient enrichment factors. In accordance with the first principle of green chemistry, it is better to prevent waste than to treat to clean up waste after it has been created. Therefore, there is a clear demand for the minimization of solvent consumption in sample preparation. Biodegradable and environmentally benign extractant phases should also be used when possible.

Nowadays, different alternatives can be found in the literature concerning the minimization of conventional LLE. These developments are included under the generic name liquid-phase microextraction (LPME). Some of the most widely used LPME techniques are single-drop microextraction (SDME), hollow fibre liquid-phase microextraction (HF-LPME) and dispersive liquid–liquid microextraction (DLLME). Recent reviews[134,135] and a book[136] dealing with this subject can be consulted for further details.

4.7.3.1 Single-Drop Microextraction

In SDME, a microdrop of extractant (1–10 µl) is used as acceptor phase of the analytes present in the sample. A microsyringe is used to hold the drop during the extraction process. There are different modes of SDME, which are schematically represented in Figure 4.2.

1. Direct-SDME (Figure 4.2(3A)): an immiscible microdrop of extractant (organic solvent or IL) is directly immersed into the bulk aqueous solution, which is continuously stirred during the extraction process.[137] Both volatile and non-volatile non-charged analytes are likely to be extracted according to the principle 'like dissolves like'. To extract ionic analytes, derivatization can be performed before (in sample) or during (in drop) the microextraction process by addition of the corresponding derivatizing agent to the sample or to the extractant phase, respectively, then forming a non-charged compound that can be extracted and enriched within the extractant phase.
2. Headspace-SDME (HS-SDME) (Figure 4.2(3B)): it is similar to HS-SPME.[138,139] A drop of extractant phase (organic solvent, IL or even an aqueous drop) is exposed from the needle of the syringe to the headspace above the sample to extract volatile or semivolatile analytes. Since the analyte mass transfer is time dependent, it is advisable to use extractant phases with low volatility (low vapour pressure and high boiling point) in order to minimize drop evaporation losses during the microextraction process. Derivatization reactions can be carried out in order to generate derivatives with suitable properties to be detected by the corresponding analytical technique.[140–142]
3. Liquid–liquid–liquid microextraction (LLLME) (Figure 4.2(3C)): an aqueous microdrop is immersed into a thin film of low density organic solvent placed over the aqueous sample.[143] This SDME mode is indicated for the extraction of ionizable analytes. The pH of the sample solution must be controlled in such a way that non-charged analytes are formed to allow their transfer into the organic phase. Then, the pH of the aqueous microdrop should allow the ionization of analytes so that they are finally extracted into it. A high degree of sample clean-up can be achieved with this SDME mode, although it requires thorough manipulation of the microdrop.

Organic solvents used in SDME must be immiscible with water and show very low water solubility when used in its immersed mode (direct-SDME), and low volatility when used in HS-SDME. Extractant phase volumes of 1–10 µl are commonly used in SDME, although nanolitre volumes of organic solvents have been reported.[144,145] In spite of the virtually solvent-free nature of the SDME technique, the possibility of using ILs and aqueous microdrops as extractants of different analytes leads to the employment of certainly clean methodologies which, moreover, offer powerful analytical advantages, such as the excellent enrichment factor that may be achieved in relatively short extraction times.

4.7.3.2 Hollow Fibre Liquid-Phase Microextraction

In hollow fibre liquid-phase microextraction (HF-LPME) the extractant phase is protected by a porous hollow fibre attached to the needle of the microsyringe.[146,147] Thus, HF-LPME can be considered more robust than immersed SDME modes (direct-SDME and LLLME), since the hollow fibre provides an improved stability to the extractant phase at high stirring rates of the sample. In addition, the use of the hollow fibre gives rise to a configuration of the extractant phase (cylindrical instead of spherical) that speeds up the extraction process.

To perform a HF-LPME, a microsyringe is used to fill the lumen of the hollow fibre with the appropriate extractant phase, as well as to hold it in place during the microextraction process. Once the microextracton process has finished, the extractant phase enriched with target analytes is retracted and incorporated into the corresponding analytical instrument for data acquisition. A new piece of hollow fibre is generally used for further extractions to avoid any possible carryover effect. Two different HF-LPME modes analogous to direct-SDME and LLLME can be used:

4. HF-LPME (Figure 4.2(4A)): Both the lumen and the pores of the hollow fibre are filled and impregnated with the corresponding extractant phase (organic solvent or ionic liquid). The mass transfer of analytes is produced through the pores of the hollow fibre.
5. HF-LLLME (Figure 4.2(4B)): This mode is used for the extraction and preconcentration of ionizable compounds. In this case, organic solvents or ILs may be used to impregnate the pores of the hydrophobic hollow fibre, while an aqueous solution is used to fill the lumen of the polymeric membrane. As in the case of LLLME, suitable pH adjustment of both the aqueous sample and the aqueous extractant phase is essential.

As a result of the controlled size of pores in the hollow fibre, this technique provides a high degree of clean-up even when complex matrices such as biological fluids are analysed by excluding macromolecules and other compounds that could interfere.[148] In addition, extractant phase volumes are commonly between 2 and 25 µl, thus involving negligible solvent consumption.

4.7.3.3 Dispersive Liquid–Liquid Microextraction

In DLLME,[149,150] a mixture of an immiscible organic solvent with higher density than water and a solvent miscible with both sample and extractant phase, known as disperser solvent, is used to extract the analytes. The injection of this mixture leads to the formation of tiny drops of extractant phase distributed along the whole sample. After centrifugation, the enriched sedimented phase is used for analysis. This LPME technique is characterized by its ease of operation, expeditiousness, and the possibility of performing several extraction processes simultaneously, hence enhancing the sample throughput. Ionic liquids have been proposed as alternative extractant phases to commonly used chlorinated organic solvents.[151] However, the addition of an antisticking agent to the sample solution is needed because of their high viscosity. In addition,

DLLME with *in situ* metathesis reaction has been proposed for the extraction and preconcentration of interesting compounds.[152] The method is based on the formation of a hydrophobic IL in the sample solution by addition of a hydrophilic IL and an ion-exchange reagent to the aqueous sample.

The use of microvolumes of extractant phase, generally in the range 1–100 μl, makes LPME a powerful and clean methodology for separation and preconcentration. LPME modes facilitate the reduction of organic solvents needed to perform the separation process, resulting in reduced toxic wastes and minimum exposure risks. Moreover, the reduction of the volume of extractant phase potentially contributes to the achievement of high enrichment factors. Thus, combination of LPME with the most appropriate analytical technique allows trace and ultratrace analysis.

Several attempts to automate LPME methodologies have been reported, including a patented system for automation of LPME presented by Kokosa, which makes use of commercial autosamplers and software to control the different steps needed to perform the microextraction process.[153] However, full automation of the whole microextraction process seems to be difficult to achieve, *e.g.* in the particular case of HF-LPME, where the fibre is commonly attached to the needle of the syringe by hand.

4.8 Membrane-Based Extraction

Membrane-based extraction can be considered today as an alternative to miniaturization and automation of traditional extraction procedures,[154] especially when it replaces the traditional LLE. The use of small amounts of organic solvents (<1 ml) or their virtual elimination are the main characteristics, as in novel microextraction techniques, for considering membrane-based extraction as a greener tool for isolation and preconcentration of analytes.[41] In comparison with SPME and SDME, membrane-based extraction has a greater capability for sample clean-up. The use of membranes for analytical extractions has increased in recent years. Different applications in biomedical, environmental, food, and industrial fields have been published. Several reviews,[154–160] books,[161] and book chapters[162–164] have been published on the use of membrane-based extraction.

In general, membranes can be considered as selective barriers between two phases: donor or feed phase and acceptor or stripping phase.[159] There are two basic types of membranes: size-exclusion and extraction membranes. In the first type, the separation occurs as a result of the sieving effect. Then, the goal in the selection of a material for size-exclusion membranes is to minimize chemical interactions between the membrane and the sample matrix or the analytes. On the other hand, membrane-based extraction techniques provide real chemical selectivity. An interaction between the membrane material and the analyte must take place.[163]

Extraction membranes can be classified according to membrane morphology (quantity, size and distribution of porous through the membrane structure) as:

Green Sample Preparation Methods

- Selective microporous membranes
- Homogeneous extraction membranes (nonporous membranes):
  1. Liquid membranes
  2. Polymeric membranes.

With the exception of the membranes called polymeric, which are non-porous, the extraction membranes are porous hydrophobic synthetic organic polymers and they become non-porous when an organic solvent fills their pores. Polypropylene (PP), PTFE, and polyvinylidene difluoride (PVDF) are the polymers most frequently used in their manufacture.

Two basic configurations of membrane extraction can be used: flat sheet (FS) and hollow-fibre (HF) membranes (Figure 4.3). The first configuration requires the implementation of an extraction module (Figure 4.3A,C) to support the membrane (usually two blocks between which the membrane is placed, with microchannels and typical volumes of 10–1000 μl).[159] FS systems can be easily miniaturized and, when operated in flow, they can be automated and linked with different analytical techniques, usually chromatography (HPLC and GC), but also capillary electrophoresis (CE), AAS, spectrophotometry and electrochemical techniques. In on-line configuration, the donor phase (sample) is pumped up to its membrane site. In the other membrane site, the organic solvent can be pumped and used in flow or static mode. According to the phase volumes, peristaltic (with large membranes, channel volumes around 1000 μl) or syringe pumps can be used with this purpose. The HF has capillary shape

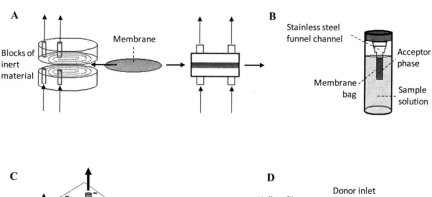

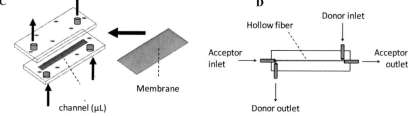

Figure 4.3 Schematic representation of membrane based extraction systems: (A) flat membrane module (1 ml); (B) bag-shaped membrane; (C) flat membrane module (10 μl); (D) on-line hollow fibre module (1.3 μl).

and usually is self-supporting (Figure 4.2(4A), (4B)), though HF units can also be used (Figure 4.3D).[157] The lumen of the HF contains the acceptor phase (typically 2–25 µl of an organic solvent). HF membrane configuration is also used in microextraction (see section 4.7.3.2). A faster mass transfer takes place in HF format as compared to FS format due to its high surface-to-volume ratio. However, the HF format only can withstand small pressure differences across the membrane. Different approaches of HF automation have been published,[154] though the in-vial configuration is the most used.

Different extraction techniques can be associated to the use of membranes: microporous membrane liquid-liquid extraction (MMLLE), supported liquid membrane extraction (SLME), and polymeric membrane extraction (PME). Membrane-assisted solvent extraction (MASE) is a PME modality with an organic acceptor phase. The procedure that combines PME with a sorbent interface is called membrane extraction with sorbent interface (MESI).[159]

In MMLLE, the microporous membrane acts as a support that allows the contact between the phases without merging them.[163] The donor is usually aqueous and the acceptor is a water-immiscible organic solvent. The acceptor phase fills the pores of hydrophobic membranes (making them non-porous), then a two-phase system is formed and one partition reaction occurs. This is equivalent to a single-step LLE and so the efficiency of this system is largely dependent on partition coefficients.[160] It is uncommon to use trapping reactions, so this technique is applied to hydrophobic, preferably uncharged, compounds such as polycyclic aromatic hydrocarbons (PAHs), chlorinated pesticides, or polychlorinated biphenyls (PCBs). MMLLE with FS (FS-MMLLE) is simpler than classical continuous LLE. Fully automated and miniaturized (1 µl) systems, both for FS-MMLLE and for HS-MMLLE, can be obtained with extracting syringe devices. More information about the automation of MLLE can be found elsewhere.[154,160] Miniaturization of MMLLE in a chip has been developed. The chip was composed of two microfabricated glass plates with a sandwiched microporous membrane.[165]

In SLME, membrane pores are filled with an organic liquid different from that of the acceptor and donor phases (typically two aqueous phases). This type of membranes allows performing extraction processes without any acceptor or donor organic phase, although the membrane must be previously impregnated in an organic solvent. This supposes introducing the membrane into certain volume of organic solvent, with the consequent generation of wastes. Solvents such n-decane or kerosene (long-chain hydrocarbons) and more polar compounds, *i.e.* dihexyl ether and dioctyl phosphate, are commonly used in these membranes. In addition, different additives can be used in order to increase the efficiency and selectivity of the extraction process. SLME can be considered chemically analogous to an extraction followed by a back-extraction in LLE.[162] Different chemical principles have been used for polar and non-polar analytes, *e.g.* simple permeation for acids and bases, carrier-mediated transport for metals and organic ions, or even immunological trapping. The kinetics of microextraction across supported liquid membrane (SLM) has been improved using electrokinetic migration.[166] High enrichment factors, and especially high selectivity, are characteristics of SLME processes. Metal ions, amino acids, surfactants, sugars, herbicides, drugs, *etc.* are among the analytes determined by means of SLME. HF-SLME configurations without automation are probably now the most used, although microfluidic devices with FS allow easy automation and miniaturization of SLME.[167] Other micromembrane configurations, *e.g.* vial devices,[168] have been proposed for automation of SLME. An important drawback of these membranes is stability. Pressure differences between both sides of the membrane must be low enough for the solvent to settle in the pores.

PME is a three-phase system that uses nonporous membranes made of silicone rubber (normally PDMS), low-density polyethylene (LDPE), or dense PP. These membranes provide a stable system because their lifetime is longer than that of the SLMs. The difference in the solubility and diffusion of different analytes into the polymer is the basis of the selectivity.[169] The fixed composition of the membranes reduces their applicability because the possibilities of obtaining chemical selectivity diminish. PME is especially troublesome for relatively polar analytes. In these cases, the addition of complex or ion-pair formers is necessary. Two modes can be used in PME, one similar to SLME (aqueous phase–polymer–aqueous phase) and other similar to MMLLE (aqueous phase–polymer–organic phase). The latter system is applied in MASE. A fully automated MASE device is commercially available. A bag-shaped membrane configuration is used (Figure 4.3B) and only 500–800 µl of organic solvent are necessary. However, a preconditioning or cleaning of the membrane is necessary in order to remove co-extractable compounds. This process can consume 1.6–50 ml of organic solvent. PME processes are slower than SLME or MMLLE because polymeric membranes have smaller diffusivity and consequently smaller transport velocity through them. A combination of MASE with PLE allows the reduction of extraction time.[170] When PME is used with a sorbent trap on-line, it gives way to MESI. This technique allows enhanced sensitivity in a simple and continuous operation. The donor phase is aqueous or gaseous and the acceptor phase is gaseous. Cryogenic or sorption traps are used. MESI is an interesting green sample preparation system in combination with gas chromatography (MESI-GC). Volatile or semivolatile organic compounds (VOCs and SVOCs) in different environmental matrices (aqueous and air samples) have been determined. MESI is compatible with portable GC for on-site field sampling.

Other green membrane extraction approaches have been proposed, such as membrane inlet mass spectrometry (MIMS). When the membrane is fitted into the inlet of a mass spectrometer, as an integral part of the sampling system, the analytes pass across the membrane depending on their solubility following a pervaporation process (adsorption–diffusion–evaporation).[159] VOCs are rapidly and directly determined without using organic solvents.

In principle, extraction-based membranes seem to fulfil the goal of GAC, *i.e.* the development of analytical procedures that generate less hazardous wastes, are safer to use and more environmentally friendly.[4] However, we should not forget that phenomena occurring in the use of hydrophobic membranes, such as soiling, demand a cleaning operation involving organic solvents, with the subsequent generation of wastes. Organic solvents are also required for regeneration of SLMs.

4.9 Surfactant-Based Sample Preparation Methods

Procedures involving surfactants (or surface-active agents) can be considered a milestone in the development of greener sample preparation methods. In comparison with organic solvents, surfactants are non-volatile, non-flammable, non-toxic, and cost-effective.

One of the most important properties of surfactants is their good capacity to solubilize water-insoluble or sparingly soluble solutes in water through binding to micelles in the aqueous solution.[171] This ability allows surfactants to replace more expensive and dangerous organic solvents. Paleologos *et al.* consider them as the solvents of the modern era, together with ILs.[172] These authors emphasize the privileged selectivity of the micellar microworld *versus* the macroenvironment of bulk organic solvents.

Surfactant micelles have attracted considerable attention in the last few years in analytical chemistry, especially as potential extracting media, and continue to have a broad appeal for different applications.[173] In this sense, some surfactant-based sample preparation methods can be outlined: cloud-point extraction (CPE), MW-assisted micellar extraction (MAME), ultrasound-assisted micellar extraction (USME), emulsification, and others such as micellar-enhanced ultrafiltration (MEUF) or reverse micelle-based extraction.

4.9.1 Surfactant-Based Extraction

Since Watanabe and Tanaka introduced CPE to preconcentrate metal ions from aqueous samples in 1978,[174] it has been extended and exploited as a versatile and simple extraction method for metal ions as well as organic compounds from environmental, biological, and clinical samples.[5,173,175]

CPE is based on cloud point and phase-separation phenomena in a surfactant aqueous solution (about 1% w/v). When a micellar solution is heated, it becomes turbid over a narrow temperature range, which is referred to as the cloud-point temperature.[176] Above the cloud-point temperature, a turbid solution is formed and separates into two isotropic phases: the 'surfactant-rich phase', composed almost totally of surfactant (micellar phase), and the 'aqueous phase', in which the surfactant concentration is approximately equal to the critical micelle concentration of the surfactant.[177]

The separation of the two phases requires appropriate experimental conditions depending on the nature of the surfactant. Thus, phase separation requires temperature changes for non-ionic and zwitterionic surfactant solutions. For ionic surfactants, changes in pH or addition of ionic salt or of organic solvent can be also necessary.[175] Then, the most important factors that can be taken into account for performing CPE are the nature and concentration of the surfactant, pH of sample solution, temperature, time to reach equilibrium, and ionic strength. For the efficient extraction of inorganic species, quantitative formation of a hydrophobic complex is an essential prerequisite.

In general, CPE is more efficient whether the surfactants or the analytes are more hydrophobic. The range of surfactant concentrations that can be used is narrow, since above the optimal range the analytical signal deteriorates and the preconcentration factor decreases, and below the optimal range accuracy and reproducibility can worsen.

pH is perhaps the most critical factor, especially for organic molecules (typically 6–9). For ionizable species, maximum extraction efficiencies are achieved at pH values where the uncharged form of the analyte prevails. For inorganic species, few differences are observed in the extraction efficiencies at different pH, with the exception of the pH-dependent reactions, where pH seems to control extraction efficiency.

Temperature seems to be a key factor in improving preconcentration efficiency and enhancement factors (typically <100 °C, 2–90 °C). CPE should be carried out at a temperature higher than the cloud-point temperature for a given time. This increases the signal by a factor as high as 3.[172] As a general principle, CPE at high temperatures cannot be used when thermolabile compounds are determined and acidic solutions are not suitable for weakly basic compounds (ionizable in low-pH solutions).

In general, CPE involves five consecutive operation steps:

1. Addition of the surfactant to the sample
2. Maintenance of suitable temperature for some time
3. Centrifugation
4. Decantation of the supernatant
5. Suitable treatment of the surfactant-rich phase.

Decantation of supernatant is the most critical step. Removal of aqueous surfactant phase from the micellar phase is usually performed by cooling the test tube, hence increasing the viscosity of micelles.[175] Total removal of water can be attained by evaporation using a neutral gas (*i.e.* nitrogen, argon, or helium). As the micelle-rich phase is viscous and cannot be injected directly into the apparatus, it should be treated by dilution. When ionic surfactants are used, steps 2 and 5 should be omitted, but additional clean-up of the extract may be necessary.

The analytical utility of CPE procedures in the analysis and isolation of organic molecules metal ions and element species has been widely accepted. For instance, the differentiation–speciation of Cr(VI) *vs* Cr(III) or As(III) *vs* arsenobetaine (AsB) and As(V) through CPE could be safely applied for microscale analytical applications.[172,173]

CPE offers a lower toxicity for the analyst and the environment than some classical extraction methods. In comparison with LLE and SPE, higher preconcentration factors with lower cost are achieved.[178] This technique is also attractive for its easy operation and rapidity (extraction times are about

Table 4.2 Some techniques mainly used for extraction of organic analytes.

Extraction technique	Extraction time/ Operating conditions	Organic extractant consumption	Level of automation/ Cost of investment	Greenness-related issues
Classical extraction techniques				
Soxhlet	6–24 h, up to boiling solvent temperature	150–500 ml	Low/low	High consumption of solvents. High risk of exposure of laboratory personnel to vapours of organic solvents. High waste generation. Soxhlet is especially time consuming. Simple techniques.
Soxtec	2–4 h, up to boiling solvent temperature	40–50 ml	High/low–moderate	
LLE	15–25 min	10–200 ml	Low/low	
Microwave-assisted extraction				
PMAE	10–30 min, up to boiling solvent temperature, closed vessel	10–40 ml	High/moderate	Less energy requirements than classical soxhlet or soxtec. Possible problems of safety when closed vessels are used
FMAE	10–30 min, open vessel	30–70 ml	High/moderate	
Ultrasound assisted extraction				
UAE	10–60 min, with ultrasonic bath	50–200 ml	Low/low–moderate	Safe procedure (atmospheric temperature and pressure). Low energy requirements. Very simple technique
Supercritical fluid extraction				
SFE	10–60 min, pressure up to 1100 psi and temperature up to 31.1 °C when CO_2 is used	2–5 ml (solid trap); 30–60 ml (liquid trap)	High/high	High energy requirements. Using solid trap only small volumes of organic solvents are necessary

Green Sample Preparation Methods

Method	Time/Conditions	Volume	Rating	Comments
Pressurized liquid extraction				
ASE	10–20 min, pressure up to 2000 psi and temperature up to 200 °C	10–40 ml	High/high	High energy requirements. More rapid procedures. When SWE is used with SPE, SPME or SBSE a virtually solvent-free technique is achieved
SWE	10–20 min, pressure up to 3000 psi and temperature up to 300 °C	10–25 ml only when a solvent is used for back-extraction	High/high	
Solid phase extraction				
SPE	5–25 min	5–15 ml (conditioning and elution)	High/moderate	Low consumption of solvent. Reduced waste generation
Microextraction techniques				
SPME	10–60 min	—	High/moderate	Virtually solvent free. Miniaturized systems
SBSE	30–200 min	—	Moderate/moderate	
SDME	5–60 min	1–10 µl	Low/low	
HF-LPME	5–60 min	2–25 µl	Low/low	
DLLME	2–10 min	10–100 µl	Low/low	
Membrane-based extraction				
MMLLE	10–90 min	1–1000 µl	High/low	Low consumption of solvent or virtually solvent free. Miniaturization. Automation. Simple procedures
SLME	10–90 min	—	High/low	
MASE	20–120 min	500–800 µl	High/moderate	
Surfactant-based extraction				
CPE	20 min, up to cloud point temperature (<100 °C)	—	High/low	Solvent free. Simple procedures
MAME	10–30 min, in closed vessel	—	Moderate/moderate	

20 min). The possibility of automation increases the green character of this technique. The combination of flow injection analysis (FIA) with CPE has been reported for both inorganic and organic analytes.[172,173]

When MAE is used in combination with the ability of the surfactant to solubilize different species, it gives place to MW-assisted micellar extraction (MAME). Since Ganzler *et al.* introduced the first application to extract organic compounds from contaminated soils in 1987,[179] MAME has been developed for different types of matrices and analytes, mainly organic compounds.[171,180] MAME entails the MW irradiation of the sample with the surfactant in a closed vessel. After extraction, samples are left to cool. Before analysis, a clean-up might be necessary and for most applications a simple filtration using a membrane syringe filter is necessary.[171] The most important parameters that can influence in the extraction process in MAME are the nature and volume of surfactant, matrix characteristics, extraction temperature, MW power, and duration of extraction. The volume of surfactant must be sufficient to ensure that the sample is entirely immersed. If the sample is fully immersed, higher surfactant volume does not influence the extraction efficiency in contrast to conventional extraction techniques.

MAME, as a combined technique, combines the advantages of MAE (see section 4.2, MW-assisted sample preparation) and CPE. In addition, considering practical and economical aspects of MAME, the facilities of operation, rapidity, low cost, and the ease of handling, makes this technique very attractive both analytically and from a green perspective.

The combination of the surfactant capacity to solubilize and ultrasound-assisted extraction (see section 4.3, ultrasound-assisted sample preparation) is referred as ultrasound-assisted micellar extraction (USME). It has been used for the extraction of organic compounds,[181,182] *e.g.* PAHs have been extracted using 10 ml of an 0.1 M surfactant solution. USME does not require the use of potentially hazardous organic solvents, which provides an important advantage over UAE.

4.9.2 Emulsification

Other surfactant-based procedures such as emulsification have been used for sample pretreatment as an interesting strategy for liposoluble matrixes. Emulsions of different samples, especially crude oils, allow the direct and rapid analysis of samples, without resorting to strong acids or large volumes of organic solvents. It can be considered an interesting approach from a green perspective, particularly when emulsification is carried out by ultrasound.

In general, mechanical or manual stirring is used for formation of emulsion as well as for maintaining its stability during analysis. In this regard, the use of ultrasound makes it possible to obtain a stable and homogeneous emulsion more quickly, and with smaller amounts of surfactant, than mechanical stirring.[183] Ultrasonic emulsification consists of two processes: (1) interfacial instability of the oil–water interface and (2) transient cavitation bubbles that generate microstreaming, high-pressure shock waves, and high local temperature during their collapse. These phenomena accelerate the mass-transfer process between two immiscible phases by generating smaller droplets of the dispersed phase.[184,185]

Ultrasonic emulsification has been mainly focused on trace element determination, but some applications have also been described for organic compounds.[186] A comparison of ultrasonic emulsification and MW-assisted digestion has been made for trace metal determination in cosmetic samples.[187] Emulsification results in a green approach because of its simplicity (minimal operations involved), rapidity, and low consumption of reagents. Only media that contain diluted acids are needed, thereby avoiding the use of concentrated mineral acids. In contrast to MAD, mild conditions (atmospheric pressure and room temperature) are used in ultrasound-assisted emulsification.

Operating conditions, organic extractant consumption, automation level, investment costs, and greenness-related issues for some techniques mainly used in the extraction of organic analytes are shown in Table 4.2.

4.10 Present State of Green Sample Preparation

In this chapter, several analytical methodologies for sample preparation have been discussed and evaluated with respect to the paradigm of green chemistry. Progress in automation, acceleration, miniaturization, and simplification along with the use of green chemicals and new materials have driven the development of many green analytical methods. The greenness reached in some of them is noteworthy. Thus, microextraction techniques such as solid-phase and liquid-phase microextraction constitute relevant examples that closely approach the premises of green chemistry. Implementation of energies such as ultrasound in laboratory operations gives rise to accelerated procedures, with low energy consumption and very safe operating conditions. For trace organic analysis, several techniques have been developed making use of fast processes, with green solvents in some cases, *e.g.* supercritical carbon dioxide, and minimum amount of chemicals. The fulfilment of green chemistry principles for inorganic trace analysis is more troublesome, since mineral acids at high concentration are generally needed, yet some digestion approaches involving small volume of acids or flow systems have a better greenness profile. Surfactants, membranes, and solid phases allow simple approaches that can diminish or even completely remove the need for organic solvents in the extraction of both organic and inorganic analytes from liquid samples.

References

1. P. T. Anastas and J. C. Warner, *Green Chemistry: Theory and Practice*, Oxford University Press, New York, 1998.
2. J. Namiesnik, *Environ. Sci. Pollut. Res.*, 1999, **6**, 243.
3. J. Namiesnik, *J. Sep. Sci.*, 2001, **24**, 151.

4. L. H. Keith, L. U. Gron and J. L. Young, *Chem. Rev.*, 2007, **107**, 2695.
5. S. Armenta, S. Garrigues and M. de la Guardia, *TrAC, Trends Anal. Chem.*, 2008, **27**, 497.
6. National Environmental Methods Index, www.nemi.gov.
7. J. Z. Buchwald, *The Creation of Scientific Effects: Heinrich Hertz and Electric Waves*, University of Chicago Press, Chicago, 1994.
8. V. Camel, *TrAC, Trends Anal. Chem.*, 2000, **19**, 229.
9. E. E. King and D. Barclay, in *Sample Preparation for Trace Element Analysis*, ed. Z. Mester and R. Sturgeon, Elsevier, Amsterdam, 2003, p. 257.
10. A. Abu-Samra, J. S. Morris and S. R. Koirtyohann, *Anal. Chem.*, 1975, **47**, 1475.
11. F. E. Smith and E. A. Arsenault, *Talanta*, 1996, **43**, 1207.
12. M. Burguera and J. L. Burguera, *Anal. Chim. Acta*, 1998, **366**, 63.
13. Q. Jin, F. Liang, H. Zhang, L. Zhao, Y. Huan and D. Song, *TrAC, Trends Anal. Chem.*, 1999, **18**, 479.
14. J. A. Nóbrega, L. C. Trevizan, G. C. L. Araújo and A. R. A. Nogueira, *Spectrochim. Acta, Part B*, 2002, **57B**, 1855.
15. J. L. Luque-García and M. D. Luque de Castro, *TrAC, Trends Anal. Chem.*, 2003, **22**, 90.
16. L. Chen, D. Song, Y. Tian, L. Ding, A. Yu and H. Zhang, *TrAC, Trends Anal. Chem.*, 2008, **27**, 151.
17. H. Matusiewicz, in *Sample Preparation for Trace Element Analysis*, ed. Z. Mester and R. Sturgeon, Elsevier, Amsterdam, 2003, p. 193.
18. M. D. Luque de Castro and J. L. Luque García, *Accelerated and Automation of Solid Sample Treatment*, Elsevier, Amsterdam, 2002, p. 179.
19. B. Griepink and G. Tölg, *Pure Appl. Chem.*, 1989, **61**, 1139.
20. A. M. Ure, L. R. P. Butler, R. O. Scott and R. Jenkins, *Pure Appl. Chem.*, 1988, **60**, 1461.
21. M. Stoeppler, *Sampling and Sample Preparation*, Springer-Verlag, Heidelberg, 1997.
22. H. Matusiewicz, *Anal. Chem.*, 1994, **66**, 751.
23. H. Matusiewicz, *Anal. Chem.*, 1999, **71**, 3145.
24. V. Carbonell, M. de la Guardia, A. Salvador, J. L. Burguera and M. Burguera, *Anal. Chim. Acta*, 1990, **238**, 417.
25. S. Baldwin, M. Deaker and W. Maher, *Analyst*, 1994, **119**, 1701.
26. J. Millos, M. Costas-Rodríguez, I. Lavilla and C. Bendicho, *Anal. Chim. Acta*, 2008, **622**, 77.
27. B. Bocca, A. Alimonti, G. Forte, F. Petrucci, C. Pirola, O. Senofonte and N. Violante, *Anal. Bioanal. Chem.*, 2003, **377**, 65.
28. L. W. Collins, S. J. Chalk and H. M. Kingston, *Anal. Chem.*, 1996, **68**, 2610.
29. M. Feinberg, C. Suard and J. Ireland-Ripert, *Chemometrics Intell. Lab. Syst.*, 1994, **22**, 37.
30. C. J. Mason, M. Edwards, P. G. Riby and G. Coe, *Analyst*, 1999, **124**, 1719.

31. M. G. A. Korn, W. P. C. dos Santos, M. Korn and S. L. C. Ferreira, *Talanta*, 2005, **65**, 710.
32. H. Engelhardt, M. Kraemer and H. Waldhoff, *Chromatographia*, 1990, **30**, 523.
33. C. H. W. Hirs, S. Moore and W. H. Stein, *J. Am. Chem. Soc.*, 1954, **76**, 6063.
34. L. Joergensen and H. N. Thestrup, *J. Chromatogr. A*, 1995, **706**, 421.
35. M. Stenberg, G. Marko-Varga and R. Oste, *Food Chem.*, 2001, **74**, 217.
36. A. Peter, G. Laus, D. Tourwe, E. Gerlo and G. van Binst, *Peptide Res.*, 1993, **6**, 48.
37. S. Armenta, J. Moros, S. Garrigues and M. de la Guardia, *Anal. Chim. Acta*, 2006, **567**, 255.
38. G. Du, H. Y. Zhao, Q. W. Zhang, G. H. Li, F. Q. Yang, Y. Wang, Y. C. Li and Y. T. Wang, *J. Chromatogr. A*, 2010, **1217**, 705.
39. C. S. Eskilsson and E. Bjorklund, *J. Chromatogr. A*, 2000, **902**, 227.
40. C. W. Huie, *Anal. Bioanal. Chem.*, 2002, **373**, 23.
41. M. Tobiszewski, A. Mechlinska, B. Zygmunt and J. Namiesnik, *TrAC, Trends Anal. Chem.*, 2009, **28**, 943.
42. J. Szpunar, V. O. Schmitt, O. F. X. Donard and R. Lobinski, *TrAC, Trends Anal. Chem.*, 1996, **15**, 181.
43. N. Saim, J. R. Dean, M. P. Abdullah and Z. Zakaria, *J. Chromatogr. A*, 1997, **791**, 361.
44. J. You, H. Zhang, H. Zhang, A. Yu, T. Xiao, Y. Wang and D. Song, *J. Chromatogr. B*, 2007, **856**, 278.
45. J. Pacheco-Arjona, P. Rodríguez-González, M. Valiente, D. Barclay and O. F. X. Donard, *Int. J. Environ. Anal. Chem.*, 2008, **88**, 923.
46. M. Monperrus, R. C. Rodríguez Martín-Doimeadios, J. Scancar, D. Amouroux and O. F. X. Donard, *Anal. Chem.*, 2003, **75**, 4095.
47. T. J. Mason and J. P. Lorimer, *Applied Sonochemistry*, Wiley-VCH, Weinheim, 2002.
48. K. S. Suslick, R. E. Cline and D. A. Hammerton, *J. Am. Chem. Soc.*, 1986, **108**, 5641.
49. T. J. Mason, *Sonochemistry*, Oxford University Press, Oxford, 1999.
50. F. Priego-Capote and M. D. Luque de Castro, *TrAC, Trends Anal. Chem.*, 2004, **23**, 644.
51. J. L. Capelo, C. Maduro and C. Vilhema, *Ultrason. Sonochem.*, 2005, **12**, 225.
52. D. S. Júnior, F. J. Krug, M. D. G. Pereira and M. Korn, *Appl. Spectrosc. Rev.*, 2006, **41**, 305.
53. M. D. Luque de Castro and F. Priego Capote, *Analytical Applications of Ultrasound*, Elsevier, Amsterdam, 2007.
54. J. L. Capelo, *Ultrasound in Chemistry. Analytical Applications*, Wiley-VCH, Weinheim, 2009.
55. C. Bendicho and I. Lavilla, in *Encyclopedia of Separation Science*, ed. I. D. Wilson, Academic Press, 2000, p. 1448.
56. K. Ashley, in *Sample Preparation for Trace Element Analysis*, ed. Z. Mester and R. Sturgeon, Elsevier, Amsterdam, 2003, p. 353.

57. J. Sánchez and E. Millán, *Quim. Anal.*, 1992, **11**, 3.
58. J. Sánchez, R. García and E. Millán, *Analusis*, 1994, **22**, 222.
59. N. Jalbani, T. G. Kazi, M. K. Jamali, M. B. Arain, H. I. Afridi, S. T. Sheerazi and R. Ansari, *J. AOAC Int.*, 2007, **90**, 1682.
60. A. Ilander and A. Väisänen, *Ultrason. Sonochem.*, 2009, **16**, 763.
61. M. B. Arain, T. G. Kazi, M. K. Jamali, N. Jalbani, H. I. Afridi, R. A. Sarfraz and A. Q. Shah, *Spectrosc. Lett.*, 2007, **40**, 861.
62. H. Teng, B. Ren, X. Tian, J. Song and H. Huang, *Z. Niangzao*, 2008, **24**, 99.
63. A. Canals and M. R. Hernández, *Anal. Bioanal. Chem.*, 2002, **374**, 1132.
64. A. Canals, A. Cuesta, L. Gras and M. R. Hernández, *Ultrason. Sonochem.*, 2002, **9**, 143.
65. C. E. Domini, L. Vidal and A. Canals, *Ultrason. Sonochem.*, 2009, **16**, 686.
66. P. Moreno, M. A. Quijano, A. M. Gutiérrez, M. C. Pérez-Conde and C. Cámara, *J. Anal. At. Spectrom.*, 2001, **16**, 1044.
67. J. L. Capelo, P. Ximénez-Embún, Y. Madrid-Albarrán and C. Cámara, *Anal. Chem.*, 2004, **76**, 233.
68. L. Amoedo, J. L. Capelo, I. Lavilla and C. Bendicho, *J. Anal. At. Spectrom.*, 1999, **14**, 1221.
69. K. Ashley, R. Andrews, L. Cavazos and M. Demange, *J. Anal. At. Spectrom.*, 2001, **16**, 1147.
70. A. Sussell and K. Ashley, *J. Environ. Monit.*, 2002, **4**, 156.
71. M. C. Yebra and S. Cancela, *Anal. Bioanal. Chem.*, 2005, **382**, 1093.
72. M. C. Yebra-Biurrun and R. M. Cespón-Romero, *Anal. Bioanal. Chem.*, 2007, **388**, 711.
73. S. M. Talebi and M. Abedi, *J. Chromatogr. A*, 2005, **1094**, 118.
74. S. Sporring, S. Bøwadt, B. Svensmark and E. Björklund, *J. Chromatogr. A*, 2005, **1090**, 1.
75. E. Psillakis, A. Ntelekos, D. Mantzavinos, E. Nikolopuolos and N. Kalogerakis, *J. Environ. Monit.*, 2003, **5**, 135.
76. R. Cornelius, J. Caruso, H. Crews and K. Heumann, *Handbook of Elemental Sepciation. Techniques and Methodology*, John Wiley & Sons, Chichester, 2003.
77. K. Ashley, *TrAC, Trends Anal. Chem.*, 1998, **17**, 366.
78. S. Rio-Segade and C. Bendicho, *J. Anal. At. Spectrom.*, 1999, **14**, 263.
79. A. Huerga, I. Lavilla and C. Bendicho, *Anal. Chim. Acta*, 2005, **534**, 121.
80. M. A. Quijano, A. M. Gutiérrez, M. C. Pérez-Conde and C. Cámara, *Talanta*, 1999, **50**, 165.
81. A. Tessier, P. G. C. Campbell and M. Bisson, *Anal. Chem.*, 1979, **51**, 844.
82. A. M. Ure, Ph. Quevauviller, H. Muntau and B. Griepinck, *Int. J. Environ. Anal. Chem.*, 1993, **51**, 135.
83. A. V. Filgueiras, I. Lavilla and C. Bendicho, *J. Environ. Monit.*, 2002, **4**, 823.
84. C. Bendicho and M. T. C. De Loos-Vollebregt, *J. Anal. At. Spectrom.*, 1991, **6**, 353.
85. N. J. Miller-Ihli, *J. Anal. At. Spectrom.*, 1989, **4**, 295.

86. V. Camel, *Analyst*, 2001, **126**, 1182.
87. M. A. McHugh and V. Krukonis, *Supercritical Fluid Extraction—Principles and Practice*, Butterworth-Heinemann, Stoneham, 1986.
88. M. L. Lee and K. E. Markides, *Analytical Supercritical Fluid Chromatography and Extraction*, Chromatography Conferences Inc., Provo, TX, 1990.
89. M. D. Luque de Castro, M. Valcárcel and M. T. Tena, *Analytical Supercritical Fluid Extraction*, Springer-Verlag, Berlin, 1994.
90. H. G. Janssen and X. Lou, in *Extraction Methods in Organic Analysis*, ed. A. J. Handley, Sheffield Academic Press, Sheffield, 1999, p. 100.
91. S. Bøwadt and S. B. Hawthorne, *J. Chromatogr. A*, 1995, **703**, 549.
92. J. R. Dean, *Extraction Methods for Environmental Analysis*, John Wiley & Sons, Weinheim, Germany, 1998.
93. S. B. Hawthorne, Y. Yang and D. J. Miller, *Anal. Chem.*, 1994, **66**, 2912.
94. M. D. Luque de Castro and M. T. Tena, *TrAC, Trends Anal. Chem.*, 1996, **15**, 32.
95. M. Koel and M. Kaljurand, *Pure Appl. Chem.*, 2006, **78**, 1993.
96. M. C. Henry and C. R. Yonker, *Anal. Chem.*, 2006, **78**, 3909.
97. H. Zhao, S. Xia and P. Ma, *J. Chem. Technol. Biotechnol.*, 2005, **80**, 1089.
98. S. Keskin, D. Kayrak-Talay, U. Akman and Ö. Hortaçsu, *J. Supercrit. Fluids*, 2007, **43**, 150.
99. W. Z. Wu, J. M. Zhang, B. X. Han, J. W. Chen, Z. M. Liu, T. Jiang, J. He and W. J. Li, *Chem. Commun.*, 2003, 1412.
100. W. Wu, W. Li, B. Han, T. Jiang, D. Shen, Z. Zhang, D. Sun and B. Wang, *J. Chem. Eng. Data*, 2004, **49**, 1597.
101. C. Turner, C. S. Eskilsson and E. Björklund, *J. Chromatogr. A*, 2002, **947**, 1.
102. M. D. Luque de Castro and M. M. Jiménez-Carmona, *TrAC, Trends Anal. Chem.*, 2000, **19**, 223.
103. M. Zhougagh, M. Valcárcel and A. Ríos, *TrAC, Trends Anal. Chem.*, 2004, **23**, 399.
104. B. E. Richter, B. A. Jones, J. L. Ezzell, N. L. Porter, N. Avdalovic and C. Pohl, *Anal. Chem.*, 1996, **68**, 1033.
105. M. M. Schantz, *Anal. Bioanal. Chem.*, 2006, **386**, 1043.
106. L. Ramos, E. M. Kristenson and U. A. T. Brinkman, *J. Chromatogr. A*, 2002, **975**, 3.
107. E. Björklund, T. Nilsson and S. Bøwadt, *TrAC, Trends Anal. Chem.*, 2000, **19**, 434.
108. E. Björklund, S. Sporring, K. Wiberg, P. Haglund and C. von Holst, *TrAC, Trends Anal. Chem.*, 2006, **25**, 318.
109. J. L. Luque-García and M. D. Luque de Castro, *TrAC, Trends Anal. Chem.*, 2004, **23**, 102.
110. J. Kronholm, K. Hartonen and M. L. Riekkola, *TrAC, Trends Anal. Chem.*, 2007, **26**, 396.
111. R. M. Smith, *J. Chromatogr. A*, 2002, **975**, 31.
112. S. B. Hawthorne and A. Kubátová, in *Sample and Sampling Preparation for Field and Laboratory*, ed. J. Pawliszyn, Elsevier, Amsterdam, 2002, p. 587.

113. C. F. Poole, *TrAC, Trends Anal. Chem.*, 2003, **22**, 362.
114. H. Braus, F. M. Middleton and G. Walton, *Anal. Chem.*, 1951, **23**, 1160.
115. M. C. Hennion, *J. Chromatogr. A*, 1999, **856**, 3.
116. C. E. Domini, D. Hristozov, B. Almagro, I. P. Román, S. Prats and A. Canals, in *Chromatographic Analysis of the Environment*, ed. L. M. L. Nollet, 2006, p. 31.
117. I. Liška, *J. Chromatogr A.*, 2000, **885**, 3.
118. V. Camel, *Spectrochim. Acta, Part B*, 2003, **58**, 1177.
119. C. F. Poole, in *Sampling and Sample Preparation for Field and Laboratory*, ed. J. Pawliszyn, Elsevier, Amsterdam, 2002, p. 341.
120. D. T. Rossi and N. Zhang, *J. Chromatogr. A*, 2000, **885**, 97.
121. J. Curyło, W. Wardencki and J. Namieśnik, *Polish J. Environ. Stud.*, 2007, **16**, 5.
122. C. L. Arthur and J. Pawliszyn, *Anal. Chem.*, 1990, **62**, 2145.
123. M. F. Alpendurada, *J. Chromatogr. A.*, 2000, **889**, 3.
124. V. Kaur, A. K. Malik and N. Verma, *J. Sep. Sci.*, 2006, **29**, 333.
125. Z. Mester, R. Sturgeon and J. Pawliszyn, *Spectrochim. Acta, Part B*, 2001, **56**, 233.
126. Z. Mester and R. Sturgeon, *Spectrochim. Acta, Part B*, 2005, **60**, 1243.
127. G. A. Mills and V. Walker, *J. Chromatogr. A*, 2000, **902**, 267.
128. J. Pawliszyn, *Solid-Phase Microextraction: Theory and Practice*, Wiley-VCH, Weinheim, 1997.
129. C. Dietz, J. Sanz and C. Cámara, *J. Chromatogr. A*, 2006, **1103**, 183.
130. E. Baltussen, P. Sandra, F. David and C. Cramers, *J. Microcolumn Sep.*, 1999, **11**, 737.
131. E. Baltussen, C. A. Cramers and P. J. F. Sandra, *Anal. Bioanal. Chem.*, 2002, **373**, 3.
132. F. David and P. Sandra, *J. Chromatogr. A*, 2007, **1152**, 54.
133. Gerstel-Twister SBSE, http://www.gerstel.com/en/twister-stir-bar-sorptive-extraction.htm.
134. F. Pena-Pereira, I. Lavilla and C. Bendicho, *Spectrochim. Acta, Part B*, 2009, **64**, 1.
135. C. Nerín, J. Salafranca, M. Aznar and R. Batlle, *Anal. Bioanal. Chem.*, 2009, **393**, 809.
136. J. M. Kokosa, A. Przyjazny and M. A. Jeannot, *Solvent Microextraction. Theory and Practice*, John Wiley & Sons, New York, 2009.
137. M. A. Jeannot and F. F. Cantwell, *Anal. Chem.*, 1997, **69**, 235.
138. A. L. Theis, A. J. Waldack, S. M. Hansen and M. A. Jeannot, *Anal. Chem.*, 2001, **73**, 5651.
139. A. Tankeviciute, R. Kazlauskas and V. Vickackaite, *Analyst*, 2001, **126**, 1674.
140. Y. C. Fiamegos and C. D. Stalikas, *Anal. Chim. Acta*, 2007, **599**, 76.
141. F. Pena-Pereira, I. Lavilla and C. Bendicho, *Anal. Chim. Acta*, 2009, **631**, 223.
142. I. Lavilla, F. Pena-Pereira, S. Gil, M. Costas and C. Bendicho, *Anal. Chim. Acta*, 2009, **647**, 112.

143. M. Ma and F. F. Cantwell, *Anal. Chem.*, 1999, **71**, 388.
144. H. F. Wu, J. H. Yen and C. C. Chin, *Anal. Chem.*, 2006, **78**, 1707.
145. B. O. Keller and L. Li, *Anal. Chem.*, 2001, **73**, 2929.
146. S. Pedersen-Bjergaard and K. E. Rasmussen, *Anal. Chem.*, 1999, **71**, 2650.
147. K. E. Rasmussen, S. Pedersen-Bjergaard, M. Krogh, H. Grefslie Ugland and T. Grønhaug, *J. Chromatogr. A*, 2000, **873**, 3.
148. E. Psillakis and N. Kalogerakis, *TrAC, Trends Anal. Chem.*, 2003, **22**, 565.
149. M. Rezaee, Y. Assadi, M. R. Milani Hosseini, E. Aghaee, F. Ahmadi and S. Berijani, *J. Chromatogr. A*, 2006, **1116**, 1.
150. C. Bosch Ojeda and F. Sánchez Rojas, *Chromatographia*, 2009, **69**, 1149.
151. M. Baghdadi and F. Shemirani, *Anal. Chim. Acta*, 2008, **613**, 56.
152. C. Yao and J. L. Anderson, *Anal. Bioanal. Chem.*, 2009, **395**, 1491.
153. J. M. Kokosa, *Automation of Liquid Phase Microextraction*, U.S. Patent 7,178,414 B1, Feb. 20, 2007.
154. T. Barri and J. A. Jönsson, *J. Chromatogr. A*, 2008, **1186**, 16.
155. J. A. Jönsson and L. Mathiasson, *TrAC, Trends Anal. Chem.*, 1999, **18**, 318.
156. J. A. Jönsson and L. Mathiasson, *TrAC, Trends Anal. Chem.*, 1999, **18**, 325.
157. J. A. Jönsson and L. Mathiasson, *J. Chromatogr. A*, 2000, **902**, 205.
158. J. A. Jönsson and L. Mathiasson, *J. Sep. Sci.*, 2001, **24**, 495.
159. N. Jakubowska, Z. Polkowska and J. Namiesnik, *Crit. Rev. Anal. Chem.*, 2005, **35**, 217.
160. K. Hylton and S. Mitra, *J. Chromatogr. A*, 2007, **1152**, 199.
161. J. A. Jönsson, *Membrane Extraction in Analytical Chemistry*, Wiley-VCH, Weinheim, 2001.
162. J. Pawliszyn, in *Sample Preparation for Field and Laboratory*, Elsevier, Amsterdam, 2002, p. 479.
163. J. A. Jönsson, in *Sampling and Sample Preparation for Field and Laboratory*, Elsevier, Amsterdam, 2002, p. 503.
164. J. A. Jönsson, in *Handbook of Membrane Separations*, CRC Press, Boca Raton, FL, 2008, p. 345.
165. Z. X. Cai, Q. Fang, H. W. Cheng and Z. L. Fang, *Anal. Chim. Acta*, 2006, **556**, 151.
166. A. Gjelstad, T. M. Andersen, K. E. Rasmussen and S. Pedersen-Bjergaard, *J. Chromatogr. A*, 2007, **1157**, 38.
167. X. Wang, C. Saridara and S. Mitra, *Anal. Chim. Acta*, 2005, **543**, 92.
168. L. Nozal, L. Arce, B. M. Simonet, A. Rios and M. Valcárcel, *Electrophoresis*, 2006, **27**, 3075.
169. L. Chimuka, E. Cukrowska and J. A. Jönsson, *Pure Appl. Chem.*, 2004, **76**, 707.
170. R. Rodil, S. Schrader and M. Moeder, *J. Chromatogr. A*, 2009, **1216**, 8851.
171. Z. Sosa, C. Padrón, C. Mahugo and J. J. Santana, *TrAC, Trends Anal. Chem.*, 2004, **23**, 469.
172. E. K. Paleologos, D. L. Giokas and M. I. Karayannis, *TrAC, Trends Anal. Chem.*, 2005, **24**, 426.
173. C. D. Stalikas, *TrAC, Trends Anal. Chem.*, 2002, **21**, 343.

174. H. Watanabe and H. Tanaka, *Talanta*, 1978, **25**, 585.
175. K. Madej, *TrAC, Trends Anal. Chem.*, 2009, **28**, 436.
176. W. L. Hinze and E. Pramauro, *Crit. Rev. Anal. Chem.*, 1993, **24**, 133.
177. G. R. Komaromy-Hiller and R. von Wandruszka, *Talanta*, 1995, **42**, 83.
178. M. de Almeida, M. A. Z. Arruda and S. L. C. Ferreira, *Appl. Spectrosc. Rev.*, 2005, **40**, 269.
179. K. Ganzler and A. Salgo, *Z. Lebensm-Unters.-Forsch*, 1987, **184**, 274.
180. F. E. Ahmed, *TrAC, Trends Anal. Chem.*, 2001, **20**, 649.
181. V. Pino, J. H. Ayala, A. M. Afonso and V. González, *Talanta*, 2001, **54**, 15.
182. Q. Fang, H. W. Yeung, H. W. Leung and C. W. Huie, *J. Chromatogr. A*, 2000, **904**, 47.
183. B. Abismaïl, J. P. Canselier, A. M. Wilhelm, H. Delmas and C. Gourdon, *Ultrason. Sonochem.*, 1999, **6**, 75.
184. A. E. Alegria, Y. Lion, T. Kondo and P. Riesz, *J. Phys. Chem.*, 1989, **93**, 4908.
185. M. Murillo, Z. Benzo, E. Marcano, C. Gomez, A. Garaboto and C. Martin, *J. Anal. At. Spectrom.*, 1999, **14**, 815.
186. M. D. Luque de Castro and F. Priego-Capote, *Talanta*, 2007, **72**, 321.
187. I. Lavilla, N. Cabaleiro, M. Costas, I. de la Calle and C. Bendicho, *Talanta*, 2009, **80**, 109.

CHAPTER 5

Miniaturization of Analytical Methods

MIREN PENA-ABAURREA AND LOURDES RAMOS

Department of Instrumental Analysis and Environmental Chemistry, Institute of Organic Chemistry, CSIC, Juan de la Cierva 3, 28006 Madrid, Spain

5.1 Miniaturization as an Alternative for Green Analytical Chemistry: Strengths and Current Limitations

Today, miniaturization is an evident trend in many analytical application fields. Miniaturization is the first requirement when trying to develop coupled (or integrated) analytical procedures, which in their turn can be considered as the first necessary step for developing completely hyphenated, (semi-)automatic, and/or unattended analytical systems. The several benefits associated with these types of analytical arrangements are evident and do not need any further explanation. These advantages partially explain the many efforts during the last two decades to miniaturize analytical processes and treatments, and the several attempts carried out to develop new analytical alternatives and techniques that contribute to achieve the hyphenation goal.

But miniaturization is both stimulated by and stimulates what can be considered another clear trend of modern analytical chemistry: the development of greener analytical methodologies and procedures that efficiently contribute to reducing the use of toxic solvents and reagents as well as the amount of wastes generated. Attempts in this direction have promoted the development of new analytical techniques for both sample preparation and instrumental analysis and the use of alternative and original analytical approaches, sometimes involving the use of existing technologies from new and different perspectives. Some of these advances have been covered in previous chapters and many more will be presented in the following ones.

The present chapter focuses on advances in the field of sample preparation for the analysis of organic components, an area in which progress has up to now been somehow more limited than in other application areas. This is especially true for the analysis of trace organic compounds, such as organic microcontaminants, in environmental and biological samples. In these fields, the variety of matrices to be analysed, the complexity of most of them, and the low levels at which the analytes should accurately be determined, has meant that, despite the extremely powerful and sophisticated separation-plus-detection instrumental techniques used for final determination of the analytes, quite often specific sample preparation protocols are still used for each particular analyte–matrix combination. Furthermore, the more complex the sample and the lower the analyte concentration, the higher the number of treatments involved in the rather conventional, although often robust and well-established, sample preparation procedures in use in this research field. Most of these analytical protocols start with an exhaustive extraction of the target compound(s) from the matrix using a conventional (but widely accepted) large-scale and time-consuming technique, such as liquid–liquid extraction (LLE) or Soxhlet extraction. This step should also effect the required trace enrichment of the analytes. However, because of the essentially non-selective nature of most of these extraction techniques, a laborious multistep procedure is often needed to remove co-extracted material and isolate the analyte(s) of interest (unless, of course, separation-plus-detection is highly selective). The several analytical steps involved in such procedures are usually carried out off-line, which make them tedious and time-consuming, prone to loss of analytes, and to contamination, because of the continual manual manipulation of the extracts. In recent years, much effort has been devoted to eliminating these drawbacks, and faster and more powerful and/or more versatile extraction techniques are now available. However, the level of integration among the several treatment steps (*i.e.* extraction, purification, and concentration), and among these and the final instrumental determination is still rather variable and certainly highly dependent on the nature of the matrix.

Today, on-line coupling (with or without automation) is a recognized feature in many areas of application which deal with gases or volatile analytes, and with a wide variety of analytes of divergent polarity in liquid samples (*e.g.* water, urine, plasma, soft drinks, and spirits). Initial problems regarding the compatibility of the various steps in terms of, for example, sample size, chemicals required, time taken, and liquid or gas flows, have been solved satisfactorily. One main benefit of using on-line (*i.e.* integrated) systems is that instead of an aliquot of a sample extract of, often, 1% or even less, the entire sample is now subjected to the final separation-plus-detection. This enables considerable (and in some cases almost proportional) reduction of the initial sample size required for the analysis.

In its turn, such miniaturization has helped to solve problems regarding the analysis of labile analytes, small (*i.e.* size-limited) samples,[1,2] the study of processes that take place in times shorter than those involved by traditional methodologies, and/or the use of powerful separation techniques with limited sensitivity due to reduced loading capacity, e.g. narrow-bore chromatographic or electrophoretic separation systems.[3,4] In some cases, developments have led to the preparation procedure being minimized, e.g. by using highly efficient and selective preconcentration sorbents based on immunoaffinity recognition; or even discarded completely, e.g. substitution of the concentration step by large-volume injection (LVI; typically 1 ml for liquid chromatography, LC, and 0.1 ml for gas chromatography, GC) and direct thermal desorption.[5–7]

For obvious reasons, the development of procedures similar to those mentioned above has been more limited for semisolid and solid samples. For these, rather complex and large-scale (off-line) approaches are still the rule rather than the exception; in other words, improvement has been rather limited (see references 8–10 and references cited therein). In this research area clearly even miniaturization of the basic processes and partial integration are, in most instances, highly demanding, if not unachieved objectives.

It is therefore possible to conclude that the generic goal of miniaturization in analytical chemistry is contributing to the development of integrated (i.e. hyphenated) systems for potential subsequent automation of the analytical process in different research areas. In the analytical chemistry field, miniaturization has also some specific goals including the development of greener and cheaper analytical processes (by reducing the required amount of reagents, solvents and wastes, the time and energy required per determination, the production costs, etc.), speeding up the analytical process (shorter analytical times, higher throughput), and the feasibility of setting up more integrated (simpler, smaller, portable, etc.) and closed (reducing the risk of decomposition, contamination or lost of the analytes, the exposure of the analyst to toxic chemicals, the amount of generated wastes, etc.) systems. Other relevant goals, such as the (potential) suitability of the miniaturized analytical system for *in situ* and continuous monitoring, the analysis of micro- and nanosamples, and the study, for the first time, of microenvironments (e.g. the analysis of the pore water in soils and sediments) can be inferred from the previous considerations.

The tremendous analytical potential of these miniaturized approaches explains the many efforts conducted in this research field in recent years. However, as previously mentioned, miniaturization is a trend observed nowadays in many scientific and technical areas. Therefore, it is important to set the scale range of the analytical approach that will be reviewed in the present chapter. Table 5.1 summarizes a simplified classification of the different scale ranges defined in analytical chemistry for the miniaturization process.

For readers who are not completely familiar with the terminology, Figure 5.1 shows a graphic description of the several possible degrees of integration in the analytical process.

- An analytical system or process is defined as **on-line** (integrated or hyphenated) when the extract (or effluent) from the sample preparation procedure is directly transferred, without intervention of the analyst, to

Table 5.1 Characteristic size range of magnitudes, volumes and sample level associated with the different degrees of miniaturization achieved in analytical chemistry.

Prefix	Typical size range	Typical volume	Sample size
Mini-	>1 mm	µl	Cellular tissue
Micro-	1 mm–1 µm	>10 nL	Macromolecule
Nano-	<1 µm	<10 nL (fL)	Molecule; ion

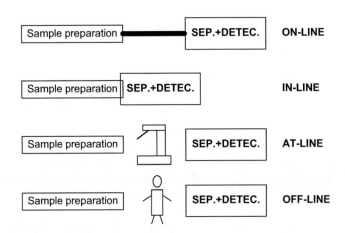

Figure 5.1 Different levels of integration in analytical systems.

the separation-plus-detection system selected for the determination of the target compound(s).
- When the final part of the sample treatment is carried out in the instrument used for final instrumental analysis, the system is said to be **in-line**.
- When the extract (or effluent) from the sample preparation procedure is transferred to the instrument selected for final determination *via* a mechanical or robotic system, the two systems are considered to be **at-line**.
- Finally, if the extract is transferred to the separation-plus-detection instrument by the analyst, the processes (or systems) involved in the analytical procedure are not coupled (or hyphenated) and the systems are defined as **off-line**.

5.2 Miniaturized Analytical Techniques for Treatment of Liquid Samples

5.2.1 Solvent-Based Miniaturized Extraction Techniques

Miniaturization of a procedure can be achieved simply by reducing the dimensions of the systems used in earlier approaches or by developing

Miniaturization of Analytical Methods

completely new set-ups or techniques. As we will show in the next section, both strategies have been explored for LLE.

5.2.1.1 In-vial Liquid–Liquid Extraction

LLE consists on the sequential extraction of an aqueous sample with an immiscible organic solvent in which the target compounds show a (much) higher solubility than in water. LLE is one of the simplest techniques for the extraction of liquid samples, does not require any special equipment, and allows several extractions to be performed in parallel. These reasons probably explain its wide acceptance in past and why it is still in use in many laboratories. However, in a typical LLE experiment, relatively large volumes of sample (typically >100 ml) are successively extracted with similar volumes of the selected organic solvent until the target compound is completely extracted from the original aqueous phase. This process obviously demands constant sample manipulation and the subsequent concentration of the organic extractant before instrumental determination to ensure proper detection. Apart from the high solvent consumption, the LLE process is prone to formation of emulsions, which lead to long analysis times and, in many instances, to analyte losses.

The goal of miniaturized LLE is to reduce the organic solvent/aqueous ratio as much as possible to enhance analyte enrichment, and (virtually) eliminate the need for subsequent concentration of the separated organic layer. As in conventional LLE, stirring and salting-out of the aqueous phase are strategies often employed to improve transfer of the analytes from the sample to the extractant. Derivatization of the target compounds either in the aqueous phase or in the organic layer has also been used with satisfactory results.

When the volumes of the aqueous phase and the extractant are small enough (*i.e.* about 1–2 ml and 500 µl, respectively), LLE can be developed in a chromatographic vial and the analytical approach is called in-vial LLE. In this case, a single extraction is carried out and, for obvious reasons, the selection of the extraction solvent and the optimization of the different experimental parameters affecting the partition process become more critical. In any case, the analytical procedure is still very simple and samples can be treated in parallel, which contributes to increasing the analytical throughput. However, probably the most interesting feature of this technique is that it allows a highly significant reduction of the organic solvent consumption, so contributing to greening the analytical process. It also helps to greatly reduce the time required per analysis as at this scale emulsions are (virtually) non-existent. This latter fact, together with the more favourable phase ratio, result in higher extraction efficiencies. Because of the higher enrichment factors achieved, concentration of the organic extract becomes unnecessary in many instances. When the extractant is clean enough to avoid compromising the final instrumental determination, this organic phase can therefore be sampled directly by conventional autosamplers and transferred without any further treatment to the separation-plus-detection system. In other words, at-line coupling between the sample preparation and the instrumental steps is possible and, because of the simplicity of the operations performed, some degree of semi-integration (*e.g.* via a robotic arm or some of the sophisticated autosamplers nowadays commercialized) is possible.

In-vial LLE has been used, although with a variable success, to enrich polychlorinated biphenyls (PCBs) from spiked aqueous samples; for the semiautomatic miniaturized LLE of anilines and pesticides (in this case using membranes to avoid co-extration of other matrix components); and for the analysis of tap, river, and residual waters. In all cases, ready-to-analysis extracts of less than 1 ml were obtained, and, not unexpectedly, the less polar the target analyte, the better the result obtained.

Unfortunately, somewhat more disappointing results have been obtained when this approach has been applied to the analysis of non-aqueous samples, such as foodstuffs. The higher complexity of these matrices leads to much lower extraction efficiencies and, owing to the limited selectivity of the process, the introduction of additional clean-up step(s) became mandatory.

5.2.1.2 Solvent Microextraction Techniques

Several new micro-LLE-based techniques that have been developed during the last 15 years, grouped under the name of solvent microextraction (SME) techniques, have recently been revised by Kokosa *et al.*[11]

The various SME techniques and working modes described in the literature can be classified according to different criteria. In this chapter we follow the terminology and classification proposed by Kokosa *et al.*,[11] which is essentially based on the number of phases involved in the extraction process and the two basic working modes: direct-immersion sampling and headspace (HS) sampling (Table 5.2).

5.2.1.2.1 Single-Drop Microextraction (SDME). This miniaturized technique was firstly introduced by Jeannot and Cantwell[12] and He and Lee[13] in

Table 5.2 Summary of most commonly used solvent microextraction techniques and working modes.

SDE technique	Definition
Direct-immersion mode	
SDME	Single-drop microextraction
HF(2)ME	Hollow fibre-protected two-phase solvent microextraction
HF(3)ME	Hollow fibre-protected three-phase solvent microextraction
DLLME	Dispersive liquid–liquid microextraction
Headspace mode	
HS-SDME	Headspace single-drop microextraction
HS-HF(2)ME	Headspace hollow fibre-protected two-phase microextraction
Dynamic modes	
In syringe	Repeated withdrawal of sample into syringe
In needle	Repeated withdrawal of sample into needle

Adapted from Ref. 11.

1997 when they simultaneously realized that a standard GC syringe could be used for SME. In SDME, the extractant phase is a single microdrop of a water-insoluble solvent (typically 1–8 µl) suspended at the tip of a syringe and immersed in an aqueous sample (~1–10 ml) contained in a vial (Table 5.3). After a preselected extraction time, the drop is withdrawn into the syringe and the enriched organic solvent directly transferred to the separation-plus-detection instrument. The experimental parameters affecting the efficiency of this SDME format are similar to those considered in LLE. Stirring of the sample, salting-out, application of temperature, and analyte derivatization (to reduce its polarity or increase its volatility) are common practices that, in general, contribute to increase the extraction efficiency and reduce the analysis time. However, to prevent drop dislodgement relatively slow stirring rates of up to 600 rpm should be used. The simplicity of the analytical procedure, which can be performed manually, or even better with a computer-controller autosampler, the extremely low solvent consumption, and the relatively high enrichment factors achieved (typically in the range 10–100) can be considered key factors contributing to the fast development of this miniaturized extraction technique and its application in different research fields.

Direct-immersion SDME has been demonstrated to be useful for the extraction of relatively non-polar and semivolatile analytes, including non-polar microcontaminants and drugs, from water samples that contain little or no particulate or dissolved matter (Table 5.3). For these types of samples, experimental results suggest that the nature of the sample has little effect on the enrichment process. However, when analysing more complex matrices such as urine,[16] previous filtration of the sample is recommended.

A recent modification of this method has been introduced by Wu et al.[31] and proved to be especially useful with size-limited biological samples. The technique is called *drop-to-drop microextraction* (DDME) and involves very small volumes of both sample (~10 µl) and organic solvent (~0.5–1.0 µl). The sample is placed in a conical-bottomed microvial, in which the drop of organic solvent suspended at the tip of a microsyringe needle is immersed. Because of this its format, the mass transfer rate of analytes into the extractant is high. Its main limitation is the relatively high limits of detection (LOD). However, it has been proved to be useful for the fast and simple extraction of drugs from blood, serum, and urine.

A three-phase SDME-based alternative was introduced in 1999 by Ma and Cantwell[32] with the name *liquid–liquid–liquid microextraction* (LLLME). This technique enabled the simultaneous enrichment and purification of polar analytes from aqueous samples and consisted of the extraction of the deionized polar analytes from the aqueous sample into a few microlitres of organic phase, which acted as an organic liquid membrane contained in a PTFE ring. The preconcentrated analytes were then back-extracted into a microdrop of aqueous receiving phase suspended in the organic phase. Next, the aqueous microdrop was withdrawn into the syringe and directly subjected to liquid chromatography (LC) or capillary electrophoresis (CE).

Table 5.3 Selected applications of solvent microextraction techniques.

Matrix (ml)	Analytes	Extraction solvent (vol., μl)	Extraction mode	Extraction time (min)	Enrichment factor	Ref.
SDME						
Spiked aqueous solution (1)	4-Methylacetophenone	n-C_8 (8)	D^a	5	30	14
Spiked aqueous solution (1)	4-Methylacetophenone + 4-Nitrotoluene + progesterone + malathion	n-C_8 (1)	D	1	380	12
Spiked aqueous solution (4)	1,2,3-Trichlorobencene	Toluene (1)	E^b	15	12	13
Spiked river water (5)	11 OCPs	n-C_6 (2)	D	5	21	15
Urine (2)	Cocaine and its metabolites	Chloroform (2)	D	6	7–17	16
LPME						
Spiked aqueous solution (0.06)	1,2,3-Trichlorobencene	Toluene (1)	D	3	27	13
River water (0.8)	PAHs	Chloroform (5)	D	20	>280	17
		Toluene (3)	E	20	60–180	
Wastewater (0.09)	10 chlorobenzenes	Isooctane (1)	D	2–3	130	18
Urine (1)	Basic drugs	Di-n-hexyl ether (20) + 0.01 M HCl (25)	D	45	60–140	19
Plasma (0.5)	Antidepressant drugs	Di-n-hexyl ether (25) + 0.01 M HCl (2)	D	45	20	20
Slurry sediment: water (4:100) (0.16 g)	Chlorobenzenes + OCPs	Toluene (3)	D	2	30–490	21
LLLME						
Spiked aqueous buffer, pH = 13 (2)	7 aromatic amines	$EtOAc^c$ (150) + sodium phosphate buffer, pH = 2.1 (2)	D	15.4	220–380	22

Matrix (volume)	Analytes	Extractant (volume)				Ref.
Cow milk + HCl 0.5 M (8)	5 phenoxiacids	n-C_8 + NaOH 0.1 M (7)	E	60	260–950	22
SLM						
Industrial wastewater (120)	7 aniline derivatives	Water, pH = 3.3 (200)	4^d	30	12–30	23
Human plasma (0.5)	Anaesthetics	n-C_6 (360)	0.018^d	25	1.5	24
Assisted membrane LLE						
Spiked aqueous solution (15)	Triazines + apolar pollutants (OCPs, PAHs)	n-C_6 (500)	E	30	3	25
Tap water (60)	5 sulfonylurea herbicides	Chloroform (960)	3^d	20	55–60	26
DLLME						
River and lake water (100)	DDT and its metabolites	Carbon tetrachloride (50)	E	2	200	27
Tap, spring and sea water^e (12)	PAHs	Toluene (14)	D	0.5	857	28
CFLME^f						
Spiked aqueous solution (80)	Bisphenol A	0.1 M sodium phosphate buffer, pH = 12 (400)	0.8^d	40	200	29
Tap, sea and mineral water (20)	Sulfonylurea herbicides	0.2 M sodium carbonate buffer, pH = 10 (50)	0.8^d	10	100	30

OCP, organochlorine pesticide; PAH, polycyclic aromatic hydrocarbons;
^{a}Dynamic extraction;
^{b}Static extraction;
^{c}Ethyl acetate;
^{d}Flow rate, ml min^{-1};
^{e}Ultrasonic-assisted DLLME;
^{f}Continuous-flow liquid membrane extraction, combination of continuous-flow LLE plus SLM.

Compared with regular SDME, this approach enables the use of higher stirring rates due to the improved stability of the drop-organic membrane. In addition, the small volume of receiving solution enables high enrichment factors (in the range 200–500) to be obtained in a rather short time (~15 min), and the complete removal of the organic phase after each extraction prevents cross-contamination. Several examples have illustrated the feasibility of the approach for the determination of a variety of polar compounds from model buffered solution (Table 5.3) but, to the best of our knowledge, no application involving real-life samples has been reported up to now.

SDME can also be accomplished by direct exposure of the drop to the headspace of the samples. In this case, the technique is referred as *headspace single-drop microextraction* (HS-SDME). HS-SDME can be applied to aqueous, non-volatile liquids, solids, and gas samples. Much higher stirring rates, without splashing, can be used, but otherwise all previous considerations apply also in this case. The technique works very efficiently for the preconcentration of volatile non-polar analytes, and has the advantage over direct-immersion SDME of providing cleaner extracts in shorter analytical times.

Irrespective of the SDME mode used, a main factor with a profound effect on both the extraction efficiency and the extraction time is the diffusion of the extracted analytes from the surface of the drop to its interior. In the previously described static SDME-based techniques, this diffusion rate can be increased by using less viscous solvents, stirring the sample or increasing the extraction temperature.[11] However, probably a more efficient alternative is the constant renewal of the solvent surface by using a dynamic approach. Two type of dynamic SDME are possible (Table 5.3). In the *in-syringe* dynamic method,[33] the aqueous sample or headspace is withdraw into the syringe needle or lumen and ejected repeatedly to perform the desired solvent enrichment. In the *in-needle* dynamic approach,[34,35] around 90% of the extraction drop is withdrawn into the syringe needle and then pushed out again repeatedly for sample exposure. The former dynamic approach is more effective when dealing with relatively pristine samples (*i.e.* without high levels of salts or major matrix components). The latter may be more useful for the analysis of relatively 'dirty' samples (*i.e.* samples containing a relatively high concentration of matrix components that could affect the subsequent instrumental analysis). It is important to highlight that in both types of dynamic approaches the use of a computer-controller autosampler for accurate and reproducible control of the syringe plunger movements is mandatory.[11]

5.2.1.2.2 Hollow Fibre-Protected Two/Three-Phase Solvent Microextraction. Hollow fibre-protected two-phase solvent microextraction (HF(2)ME), was first introduced by He and Lee in 1997[13] under the name of *liquid-phase microextraction*. In its simplest version, the technique involves a small-diameter microporous polypropylene tube (the hollow fibre), usually sealed at one end, to contain the organic extracting solvent. The open end of the hollow fibre is attached to a syringe needle containing the selected extraction solvent and is used to fill the fibre with the organic solvent. The fibre is then immersed in the vial containing the aqueous sample to be studied. Enrichment is achieved by migration of the analytes through the fibre. Once the extraction time is completed, the solvent is withdrawn with the syringe and transferred to the instrument selected for final determination of the analytes. In practice, HFME can be considered as a liquid–liquid membrane extraction in which the porous polymer effectively protects the extraction solvent from contamination with particulate matter and soluble polymeric material, such as humic acids and proteins.[11] Consequently, HF(2)ME is more appropriate than SDME for the analysis of 'dirty' aqueous samples. Other advantages of HF(2)ME over SDME are the use of larger extractant volumes (in the 4–20 μl range), which leads to higher extraction efficiencies, and the fact that the solvent cannot be dislodged, which allows the use of higher stirring rates. On the other hand, HF(2)ME usually involves longer extraction times than SDME (20–60 min *vs* 5–15 min with SDME), and, at least if LVI is used, only a fraction of the organic extractant is transferred to the instrument selected for final determination. When used in combination with LC, it also requires a previous solvent exchange. Although this technique can be adapted for use with an autosampler,[36] probably its main limitation is that each individual hollow fibre must be carefully sized and prepared before use.[11]

The three phases involved in HF(3)ME are the aqueous sample investigated, the water-immiscible organic solvent that fills the pores of the hollow fibre polymer before this is attached to the syringe needle, and an aqueous acceptor phase that is placed in the lumen of the fibre with the help of the syringe.[37] HF(3)ME is carried out similarly to HF(2)ME but, since the final acceptor solution is aqueous, the technique is used to extract water-soluble analytes from aqueous matrices. For obvious reasons, in HF(3)ME the pH of both the aqueous sample and the acceptor phase are key parameters controlling the efficiently of the extraction process.

HF(3)ME shares the most obvious shortcomings of HF(2)ME, namely its relatively long extraction times, difficulty of complete automation, and intensive manual preparation of the fibre before use. However, despite these limitations, like HF(2)ME, HF(3)ME has proved to be useful for the analysis of a variety of compounds in aqueous samples of limited size, for which it allows high enrichment factors in relatively short times, especially when the technique is modified to favour the migration of analytes through the porous fibre, for instance by applying a potential difference between the two phases.[38] This latter modification is known as *electromembrane extraction* (EME).

Interestingly, the fibre filled with solvent can also be sealed at both ends. In this case, it can be placed directly into the stirred solution for extraction and retrieved after a preselected extraction time. The enriched solvent is then removed by polymer puncture with a chromatographic syringe. This approach is also called *solvent bar microextraction*[39] and can be use as a two- or three-phase system.

HF(2/3)ME can be used in the static versions previously described, or using dynamic (*i.e.* in syringe or in-needle) approaches similar to those described in the SDME section. In terms of method development, their corresponding dynamic alternatives are more demanding than the static approaches because of the need for careful optimization of a larger number of experimental parameters affecting the efficiency of the extraction process, including (apart from those described in the static approach) the speed of withdrawal and ejection of the sample, the duration of the static extraction step in between these two actions, and the amount of sample aspirated. However, higher enrichment factors in shorter extraction times have been reported in all instances for these dynamic approaches.

In Table 5.3, the experimental conditions used and the typical enrichment factors achieved with the most common SME techniques are summarized and compared on the basis of selected examples.

5.2.1.2.3 Dispersive Liquid–Liquid Microextraction. *Dispersive liquid–liquid microextraction* (DLLME) was introduced in 2006 by Assadi's group[40] and can be considered a modification of miniaturized LLE. The technique consists of the rapid injection into the studied aqueous sample (up to 10 ml) of a relatively small amount of a water-immiscible extraction solvent (typically 10–50 µl) dissolved in 0.5–2 ml of a water-soluble solvent with the help of a syringe. The rapid injection of the mixture of organic solvents into the water efficiently disperses the water-immiscible solvent in the aqueous mass as small microdrops into which the target analytes are rapidly extracted. The enriched organic phase is then separated from the aqueous phase by centrifugation, or frozen (depending on its density) and directly subjected to instrumental analysis, typically by GC.

The feasibility of direct DLLME has been demonstrated for the accurate determination of non-polar compounds, including trace non-polar microcontaminant families such as polycyclic aromatic hydrocarbons (PAHs), chlorobenzenes, and trihalomethanes for which enrichment factors in the 200–900 range have been reported (Table 5.3). The analysis of polar analytes demands previous pH adjustment and/or *in situ* derivatization of the polar analytes to improve the extraction efficiency. The derivatization agent can be directly added to the sample or dispersed together with the extraction solvent.

The several manual manipulations involved in DLLME made the technique difficult to automate and the use of internal standards and surrogates even more necessary than for the SME-based techniques described earlier.

5.2.2 Sorption-Based Miniaturized Extraction Techniques

Many pretreatment techniques currently used for clinical or environmental analysis of fluid (gaseous or liquid) samples are based on trapping the target compounds on, or in, a suitable sorbent. The amount of sorbent that has to be used, *i.e.* the capacity, is determined by the amount of analyte(s), the level of matrix interferences, and the nature of the interactions between the analyte(s) and interferences and the sorbent. Depending on the characteristics of the sorbent, analyte retention is governed by adsorption, *i.e.* by real chemical interaction between sorbent and analytes, by absorption, *i.e.* by partition of the analytes between sorbent and sample, by ionic interactions, and/or by a mixed retention mechanism. Frequently used sorbents include alkyl-bonded silicas, extremely hydrophobic styrene–divinylbenzene copolymers, and polydimethylsiloxane (PDMS) for sorption and enrichment purposes; Tenax; carbon for the (more or less) selective trapping of highly polar or planar compounds; ion exchangers; mixed-mode materials, *e.g.* cation and anion exchangers mixed with C18-bonded silica for the simultaneous enrichment of both highly polar and non-polar compounds; and the class-selective immunosorbents (ISPEs) and molecular imprinted polymers (MIPs). For detailed overviews of the relevant characteristics of these and other types of sorbent, and the most commonly used formats, the reader is referred to reviews of a more specialist nature.[41,42]

After preconcentration, analyte desorption can be accomplished by elution with a small volume of an appropriate solvent, either in a vial or at an appropriate interface, which is then partly or completely transferred to the instrument selected for final determination; or by thermal desorption, typically in the injection port of the instrument. Of course, thermal desorption offers the advantage of no dilution but it is obviously limited to semivolatile thermally stable analytes.

In this section we have reviewed the most relevant miniaturized sorption-based techniques in use on the basis of selected applications dealing with the analysis of fluid and liquid real-life samples.

5.2.2.1 Solid-Phase Extraction

Today, *solid-phase extraction* (SPE) is the most widely used technique for the clean-up and enrichment of analytes from biofluids and environmental aqueous samples. In recent decades, a large variety of applications involving sorbents with increasingly improved selectivity and loading capacity, and even tailored-designed ISPEs and MIPs, have been described. Many of these sorptive phases are now commercially available, contributing to the wider use of the technique. Although research in sorbent development continues, contributing to simplification of sample preparation procedures and greening the analytical process, probably the main achievement in this context was the introduction of on-line SPE-based procedures, with SPE-LC and SPE-GC as its most representative and relevant examples. The key factor which made these hyphenated systems possible was the miniaturization of the SPE process. Reducing the size of the conventional 1–6 ml SPE syringe barrels to the 10 mm × 1–2 mm internal diameter of the so-called Prospeckt-type cartridges used in the hyphenated systems led to a reduction of sample volumes from 0.5–1.0 L to less than 50–100 ml. Quite often, even 5–10 ml turns out to be sufficient to obtain similar LOD of 0.01–0.1 µg L^{-1} with SPE-LC and 1000-fold lower with SPE-GC, which previously required 100-fold larger volumes. The relatively small particle size of the sorbent packed in the miniaturized SPE cartridge ($\sim$40 µm) results in high retention efficiencies and adequate breakthrough volumes despite the small amount of sorbent used. More importantly, quantitative elution of the analytes can be achieved with 50–100 μl of the appropriate solvent, *i.e.* with a volume small enough to allow complete transfer to the instrument selected for final determination.

Prospeckt-type cartridges are typically mounted in holders similar to those used for the LC precolumns and integrated in systems containing 4-, 6- or 10-port valve systems, depending on the complexity of the operations to be performed. LC pumps are used to pump the sample and solvent(s) through the system. Automation and computer control of the various operations can easily be achieved using programmable valves.[43]

As in any other SPE approach, in principle, two different working modes are possible: the sorbent can be used to retain either the interfering matrix components or the target compounds. The former working mode allows purification of the analytes but without concentration. The latter would ideally allow performing both operations, *i.e.* preconcentration plus clean-up of the studied analytes, in a single step. In practice, the SPE process is not as selective (or samples as clean) as would be desirable, and complete removal of potentially interfering matrix components is only accomplished after an extra purification step. In the on-line systems, this and other operations, such drying of the sorbent, can be performed by simply changing the position of the valves.

For readers who are not completely familiar with the valve-based systems used for miniaturized SPE, Figure 5.2 shows the basic set-up required for on-line SPE with two cartridges.

Figure 5.2A shows the valves position during sample loading. In this case, the sample, previously loaded in the injection loop, is pumped through both SPE cartridges. Analytes will be selectively retained in any of the cartridges according to their affinity for the respective sorbents; or alternatively eliminated via the waste port. Then, changing the position of valves 1 and 3 and using a micro-LC pump, the analytes previously retained in cartridge 1 can be desorbed and directly transfer to the instrument selected to perform the analytical determination. When this step is completed, a simple change of the position of valves 2 and 3 make sit possible to proceed with desorption of the analytes retained in cartridge 2, which can then be quantitatively transferred to the analytical system. The number of valves and configuration of the system would obviously depend on the goal of the analysis. As an example, a similar valve-based approach was used to study the so-called fast adsorption of pesticides in soils and sediments.[44] In this set-up, the injection valve was slightly modified to allow direct injection of the contaminated soil/sediment slurry and a filter was incorporated for retention of the solid particles, while pesticides remaining in solution were concentrated in an SPE cartridge. Both pesticide fractions were separately collected for independent instrumental analysis. That is, a single injection of the slurry in the system provided simultaneous information on the pesticide concentration in both phases in less time than the conventional batch approaches used at that moment, so allowing a more accurate evaluation of the adsorption isotherms at early steps of the contamination process.[45]

As a further illustration of the potential of this type of on-line SPE system, Table 5.4 summarizes the several steps involved in an SPE-LC analysis.

Miniaturization of Analytical Methods

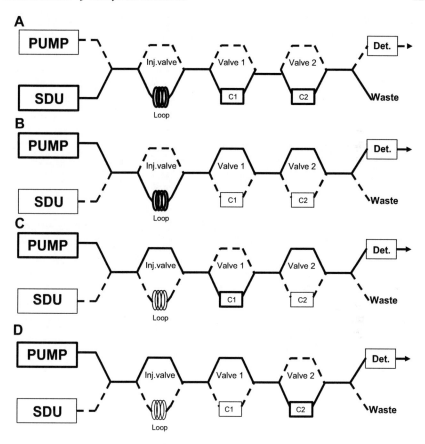

Figure 5.2 Schematic of the basic valve set-up required for miniaturized on-line SPE with two cartridges (C1 and C2). (A) Configuration for sample loading, (B) test of the tube and injection loop, (C) cartridge 1 elution, and (D) cartridge 2 elution. SDU, solvent delivery unit; Det., detector.

Table 5.4 Different steps in an SPE-LC analysis; typical solvent volumes, flow rates and times involved in each case when using a 10 mm × 2 mm SPE cartridge (C18, 40 µm) are indicated.

Step	Solvent (ml)	Flow (ml min^{-1})	Time (s)
Activation	Methanol (2)	5	24
Equilibration	Water (2)	5	24
Equilibration	Buffer (2)	5	24
Sample loading + washing (2.5 ml buffer)	Serum (0,5)	1	180
Elution	Mobile phase	0.5	45
Complete process			*297*

The small volumes of solvents required for sorbent activation and equilibration, combined with the relatively high flow rates used (typically, 5 ml min^{-1}), make it possible to complete these steps in much shorter times (~1 min) than those required for conventional SPE. Sample loading, in this case 500 μl of serum, should be carried out at slower flow rates to allow proper interaction of the analytes with the sorbent. For this step, a micro-LC pump must be used to ensure adequate flow control (typically in the 10–200 μl min^{-1} range). The subsequent washing of more weakly retained matrix components out of the cartridge can be performed at intermediate flow rates of approximately 1 ml min^{-1}. Despite the relatively slow flow rates used in the loading and washing steps, both processes are completed in quite a short time (~3 min) because of the small volume of sample and cleaning solvent pumped. Elution of the analytes from the cartridge is typically carried out at 100–500 μl min^{-1} to ensure quantitative desorption of the studied compounds in a minimum solvent volume and its transfer to the analytical system as a narrow band for optimal LC separation. In most instances, 50–75 μl suffices for the quantitative desorption of analytes from a miniaturized SPE cartridge, so this step could be completed in 30–50 s, depending on the application. It is therefore possible to conclude that the average time required for complete sample preparation is around 5 min, which is significantly shorter than the time required for conventional (*i.e.* regular size and off-line) SPE.

The small volumes and short analytical times involved in on-line miniaturized SPE also means that method development and optimization with these systems are, in principle, faster than those needed in conventional approaches. The experimental parameters to consider during method development in miniaturized SPE are the same as for conventional SPE, namely the nature and amount of sorbent, the nature of the solvents used in the different SPE steps, and, in particular, their flow rates. Also, the main reasons for low analyte recoveries are reduced sorbent capacity or too strong retention; slow kinetics of the sorption process (or, to put it another way, sample and/or solvent flow rates that are too high); and possible adsorption of the analytes in the tube used to connect the different parts of the system. On the other hand, in these closed systems the risk of analyte degradation and oxidation is greatly reduced as compared to the (open) conventional approaches.

Assuming that automation of on-line SPE-based systems can be considered as an achieved goal, development in SPE-LC is at present mainly orientated to the use of sorbents with (1) higher loading capacities that contribute to improving the present LODs; (2) new functionalities to extend this type of routine analysis to more polar analytes; and (3) more selective sorbents, such as ISPEs and MIPs, which contribute to reducing the risk of interference when less selective detectors are used as alternative to mass spectrometry (MS). It should be added that next to the on-line set-up, replacing the SPE–LC part by a single short column, SSC (1–2 cm length), and using MS[46] and especially MS–MS[47] detection has efficiently contributed to further reduction of both the analytical times and the LODs. As an example, it can be mentioned that this type of approach has facilitated the real-time study of analyte degradation at the trace level,[47,48] with LC run times of, frequently, only some 3 min. Finally, this progressive reduction of the sample size required to perform the analysis has made possible, in some cases, the direct injection of the aqueous sample[49] or of the aqueous extract obtained from fruits and vegetables.[50] In the former, 4 ml sufficed to achieve LODs of 0.01–0.1 µg L^{-1}. In the latter, the injection of the 3 ml extract provided LODs in the 0.5–2 µg kg^{-1} range for some particular pesticides, which can be appropriate for their fast screening. However, the use of highly selective and sensitive detectors, such MS-MS, then becomes mandatory.

In principle, the on-line combination of SPE and GC could be considered more difficult than SPE-LC because of incompatibility of the solvents and flow rates used in both techniques. However, in practice, SPE-GC experienced a development parallel and similar to that SPE-LC. Experience has shown that with SPE cartridges as small as 10 mm × 1 mm internal diameter, water samples of up to 10 ml suffice to reach LODs of 20–50 ng L^{-1} in the full-scan MS acquisition mode for a wide variety of semivolatile microcontaminants.[51] The better performance of the GC compared with the LC-based approach is mainly because of the superior detection/identification performance of GC-MS. Actually, if GC-MS-MS was used instead of GC-MS, LODs for several pesticides in surface water were 0.01–4 ng L^{-1} for 10 ml samples[52] or, alternatively, 0.2 µg L^{-1} for 0.1 ml samples.[53] These low LODs and the progressive reduction of the sample volume required have promoted an important development in the field of LVI in GC. Some of the novel LVI interfaces[5,6] allow direct water injection in GC, and the introduction of up to 10 ml of solvent using slow injection or multiple fast injections in a packed liner. Nevertheless, the application field of these modifications depends strongly on the type and concentration of the interferences present in the sample.

For a detailed discussion on the feasibility of SPE-GC combined with different detectors for the accurate determination of trace compounds in aqueous samples, and on the possibility of on-line coupling of SPE with other modern separation techniques, including microbore LC and CE, the reader is referred to a more specialized review.[54]

As regards other SPE formats, the so-called Empore discs (discs 2–3 mm in diameter and 0.5 mm thick cut from the larger original discs) have shown to be attractive alternatives to conventional cartridges, both for on-line SPE-LC[55] and SPE-GC.[56] Resin discs (0.7 mm diameter) have also been mounted inside the removable needle chamber of a 50 µl Hamilton gas-tight syringe and proved to be a valuable miniaturized automated alternative that enabled the efficient preconcentration of substituted benzenes from a volume of water as small as 2.5 ml. This method provided recoveries higher than 90% at the 10 ng ml^{-1} level with GC-FID and required only 5 µl of acetonitrile for desorption.[57] A recently introduced modification to this approach proposes the use of a sorbent chamber (or cartridge) placed at the top of the syringe needle to yield the so-called *microextration in a packed syringe* (MEPS) technique. This miniaturized technique, through successive withdrawal and ejection of the aqueous solvent, allows the preconcentration of the analytes in the aqueous sample on the selected sorbent. As in other SPE approaches, a washing step (typically with 50 μl of water) can easily be incorporated to remove any undesirable matrix component. Then, analytes are eluted with an appropriate amount of solvent (~20–50 μl) and transferred to the GC/LC port. MEPS applications includes, for instance, the determination of PAHs in water[58] and of drugs in blood.[59]

Finally, the 96-well plates, very popular in clinical research and for biological applications, have up to now hardly been used in environmental studies, despite the high throughput and low LODs that can be achieved when they are combined with an appropriate separation-plus-detection system, as demonstrated for the trace-level determination of alachlor in water and vegetables, with an LOD of $0.4\,\mu g\,L^{-1}$ by GC-MS.[60]

5.2.2.2 Solid-Phase Microextraction

Solid-phase microextraction (SPME) was introduced in 1990 by Pawliszyn's group[61] as a solvent-free preconcentration technique in which analytes are adsorbed on to a fused-silica fibre coated with an appropriate sorbent layer by simple exposure of the fibre for a preselected time to the gas or liquid sample. The preconcentrated analytes are then desorbed into a suitable instrument for separation-plus-detection. The SPME device is extremely simple (Figure 5.3). It consists of a syringe-like metallic body equipped with a needle that houses the SPME fibre and a plunger that allows the fibre to be retracted or pulled out of its protective holder (*i.e.* the syringe needle) for analyte sorption and desorption. The fibre position is adjusted with a screw placed in the syringe body. Other SPME formats, such as in-tube SPME[62,63] have achieved only rather limited success.

In general, analyte preconcentration is achieved by exposure of the fibre to the HS of a 1–10 ml vial containing the gaseous, liquid or solid sample; or by direct immersion of the fibre in the liquid sample (or extract) in which the analytes are dissolved. Figure 5.4 shows an overview of the most common sorption and desorption approaches for SPME as well as some common

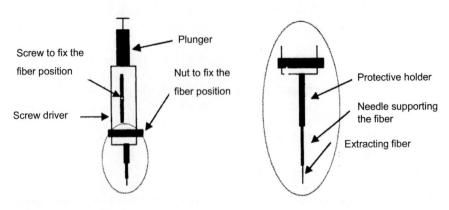

Figure 5.3 SPME device.

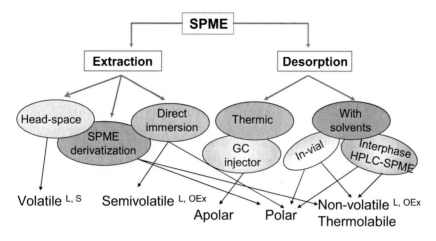

Figure 5.4 Different working modes for SPME and fibre desorption and their corresponding fields of application.

analytical strategies adopted to increase either the selectivity or the application range of this miniaturized technique. In most instances, SPME is used for preconcentration but other applications, such as purification, phase exchange, and field sampling are also possible. This chapter focuses on the first of these aims since the others can be considered particular cases of that one.

SPME is an equilibrium technique. In the HS approach, the volatile analytes are displaced from their original sample surface–HS equilibrium to the fibre surface in which they are progressively preconcentrated. This working mode has been applied for the extraction at ambient temperature of compounds with Henry's constants above 90 atm cm^3/mol from liquid ($\sim$5 ml) or solid (a few mg) samples contained in a closed vial (or recipient). Extraction times as short as 1 min suffice for quantitative SPME of benzene, toluene, ethylbenzene, and xylenes (BTEX) from water.[64] For less volatile compounds, the extraction time can efficiently be shortened by reduction of the HS volume, heating and salting-out of the sample, and agitation of the liquid sample and the HS. However, one should be aware that these treatments can also promote desorption of other, less volatile, matrix components from the sample, which will then compete with the target compounds for sorption in the fibre. Careful optimization of the several experimental parameters affecting the SPME process is consequently mandatory. In any case, the selectivity of the HS-SPME process results in rather clean (*i.e.* simple) chromatograms, which has led this approach to be sometimes preferred for indirect determination of non-volatile analytes which are previously transformed on to more volatile derivatives in the extraction vial.

In principle, direct immersion of the fibre in a liquid sample should force the analyte partition between these two phases. In practice, the fibre is similarly exposed to all matrix components which, depending of their affinity for the selected sorbent, will compete with the target analytes for the active points on the fibre. Therefore, direct fibre immersion makes it possible to extend the application field of SPME to less volatile compounds. However, as compared to HS-SPME, selectivity is somewhat sacrificed and method optimization is usually more demanding because of the higher possibility of a matrix effect.

As in any other extraction technique based on the sorption of an analyte on a sorbent, one of the main parameters affecting the efficiency of the SPME process is the nature and amount of sorbent. Several fibre coatings are now commercially available, including the non-polar PDMS, semipolar polydimethyl siloxane–divinylbenzene (PDMS–DVB), and polar polyacrylate (PA), Carbowax–divinylbenzene (CW–DVB) liquid-like phases, the coated porous particle phase polydimethyl siloxane–Carboxen (PDMS–Carboxen), poly(3-methylthiophene), Nafion and, less frequently, carbon nanotubes,[65] MIPs,[66] or simply anodized metals (reference 67 and references therein), which contributes to the expansion of the range of analyte classes that can be successfully analysed (Table 5.5). This variety of sorbents results in diverse retention mechanisms depending on the nature of the extracted analytes, which can be adsorbed, absorbed or react with the fibre coating.

In most applications, the analyte partitions into the stationary phase until plateau conditions are reached, which typically takes 2–60 min (with higher values for higher molecular weight analytes). As previously mentioned, the process can be aided by salting-out, sample agitation, pH adjustment and/or heating, and matrix effects can be avoided by using the standard addition procedure for quantitation or, less frequently, protective membranes to prevent the adsorption of matrix components on the fibre.[76] At equilibrium, the amount of analyte sorbed in the fibre coating is directly related to its concentration in the sample according to the following equation:

$$n = K_{fs} V_f V_s C_o / (K_{fs} V_f + V_s)$$

where n is the number of moles of compound retained by in the stationary phase, K_{fs} is the partition coefficient of the compound between the station at phase and the sample, V_f is the volume of the stationary phase, V_s is the volume of the sample, and C_o is the initial concentration of the compound in the sample. The relationship between the amount of analyte preconcentrated in the fibre and that on the original sample is consequently linear, and SPME can provide quantitative data. Moreover, when the volume of the sample is very large compared to that of the fibre, the previous equation is simplified to

$$nV_s = K_{fs} V_f C_o$$

which justifies the use of SPME in field sampling.

SPME allows to achive LODs at low ng L^{-1} levels for both volatile[70] and semivolatile[69,73] analytes when selective detection is used, e.g. SIM-MS or AED.[72] Increasing fibre thickness, typically in the 7–100 μm range, helps to increase the sensitivity because of the improved partitioning ratio but it also

Table 5.5 SPME analytical approaches and typical applications.

Sample ($ml\,mg^{-1}$)	Analyte	SPME fibre (thickness, μm)	Fibre description	Sample pretreatment	Time (min)	Ref.
Headspace						
Tap and swimming-pool (10)	Haloacetic acids	PDMS–Carboxen (75)	Partially cross-linked	Derivatization + ion strength adjustment	35	68
Industrial harbour water (10)	PCBs	PDMS (100)	Non-bonded	–	30	69
Human urine (2)	BTEX	PDMS (100)	Non-bonded	–	30	70
Slurry plant:water (1:3) (5000)	Organophosphorus pesticides	PDMS (100)	Non-bonded	Homogenization	90	71
Waste oil (0.5)	PCBs	PDMS–DVB (65)	Partially cross-linked	Acid digestion + LLE + water dilution	10	72
Direct immersion						
Surface water (3)	Herbicides	CW–DVB (65)	Partially cross-linked	–	30	73
Industrial wastewater (5.3)	Industrial organic pollutants	PA (85)	Partially cross-linked	pH adjustment	30	74
Fruit and fruit juice (3)	Organophosphorous pesticides	PDMS (100)	Non-bonded	Slurry soil:water (1:100)	20	75

BTEX, benzene, toluene, ethylbenzene, and xylenes; PCB, polychlorinated biphenyls.

increases equilibrium times and sometimes results in problems in achieving complete desorption. Strategies involving derivatization of the analytes in the aqueous phase, combined with SPME, have extended the range of application to very polar[77] or ionic substances.[78,79] Applications involving SPME with on-fibre derivatization require conversion of the analytes after extraction by applying the reagent as a gas and are, as far as we know, still scarce in the literature. The relatively high relative standard deviation (RSD) values reported up to now when using this approach (*e.g.* 10–35% for chemical warfare agents at 1–20 µg ml^{-1} levels, n = 6)[80] can be regarded as an indicator of conditions which are difficult to control. Generally speaking, although the analysis of aqueous samples can be accomplished with little or no pretreatment (Table 5.5), SPME of target compounds from more complex (solid) matrices typically requires a previous separation of the analytes from the main matrix components,[81] usually involves longer extraction times and is frequently less exhaustive than for liquid samples because of the less favourable extraction conditions.

Desorption of analytes from the SPME fibre in which they have been preconcentrated is most frequently accomplished by direct thermal desorption in the GC injection port for subsequent GC separation and detection (Figure 5.3). In this approach, the injection conditions should ensure complete analyte volatilization from the fibre and introduction in the GC column. The only practical limitation is the working temperature, which is determined by the nature of the fibre stationary phase (typically in the 260–300 °C range, depending on the coating). Nowadays, a number of commercial GC autosamplers allow complete automation of the SPME process.

More polar or thermolabile analytes can be manually extracted by immersion of the fibre in a small amount of solvent contained in a vial. In this case, after a preselected extraction time, a fraction of the enriched solvent is typically injected into an LC or CE system for separation and analyte detection. This approach requires the careful optimization of the several parameters affecting the solid–liquid extraction process, such as the nature and volume of the extraction solvent, extraction time and, when required, solvent agitation. One should also be aware that this approach is not applicable to fibres with non-bonded phases because they tend to disintegrate and dissolve in contact with organic solvents. Several interfaces have also been developed to allow automation of this liquid extraction process and direct transfer of the complete liquid phase to the LC system (see reference 82 and references therein). However, the practical application of this configuration appears to be somewhat more limited than that achieved by the GC-based approaches.

It is evident that the small size of the fibre is mainly responsible for both the advantages and the main shortcomings of the technique. Most of these have been discussed in detail in this section, but there is a particular type of analysis for which the small size of the needle is the key feature making SPME the only applicable technique. This approach is called *non-depletive SPME* (nd-SPME) and has been used, for instance, to determine pollutants dissolved in pore water with high precision.[83,84] The miniaturized fibre design was also the key aspect in the determination of both the free and total internal amounts of chlorfenvinphos in laboratory- and field-exposed small insects (*Trybliograpa rapae*, up to 3 mm long) extracted with only 200 μl of the selected solvent.[1] LODs 10 times lower than reported for solvent-based extraction procedures (*i.e.* <0.5 ng), were obtained after 45 min SPME, thermal desorption, and GC-ECD analysis.

5.2.2.3 Stir bar-Sorptive Extraction

One of the limitations of SPME mentioned above, the relatively small volume of bound stationary phase, prompted the development of a new miniaturized extraction technique, *stir bar-sorptive extraction* (SBSE), introduced in 1999 by Sandra's group[85] and marketed commercially as the Twister. In a typical SBSE experiment, a magnetic stir bar coated with 55 or 219 μl PDMS (corresponding to magnets 10 and 40 mm long, respectively) is rotated in an aqueous sample (or extract) for a selected, but often fairly long, extraction time. SBSE of the HS of a gas, liquid or solid sample contained in a sealed vial is also possible, although less frequently used. The magnetic stir bar can also be inserted into a short length of PDMS or silicon tubing. In any case, because the surface area of the stir bar is greater than that of the SPME fibre and the volume of the adsorbent is at least a factor of 100 larger, there is a higher phase ratio than in SPME and hence a higher extraction efficiency, which results in lower LODs. After the extraction, the stir bar is removed, often manually, and transferred to the injection port of a GC for thermal desorption,[86] or into a solvent for LC analysis.[87,88] A novel desorption unit enables fully automated analysis of 98 or 196 PDMS-coated stir bars.[89]

The similarities between SPME and SBSE could easily make conclude that all working modes described for SPME (Figure 5.3) are also possible for SBSE. However, it is important to note that the still rather limited variety of coating materials available for SBSE limits the practicability of the technique. Anyway, SBSE has been shown to be a valuable simple, green, and miniaturized analytical alternative for many applications and, in some of them, has proved to be superior to SPME. When combined with a selective GC detector such as MS and using sample volumes of typically 10 ml the technique is feasible for analysis of compounds ranging from non-polar PAHs[87] to some organotin compounds[90] in water, dicarboximide-type fungicides in wine,[91] and additives in beverage and sauce samples[92] at the μg L^{-1} level. However, and similarly to SPME, application to more complex samples such as biological fluids[93,94] or solid samples[89] can only be accomplished after a pretreatment step which effects appropriate isolation of the target compounds from the matrix.

Compounds preconcentrated on the stir bar can also be extracted with a small volume of solvent (*e.g.* 500 μl) in a vial.[88,95] However, the obvious drawback is more manipulation and dilution of the analytes, *i.e.* loss of analyte detectability, because only a fraction of the extract is injected typically into the LC system.

5.3 Miniaturized Analytical Techniques for Treatment of Solid Samples

Most of the previously reviewed techniques cannot be directly applied to semi-solid environmental and biological matrices, *i.e.* to samples containing high amount of lipids, proteins, or organic matter. In general, the analysis of these types of samples requires the initial extraction of the target compound(s) from the complex matrix in which they are entrapped. The non-selective character of most of the exhaustive extraction procedures used in this step makes subsequent purification and/or fractionation of the studied analytes from the co-extrated material mandatory to ensure the accurate instrumental determination of the target compounds. For these subsequent clean-up steps, techniques and analytical procedures similar to those described for liquid samples in previous sections are typically used.

This section reviews modern techniques that have already been demonstrated to provide extraction efficiencies similar to those of other conventional (*i.e.* large scale) and widely accepted extraction techniques but that involve a much smaller sample size (*i.e.* <0.5 g). Because one of the main goals of green sample treatment is the effective reduction of reagent consumption, techniques and analytical strategies allowing an enhanced selectivity, so contributing to the simplification (or even elimination) of subsequent clean-up treatment(s) of the generated extracts before instrumental analysis, will receive special attention.

5.3.1 Matrix Solid-Phase Dispersion

Matrix solid-phase dispersion (MSPD) was introduced in 1989 by Barker *et al.*[96] as a process that allowed the disruption of the structure of a solid matrix and its extraction in a single treatment. In MSPD, the semisolid matrix is blended with an appropriate sorbent until a dry, homogeneous mixture is obtained. This process results in the homogenous distribution of the matrix components on the sorbent surface and, in practice, can be considered a solid–solid extraction. The mixture obtained is then packed into a column (or syringe barrel) from which the analytes of interest are eluted with a suitable solvent. MSPD is also applicable to liquid or viscous matrices by simply mixing of the sample and the sorbent, and subsequent sedimentation and homogenous packing of the slurry into a column. Proper selection of sorbent and eluent can effect specific retention of impurities on the sorbent and selective elution of the target compounds. These combined clean-up effects frequently enable direct analysis of the collected extract. As an illustrative example of the efficiency of the MSPD process, Figure 5.5 compares the total-ion GC-MS chromatograms obtained with different sorbent–eluent combinations during the analysis of selected pesticides in single insects (40 mg *Porcellio scaber*).[2] MSPD does not require special equipment and it is feasible for field application.

The quoted features and the relative simplicity of MSPD are responsible for the wide and rapid acceptance of the technique in many application fields where

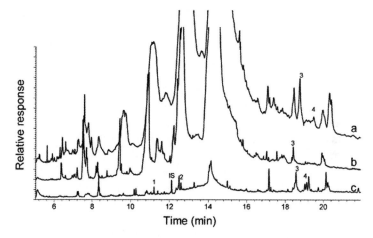

Figure 5.5 Comparison of total-ion GC-MS chromatograms of extracts obtained by MSPD from insects with 100 μl samples. (a) Ethyl acetate from a C8-bonded silica/sample mixture; (b) ethyl acetate from C8/sample mixture and washing before extraction; (c) n-hexane from silica/sample mixture. Peak assignment: (1) diazinon, (2) malathion, (3) permethrin, (4) cyfluthrin, (IS) parathionmethyl.

it has been demonstrated to be a valuable alternative to more classical exhaustive techniques, as highlighted in several reviews.[97–99]

Miniaturization of MSPD can be achieved simply by reducing the amount of sample subjected to the analysis, with corresponding reduction of the amount of sorbent used for dispersion. This simple operation makes it possible to reduce the tens of grams of sorbent and several hundreds of millilitres of solvent required for sample preparation by most conventional methods to around 1 g of sorbent or less and 5–20 ml of solvent in miniaturized MSPD. Apart from the inherent benefit when only a small amount of sample is available,[100,101] sample preparation is typically completed in less than 1 h with minimum sample manipulation, which sharply contrasts with the several hours and several treatment steps required by classical large-scale procedures. The feasibility of miniaturized MSPD for the quantitative extraction and simultaneous purification of a variety of analytes has been demonstrated in a number of application studies,[98,99] in which the initial sample reduction has easily been compensated by the high sensitivity provided by modern instrumentation. However, rather surprisingly, miniaturized MSPD is still far from being considered a common practice in laboratories.

Miniaturized MSPD has been shown to be a valuable alternative for the fast and accurate determination of relative abundant components, such as essential oils in herbs[102] and polyphenols and organic acids in tobacco;[103] and also trace analytes, including environmentally relevant PAHs in soil[104] and different classes of pesticides in non-fatty matrices, such as juice,[105] fruits,[106–108] and cereals.[106] For these latter types of matrices, and despite the different chemical structures and polarities of the target compounds, recoveries above 80% have frequently been reported. When a selective and sensitive technique is used for final determination (*i.e.* GC-MS[105,107] or LC-MS[106]), LODs in the low μg/g range are easily obtained even without extra treatment of the collected extracts. As an illustrative example of the potential of the approach for the development of complete on-line (or at-line) processes, using a Prospect-type set-up, Kristenson *et al.*[107] proved that 25 mg of fruit and 100 μl of ethyl acetate sufficed for accurate extraction (recoveries in the 83–118% range and RSDs below 13%) of a variety of triazines and organophosphorus pesticides at the maximum residue limits (MRL) typically set in European Union legislation (LODs in the 4–90 μg/g range).

When dealing with more complex samples, *e.g.* for pesticide determination in insects, more careful selection of the sorbent–solvent combination is required to perform the necessary clean-up without affecting the performance of the method (see Figure 5.5 for an example[2]). The sequential elution of closely related impurities and analytes and/or performing some extra in-line purification by packing an appropriate sorbent in the bottom of the extraction column are also successful approaches for the one-step analysis of such samples.[101]

The analysis of trace compounds in fat-containing matrices represents a particularly difficult analysis case. In this type of determination, apolar sorbents, such as C18, are usually preferred for sample dispersion and mixtures of non-polar and medium-polar solvents are used for the sequential extraction of interfering matrix components and target compounds from the column. However, this analytical strategy does not always result in the desirable complete elimination of the lipid residues and extra treatment of the collected extracts is frequently required.[109,110] As an alternative to the subsequent off-line treatment of the MSPD extracts, in-line (or on-line) packing of more polar sorbents, such as Florisil,[104,111,112] strong ion exchangers,[113] and silica modified with sulfuric acid,[114] have been proposed for complete fat removal, so yielding ready-to-analyse extracts.

Rather unexpectedly, the use of dispersant sorbents alternative to the typical C18 for fat removal in these types of determinations is still quite rarely considered. However, carbon has been proved to be an efficient alternative for lipid retention during the analysis of dithiocarbamate pesticides and their main metabolites in avocado and nuts,[106] although in this study the widely variable structures and polarities of the analytes studied resulted in rather wide recoveries (11–96%). During the analysis of ethylene bisdithiocarbamate metabolites in almond,[115] lipids were hydrolysed and removed from the column with 0.02 mol L^{-1} NaOH. Remaining traces of fat were selectively retained on alumina (in a second on-line column) during the extraction step. In this study, satisfactory recoveries (76–85%), RSDs lower than 12%, and LODs of 50–70 ng/g were obtained although only 200 mg of sample was used, and LC-DAD VU was selected for final determination. In this case, sand was used as sorbent support for MSPD to avoid column clogging. Satisfactory results (*i.e.* recoveries in the 81–130% range with RSDs of 2–12%) were also obtained when using silica modified with sulfuric acid as MSPD sorbent for fat removal of fatty animal foodstuffs containing up to 45% of lipids (w/w on a dried basis).[114]

Ready-to-analyse extracts were obtained when an extra layer of modified silica was packed at the bottom of the extraction column to ensure in-line removal of remaining lipidic traces. LODs as low as 0.09–3 pg/g were obtained for all 23 PCB congeners investigated using GC-microECD for final determination, which proved the feasibility of the proposed procedure for accurate determination of these trace lipophilic pollutants even although subsamples as small as 100 mg of heterogeneous matrices were used for the analysis.

5.3.2 Enhanced Fluid/Solvent Extraction Techniques

Extraction efficiency during the preparation of semisolid matrices can be enhanced by heating or shaking the sample or by using as extractant a fluid or solvent with a higher diffusion rate. The latter is the basis of supercritical fluid extraction (SFE) and subcritical water extraction (SWE); the former approaches are used in pressurized liquid extraction (PLE), microwave-assisted extraction (MAE), and ultrasonic-assisted extraction.

The main benefices and limitations of SFE as an essentially solvent-free, *i.e.* green, and in many instances miniaturized, analytical extraction technique have been discussed in previous chapters and will not be repeated here. The readers are encouraged to read Chapters 3 and 4 to obtain a complete vision of the applicability of this particular technique in the different research areas.

5.3.2.1 Pressurized Liquid Extraction

Since its introduction in the mid 1990s[116] and rapid acceptance as a U.S. Environmental Protection Agency (EPA) method,[117] PLE has undergone rapid development and it is nowadays a widely accepted extraction technique in many research fields (including procedures in which water is used as the extractant, *e.g.* SWE).[118] As previously explained in Chapter 4, in PLE, the sample, typically dispersed in a drying or inert sorbent such as sodium sulfate, Hydromatrix, or diatomaceous earth, is packed in a stainless steel cell and, once inserted in a closed flow-through system, extracted with the selected solvent at temperatures above its atmospheric boiling point. Because the solvent must be kept liquid during extraction, relatively high pressures are also applied. Its well-documented efficiency, rapidity, and moderate solvent consumption are recognized as the main merits of this essentially analyte- and matrix-independent technique and the reasons for its widespread application. The feasibility of the approach for on-line or in-line coupling with some of the techniques reviewed in previous sections for subsequent fractionation and/or enrichment of the extracted analytes, an aspect particularly relevant during SWE of less polar compounds, is another valuable feature of the technique.

Despite its many attractive characteristics, the number of studies dealing with miniaturized PLE has, until now, been rather limited, probably because of the relatively large size of the smaller extraction cells of commercial systems (*i.e.* at least 11 ml). Some authors[119] have demonstrated, using such a PLE systems that, for instance, 20 mg of freeze-dried bacterial cells or 100 mg of soil sufficed for the accurate determination of phenols. However, the large dimensions of the 11 ml cell as compared to that of the sample obliged them to fill the rest of the extraction cell with an inert sorbent. In other cases, the remaining space has been used to pack clean-up sorbents to perform in-line purification of the PLE extracts.[120,121] Irrespective of the approach used, these types of arrangements result in the use of amounts of sorbents and solvents similar to those of conventional PLE applications although a much smaller sample is analysed. In other words, sample reduction does not result in the desirable greening of the analytical process.

At present, the only way to solve this problem is to design a home-made miniaturized PLE system.[122–125] This type of set-up and a heatable 10 mm × 3.0 mm internal diameter stainless steel extraction cell enabled quantitative extraction of the 16 EPA PAHs (recoveries, 90–110%) from 50 mg soil with only 100 µl of toluene. Direct injection of 50 µl of this raw extract into a programmable temperature vapouriser (PTV) system containing the so-called ATAS-A sorbent, enabled in-line clean-up before GC-SIM-MS analysis and LODs as low as 2–9 ng/g soil for a large majority of the target compounds. Although a small amount of a very heterogeneous sample was used, the RSDs of 2–15% were similar to those found when using traditional methods for this type of determination.[124] The approach allowed complete sample preparation in 10 min, minimized reagent consumption, and, because of the small volume of solvent, showed a better potential for hyphenation and automation than with commercial systems. The same arrangement was subsequently used for PLE of chloroanilines from soil. Again, acceptable recoveries were obtained for most of the studied analytes (36–109%) and satisfactory RSDs (8–13%) were reported even although only 50 mg of sample was used for the determination. LODs as low as 5–50 ng/g were obtained when 20 µl of acetone:hexane (1:1, v/v) out of the 100 µl used for the extraction was injected in the GC-MS system.

The main limitation of this set-up—the lack of flexibility as regards the dimensions of the extraction cell—was overcome by using a large, heatable oven, e.g. that of a GC[122] or a specifically designed miniaturized oven.[124–126] The former solution has typically been adopted for SWE because of the need to insert a relatively large coil for heating the water before the extractant reaches the extraction cell.[122,123] The latter has been proved to be particularly interesting when the goal is the in-cell purification of the target compounds in order to obtain ready-to-analyse extracts with minimum time and reagent consumption.[125]

5.3.2.2 Microwave-Assisted Extraction

As with PLE, no miniaturized MAE system is commercially available; consequently, very few studies dealing with miniaturized MAE are reported in the literature.[127,128] The first attempt to develop miniaturized MAE was reported by Cresswell and Haswell.[127] In this study, a sediment slurry was pumped through a wide (1/8 in, 3 mm) PTFE tubing installed inside the MAE system at 0.75 ml min^{-1}. After passing through the microwave cavity, the slurry was in-line filtered to separate the solid particles from the liquid fraction. Analytes in the aqueous phase were then on-line preconcentrated in a C18 cartridge. The method was applied to the analysis of PAHs in certified sediment although the results were far from satisfactory (RSDs in the range 22–50%). More convincing results were reported by Ericsson and Colmsjo[128] who, using an essentially similar approach, proposed inserting a preheating column in front of the extraction cell in the microwave cavity and the back-elution of the target compounds from the 10 mm × 2 mm PLRP-S SPE disposable cartridge used for on-line SPE of the extracted analytes. Using this configuration the authors demonstrated the feasibility of dynamic MAE coupled on-line with SPE for accurate determination of PAHs in a reference sediment (recoveries 88–104%, RSDs 1–10%) using only 60 mg of sample, 400 μl of methyl tert-butyl ether (MTDE) for back-extraction of the analytes from the SPE cartridge and GC-PID for final determination.

5.3.2.3 Ultrasonic-Assisted Extraction

The use of ultrasound for analytical applications is relatively recent. Nevertheless, some applications have already demonstrated the potential of sonication for the miniaturized, rapid, relatively inexpensive and quantitative extraction of analytes of different nature. Early papers in this field focused on the dynamic extraction and on-line purification of, for example, Cr(VI) in soil[129] and organophosphate esters in air filters,[130] using a stainless steel extraction cell placed either in an ultrasonic bath[130] or in a water bath close to an ultrasonic probe.[129] In both cases, a dynamic approach was preferred because the continuous transfer of the analytes from the matrix to the extractant solvent reduced the risk of analyte degradation by the high temperatures and pressures generated by the cavitation process compared with the static mode. As an example of the typical results obtained, this arrangement allowed quantitative extraction of the organophosphate esters preconcentrated from air on 25 mm binder-free A/E borosilicate glass fibre filter in 3 min with only 600 μl hexane:methyl tert-buthyl ether (7:3) and with RSDs below 8%.

Interestingly, as an alternative to the more frequent static ultrasonic bath extraction of relatively small samples (*i.e.* <0.5 g) followed by off-line column purification of the slurries,[131,132] Albero *et al.*[133] have recently demonstrated that the speed and efficiency of SPE of pesticides from juice can be improved by placing the SPE cartridge inside an ultrasonic bath for a preselected time. On the base of this observation, Ramos *et al.*[134] proposed the so-called *ultrasonic-assisted matrix solid-phase dispersion* (UA-MSPD) as an alternative sample preparation technique for the fast extraction and purification of analytes in a single, and if required miniaturized, step. The feasibility of the approach was illustrated for the simultaneous extraction and cleaning-up of selected triazines and organophosphorus pesticides from fruits. Complete sample preparation was accomplished in only 1 min by direct immersion of the MSPD mixture (*i.e.* 100 mg of fruit peel dispersed in a similar amount of C8 and wetted with the extraction solvent), in a sonoreactor at 50% amplitude. Recoveries above 80% and RSDs in general better than 12% were obtained for the target compounds at spiking levels similar to those set as MRLs in current EU legislation.

In our opinion, all these preliminary results indicate that sonication, alone or in combination with other previously mentioned techniques, might become an interesting analytical alternative to other more conventional leaching procedures.

5.4 Analytical Micro-Systems: From Lab-on-a-Valve to μ-TAS

Since its introduction in 1975,[135] *flow injection (FI)* has made rapid progress as an analytical concept, rather than a technique, that effectively contributes to improving the rapidity, robustness, and reliability of many (relatively simple) sample preparation operations for which it allows complete automation.[136] The FI concept has evolved through two new generations, namely *sequential injection* (SI), referred to as second-generation FI,[137] and the so-called laboratory-on-valve (*lab-on-valve*, LOV), referred to as third-generation FI.[138] The key aspect to developing SI systems was the replacement of the continuous flow used in FI by programmable flow, which allows the digitally controlled displacement of liquids, gas, and/or beads, by stopping, reversing, and accelerating flow rates.[136] Subsequent downscaling was achieved by integrating the SI principle on a LOV platform.

The basic component of a LOV microsystem is a transparent, one-piece structure made of Perspex and mounted on top of a six-port valve. The system is designed to include connecting ports, working channels and a flow-through cell, through which other individual ports are connected. This central cell is also connected with a propulsion unit, typically a syringe pump, necessary to circulate liquids through the SI-LOV system. In principle, this basic structure allows a number of chemical and physical processes, including fluidic and microcarrier bead control, sample dilution, homogeneous reaction, liquid–solid interaction and analyte preconcentration, and real-time monitoring of various reaction processes via in-cell detection with optical fibres.[136,139] Auxiliary units (*e.g.* holding and mixing coils, T connections, and auxiliary pumps and valves) can also be incorporated in this basic set-up, so increasing the versatility of its potential applications.[136,139]

One of the main features of SI-LOV as compared to other flow systems is the miniaturization of the flow channels. While the latter operates on the millilitre scale, SI-LOV allows downscaling the sample and reagent volumes to the 10–20 μl range, and effectively contributes to minimizing waste generation (typically, only 100–200 μl per assay).[140] (For a deeper discussion regarding extra practical benefits deriving from proper choice of the LOV channel dimensions, the reader is encouraged to consult a more specialized reviews.[136,140]) These figures make evident the potential of SI-LOV (either with or without bead injection) for the accurate, automated, and green handling of minute samples. The feasibility of the approach for the on-line separation/ preconcentration of selected analytes was first illustrated for metals, for which LODs in the ultratrace range have been reported when the LOV system is combined with highly selective and sensitive spectroscopic or MS-based detectors (see reference 136 and references therein). In recent years, a number of applied studies have demonstrated the potential of the LOV concept also in the bioanalytical field. Here, the continuous monitoring of relatively abundant compounds, such as glucose, lactate, glycerol, or ethanol, has been used for the continuous monitoring of cell cultures or enzymatic reactions (see reference 139 and references therein). The incorporation of protein-coated beads in SI-LOV has been demonstrated to be a successful alternative tool for the evaluation of antibody–protein interactions. This approach has also been used for the selective determination of target analytes (antibody or protein). The reported results illustrate the potential of this approach for automated microscale affinity chromatography. Apart from the reduction of the sample and solvent volume and the reduced number of beads used to pack the microcolumn, beads can be discarded after each experiment, which contribute to increased accuracy of determination by avoiding cross-contamination or ghost peaks due to progressive column degradation. Again, the small volume of eluent typically obtained from LOV-based systems simplifies their direct coupling with sensitive detectors and, if required, with powerful (and essentially miniaturized) separation techniques, namely CE and LC, for accurate determination of minor compounds in complex mixtures.

The feasibility of further miniaturization of systems, including complete integration of the several analytical steps (*i.e.* sample preparation, analyte separation, and detection) in a single monolithic device has been an active subject of research during the last 15 years.[141] This analytical concept yielded the idea of the so-called *micrototal analytical systems* (μ-TAS), also named lab-on-a-chip, and the downscaling of volumes involved to the nanolitre range. The various interesting achievements reported in this field since its introduction, and especially during the last decade, should certainly be considered as a proof of concept. Nowadays, a number of applications, typically involving CE as separation technique and either electrochemical, fluorescence or chemiluminiscence detection, has been reported for the determination of both ions and organic analytes in liquid matrices with LODs in the micromolar range.[142] But the treatment of solid samples is still a challenge: it demands a previous (off-chip) extraction step and, because of the complexity of the matrix, immunoassay-based determination. Nevertheless, using this analytical strategy, results in agreement with those obtained with reference methods have been reported, for instance for the determination of botulinum neurotoxin A[143] and folic acid[144] in infant formula.

Although further development is still required to overcome some of the shortcomings of present μ-TAS systems, including those associated with sample introduction, slow sample transport, and the micromachining of more appropriate interfaces between the different components,[145] the promising results obtained up to now mean that new achievements can also be expected in this research field. In future years further developments will probably yield new generations of μ-TAS, allowing more sophisticated, integrated, and greener on-site analytical determinations.

Acknowledgments

The authors thank MICINN (AGL2009-11909) for financial support. MPA also thanks MICINN for her predoctoral grant.

References

1. A. Alix, D. Collot, J. Nénon and J. Anger, *Anal. Chem.*, 2001, **73**, 3107.
2. E. M. Kristenson, S. Shahmiri, C. J. Slooten, J. J. Vreuls and U. A. T. Brinkman, *Chromatographia*, 2004, **59**, 315.
3. E. Schoenzetter, V. Pichon, D. Thiebaut, A. Fernández-Alba and M. C. Hennion, *J. Microcol. Sep.*, 2000, **12**, 316.
4. N. C. van de Merbel, F. M. Lagerwerf, H. Lingeman and U. A. T. Brinkman, *Int. J. Environ. Anal. Chem.*, 1994, **54**, 105.
5. M. Pérez, J. Alario, A. Vázquez and J. Villén, *J. Microcol. Sep.*, 1999, **11**, 582.
6. J. M. Cortés, R. M. Toledano, J. Villén and A. Vázquez, *J. Agric. Food Chem.*, 2008, **56**, 5544.
7. J. A. de Koning, P. Blokker, P. Jüngel, G. Alkema and U. A. Th. Brinkman, *Chromatographia*, 2002, **56**, 185.
8. T. Hyötyläinen and M. L. Riekkola, *Anal. Bioanal. Chem.*, 2004, **378**, 1962.
9. J. L. Luque-García and M. D. Luque de Castro, *J. Chromatogr. A*, 2003, **998**, 21.
10. M. Papagiannopoulos, B. Zimmermann, A. Mellenthin, M. Krappe, G. Maio and R. Galensa, *J. Chromatogr. A*, 2002, **958**, 9.
11. J. M. Kokosa, A. Przyjazny and M. A. Jennot, *Solvent Microextraction. Theory and Practice*, John Wiley & Sons, New York, 2009.
12. M. A. Jeannot and F. F. Cantwell, *Anal. Chem.*, 1997, **69**, 235.
13. Y. He and H. K. Lee, *Anal. Chem.*, 1997, **69**, 4634.
14. M. A. Jeannot and F. F. Cantwell, *Anal. Chem.*, 1996, **68**, 2236.
15. L. S. de Jager and A. R. J. Andrews, *Analyst*, 2000, **125**, 1943.
16. L. S. de Jager and A. R. J. Andrews, *J. Chromatogr. A*, 2001, **911**, 97.
17. L. Hou and H. K. Lee, *J. Chromatogr. A*, 2002, **976**, 377.
18. Y. Wang, Y. C. Kwok, Y. He and H. K. Lee, *Anal. Chem.*, 1998, **70**, 4610.
19. T. S. Ho, S. Pedersen-Bjergaard and K. E. Rasmussen, *J. Chromatogr. A*, 2002, **963**, 3.
20. S. Andersen, T. Halvorsen, S. Pedersen-Bjergaard and K. Rasmussen, *J. Chromatogr. A*, 2002, **963**, 303.
21. L. Hou, G. Shen and H. K. Lee, *J. Chromatogr. A*, 2003, **985**, 107.

22. L. Zhu, C. B. Tay and H. K. Lee, *J. Chromatogr. A*, 2002, **963**, 231.
23. J. Norberg, Å. Zander and J. Å. Jönsson, *Chromatographia*, 1997, **46**, 483.
24. Y. Shen, J. Å. Jönsson and L. Mathiasson, *Anal. Chem.*, 1998, **70**, 946.
25. J. J. Vreuls, E. Romijn and U. A. T. Brinkman, *J. Microcol. Sep.*, 1998, **10**, 581.
26. Q. Zhou, J. Liu, Y. Cai, G. Liu and G. Jiang, *Microchem. J.*, 2003, **74**, 157.
27. Q. Zhou, L. Pang and J. Xiao, *J. Chromatogr. A*, 2009, **1216**, 6680.
28. A. Saleh, Y. Yamini, M. Faraji, M. Rezaee and M. Ghambarian, *J Chromatogr. A*, 2009, **1216**, 6673.
29. J. Liu, J. Chao, M. Wen and G. Jiang, *J. Sep. Sci.*, 2001, **24**, 874.
30. J. Chao, J. Liu, M. Wen, J. Liu, Y. Cai and G. Jiang, *J. Chromatogr. A*, 2002, **955**, 183.
31. H. F. Wu, J. H. Yen and C. C. Chin, *Anal. Chem.*, 2006, **78**, 1707.
32. M. Ma and F. F. Cantwell, *Anal. Chem.*, 1999, **71**, 388.
33. G. Shen and H. K. Lee, *Anal. Chem.*, 2003, **75**, 98.
34. G. Ouyang, W. Zhao and J. Pawliszyn, *J. Chromatogr. A*, 2007, **1138**, 47.
35. J. M. Kokosa, A. Przyjazny and R. Jones, Paper 1680-4, presented at *PittCon 2007*, Chicago, 2007.
36. G. Ouyang and J. Pawliszyn, *Anal. Chem.*, 2006, **78**, 5783.
37. S. Pedersen-Bjergaard and K. E. Rasmussen, *Anal. Chem.*, 1999, **71**, 2650.
38. A. Gjelsad, S. Andersen, K. E. Rasmussen and S. Pedersen-Bjergaard, *J. Chromatogr. A*, 2007, **1157**, 38.
39. G. Jiang and H. K. Lee, *Anal. Chem.*, 2004, **76**, 5591.
40. M. Rezaee, Y. Assadi, M. R. Milani-Hosseini, E. Aghaee, F. Ahmadi and S. Berijani, *J. Chromatogr. A*, 2006, **1116**, 1.
41. N. Fontanals, R. M. Marcé and F. Borrull, *J. Chromatogr. A*, 2007, **1152**, 14.
42. F. G. Tamayo, E. Turiel and A. Martin-Esteban, *J. Chromatogr. A*, 2007, **1152**, 32.
43. H. Bagheri, E. R. Brouwer, R. T. Ghijsen and U. A. Th. Brinkman, *J. Chromatogr. A*, 1993, **647**, 121.
44. L. Ramos, J. J. Vreuls, U. A. T. Brinkman and L. E. Sojo, *Environ. Sci. Technol.*, 1999, **33**, 3254.
45. L. Ramos, L. E. Sojo, J. J. Vreuls and U. A. Th. Brinkman, *Environ. Sci. Technol.*, 2000, **34**, 1049.
46. A. Capiello, A. Berloni, G. Famiglini, F. Mangani and P. Palma, *Anal. Chem.*, 2001, **73**, 298.
47. A. C. Hogenboom, R. J. C. A. Steen, W. M. A. Niessen and U. A. Th. Brinkman, *Chromatographia*, 1998, **48**, 475.
48. A. C. Hogenboom, W. M. A. Niessen and U. A. T. Brinkman, *Rapid Commun. Mass Spectrom.*, 2000, **14**, 1914.
49. A. C. Hogenboom, P. Speksnijder, R. J. Vreeken, W. M. A. Niessen and U. A. Th. Brinkman, *J. Chromatogr. A*, 1997, **777**, 81.
50. A. C. Hogenboom, M. P. Hofman, S. J. Kok, W. M. A. Niessen and U. A. Th. Brinkman, *J. Chromatogr. A*, 2000, **892**, 379.

51. T. Hankemeier, S. P. J. van Leeuwen, J. J. Vreuls and U. A. Th. Brinkman, *J. Chromatogr. A*, 1998, **811**, 117.
52. K. K. Verma, A. J. H. Louter, A. Jain, E. Pocurull, J. J. Vreuls and U. A. Th. Brinkman, *Chromatogrphia*, 1997, **44**, 372.
53. A. J. H. Louter, J. van Doornmalen, J. J. Vreuls and U. A. Th. Brinkman, *J. High Resol. Chromatogr.*, 1996, **19**, 679.
54. L. Ramos, J. J. Ramos and U. A. Th. Brinkman, *Anal. Bioanal. Chem.*, 2005, **318**, 219.
55. E. R. Brouwer, H. Lingeman and U. A. Th. Brinkman, *Chromatographia*, 1990, **29**, 415.
56. P. J. M. Kwakman, J. J. Vreuls, U. A. Th. Brinkman and R. T. Ghijsen, *Chromatographia*, 1992, **34**, 41.
57. J. S. Fritz and J. J. Masso, *J. Chromatogr. A*, 2001, **909**, 79.
58. A. El-Beaqqali, A. Kussak and M. Abdel-Rehim, *J. Chromatogr. A*, 2006, **1114**, 234.
59. M. Abdel-Rehim, *LC-GC Eur.*, 2009, **22**, 8.
60. J. A. Gabaldon, J. M. Cascales, A. Maquieira and R. Puchades, *J. Chromatogr. A*, 2002, **963**, 125.
61. C. Arthur and J. Pawliszyn, *Anal. Chem.*, 1990, **62**, 2145.
62. Y. Gou, R. Eisert and J. Pawliszyn, *J. Chromatogr. A*, 2000, **873**, 137.
63. H. Bagheri and A. Salemi, *Chromatographia*, 2004, **59**, 501.
64. Z. Zhang and P. J., *Anal. Chem.*, 1993, **65**, 1843.
65. Q. Li, X. Ma, D. Yuan and J. Chen, *J. Chromatogr. A*, 2010, **1217**, 2191.
66. E. H. M. Koster, C. Crescenzi, W. den Hoedt, K. Ensing and G. J. de Jong, *Anal. Chem.*, 2001, **73**, 3140.
67. D. J. Djozan and L. Abdollahi, *Chromatographia*, 2003, **57**, 799.
68. Y. Y. Shu, S. S. Wang, M. Tardif and Y. Huang, *J. Chromatogr. A*, 2003, **1008**, 1.
69. K. Li and M. Fingas, *Anal. Chem.*, 1998, **70**, 2510.
70. S. Fustinoni, R. Giampiccolo, S. Pulvirenti, M. Buratti and A. Colombi, *J. Chromatogr. A*, 1999, **723**, 105.
71. W. Chen, K. F. Poon and M. H. W. Lam, *Environ. Sci. Technol.*, 1998, **32**, 3816.
72. M. Ramil-Criado, P. I. Rodríguez and T. R. Cela, *J. Chromatogr. A*, 2004, **1056**, 187.
73. F. Hernández, J. Beltrán, F. J. López and J. V. Gaspar, *Anal. Chem.*, 2000, **72**, 2313.
74. C. Grote, K. Levsen and G. Wünsch, *Anal. Chem.*, 1999, **71**, 4513.
75. A. L. Simplício and L. V. Boas, *J. Chromatogr. A*, 1999, **833**, 35.
76. H. Lord and J. Pawliszyn, *J. Chromatogr. A*, 2000, **885**, 153.
77. M. N. Sarrión, F. J. Santos and M. T. Galcerán, *Anal. Chem.*, 2000, **72**, 4865.
78. Y. Cai and J. M. Bayona, *J. Chromatogr. A*, 1995, **696**, 113.
79. Z. Mester and J. Pawliszyn, *J. Chromatogr. A*, 2000, **873**, 129.
80. M. T. Sng and W. F. Ng, *J. Chromatogr. A*, 1999, **832**, 173.
81. A. Bouaid, L. Ramos, M. J. González, P. Fernández and C. Cámara, *J. Chromatogr. A*, 2001, **939**, 13.

82. J. R. Dean, *Extraction Techniques in Analytical Sciences*, John Wiley & Sons, Chichester, 2009.
83. P. Mayer, W. H. J. Vaes, F. Wijnker, K. C. H. M. Legierse, R. H. Kraaij, J. Tolls and J. L. M. Hermens, *Environ. Sci. Technol.*, 2000, **34**, 5177.
84. A. Rico-Rico, S. T. J. Droge, D. Widmer and J. L. M. Hermens, *J. Chromatogr. A*, 2009, **1216**, 2996.
85. E. Baltussen, P. Sandra, F. David and C. Cramers, *J. Microcolumn Sep.*, 1999, **11**, 737.
86. E. Baltussen, F. David, P. Sandra and C. Cramers, *Anal. Chem.*, 1999, **71**, 5213.
87. P. Popp, C. Bauer and L. Wennrich, *Anal. Chim. Acta*, 2001, **436**, 1.
88. A. R. M. Silva and J. M. F. Nogueira, *Anal. Bioanal. Chem.*, 2010, **396**, 1853.
89. P. Sandra, B. Tienpont and F. David, *J. Chromatogr. A*, 2003, **1000**, 299.
90. J. Vercauteren, C. Pérès, C. Devos, P. Sandra, F. Vanhaeke and L. Moens, *Anal. Chem.*, 2001, **73**, 1509.
91. P. Sandra, B. Tienpont, J. Vercammen, A. Tredoux, T. Sandra and J. David, *J. Chromatogr. A*, 2001, **928**, 117.
92. N. Ochiai, K. Sasamoto, M. Takino, S. Yamashita, S. Daishima, A. C. Heiden and A. Hoffmann, *Anal. Bioanal. Chem.*, 2002, **373**, 56.
93. B. Tienpont, F. David, K. Desmet and P. Sandra, *Anal. Bioanal. Chem.*, 2002, **373**, 46.
94. T. Benijts, J. Vercammen, R. Dams, H. Pham-Tuan, W. Lambert and P. Sandra, *J. Chromatogr. B*, 2001, **755**, 137.
95. C. Blasco, G. Font and Y. Picó, *J. Chromatogr. A*, 2002, **970**, 210.
96. S. A. Barker, A. R. Long and C. R. Short, *J. Chromatogr.*, 1989, **475**, 353.
97. S. A. Barker, *J. Chromatogr. A*, 2000, **880**, 63.
98. S. A. Barker, *J. Chromatogr. A*, 2000, **885**, 115.
99. E. M. Kristenson, L. Ramos and U. A. Th. Brinkman, *TrAC, Trends Anal. Chem.*, 2006, **25**, 96.
100. E. M. Kristenson, S. Shahmiri, C. J. Slooten, J. J. Vreuls and U. A. Th. Brinkman, *Chromatographia*, 2004, **59**, 315.
101. B. Morzycka, *J. Chromatogr. A*, 2002, **982**, 267.
102. A. L. Dawidowicz and E. Rado, *J. Pharm. Biomed. Anal.*, 2010, **52**, 79.
103. G. Xiang, L. Yang, X. Zhang, H. Yang, Z. Ren and M. Miao, *Chromatographia*, 2009, **70**, 1007.
104. M. T. Pena, M. C. Casais, M. C. Mejuto and R. Cela, *J. Chromatogr. A*, 2007, **1165**, 32.
105. B. Albero, C. Sánchez-Brunete, A. Donoso and J. Tadeo, *J. Chromatogr. A*, 2004, **1043**, 127.
106. C. Blasco, G. Font and Y. Picó, *J. Chromatogr. A.*, 2004, **1028**, 267.
107. E. M. Kristenson, E. G. J. Haverkate, C. J. Slooten, L. Ramos, R. J. J. Vreuls and U. A. Th. Brinkman, *J. Chromatogr. A*, 2001, **917**, 277.
108. J. J. Ramos, M. J. González and L. Ramos, *J. Chromatogr. A*, 2009, **1216**, 7307.
109. A. R. Long, L. C. Hsieh, M. S. Malbrough, C. R. Short and S. A. Barker, *J. Assoc. Off. Anal. Chem. Int.*, 1990, **73**, 379.

110. M. Zhao, F. van der Wielen and P. de Vooght, *J. Chromatogr. A*, 1999, **837**, 129.
111. A. R. Long, M. M. Soliman and S. A. Barker, *J. Assoc. Off. Anal. Chem. Int.*, 1991, **74**, 493.
112. H. M. Lott and S. A. Barker, *J. Assoc. Off. Anal. Chem. Int.*, 1993, **76**, 67.
113. J. Tolls, R. Samperi and A. di Corcia, *Environ. Sci. Technol.*, 2003, **37**, 314.
114. J. J. Ramos, M. J. González and L. Ramos, *J. Sep. Sci.*, 2004, **27**, 595.
115. R. M. Garcinuño, L. Ramos, P. Fernández-Hernando and C. Cámara, *J. Chromatogr. A*, 2004, **1041**, 35.
116. B. E. Richter, J. L. Ezzell, D. Felix, K. A. Roberts and D. W. Later, *Amer. Lab.*, 1995, **27**, 24.
117. EPA SW-846 Method 3545, *Pressurized Fluid Extraction (PFE)*, www.epa.gov/osw/hazard/testmethods/sw846/pdfs/3545a.pdf.
118. L. Ramos, E. M. Kristenson and U. A. Th. Brinkman, *J. Chromatogr. A*, 2002, **975**, 3.
119. J. Pörschmann, J. Plugge and R. Toth, *J. Chromatogr. A*, 2001, **909**, 95.
120. S. Sporring and E. Björklund, *J. Chromatogr. A*, 2004, **1040**, 155.
121. I. Kania-Korwel, H. Zhao, K. Norstrom, X. Li, K. C. Hornbuckle and H.-J. Lehmler, *J. Chromatogr. A*, 2008, **1214**, 37.
122. T. Hyötyläinen, T. Andersson, K. Hartonen, K. Kuosmanen and M. L. Riekkola, *Anal. Chem.*, 2000, **72**, 3070.
123. K. Kuosmanen, T. Hyötyläinen, K. Hartonen and M. L. Riekkola, *J. Chromatogr. A*, 2001, **943**, 113.
124. L. Ramos, J. J. Vreuls and U. A. Th. Brinkman, *J. Chromatogr. A*, 2000, **891**, 275.
125. J. J. Ramos, C. Dietz, M. J. González and L. Ramos, *J. Chromatogr. A*, 2007, **1152**, 254.
126. P. Richter, B. Sepúlveda, R. Oliva, K. Calderón and R. Seguel, *J. Chromatogr. A*, 2003, **994**, 169.
127. S. L. Cresswell and S. J. Haswell, *Analyst*, 1999, **124**, 1361.
128. M. Ericsson and A. Colmsjö, *J. Chromatogr. A*, 2002, **964**, 11.
129. J. L. Luque-García and M. D. Luque de Castro, *Analyst*, 2002, **127**, 1115.
130. C. Sanchez, M. Ericsson, H. Carlsson, A. Colmsjö and E. Dyremark, *J. Chromatogr. A*, 2002, **957**, 227.
131. S. Ozcan, A. Tor and M. E. Aydin, *Clean*, 2009, **37**, 811.
132. S. Ozcan, A. Tor and M. E. Aydin, *Anal. Chim. Acta*, 2009, **640**, 52.
133. B. Albero, C. Sanchez-Brunete and J. L. Tadeo, *J. Agric. Food Chem.*, 2003, **51**, 6915.
134. J. J. Ramos, R. Rial-Otero, L. Ramos and J. L. Capelo, *J. Chromatogr. A*, 2008, **1212**, 145.
135. J. Ruzicka and E. H. Hansen, *Anal. Chim. Acta*, 1975, **78**, 145.
136. J. Wang and E. H. Hansen, *TrAC, Trends Anal. Chem.*, 2003, **22**, 225.
137. J. Ruzicka and G. D. Marshall, *Anal. Chim. Acta*, 1990, **237**, 329.
138. J. Ruzicka, *Analyst*, 2000, **125**, 1053.

139. M. D. Luque de Castro, J. Ruiz-Jimenez and J. A. Perez-Serradilla, *TrAC, Trends Anal. Chem.*, 2008, **27**, 200.
140. P. Solich, M. Polasek, J. Klimundova and J. Ruzicka, *TrAC, Trends Anal. Chem.*, 2003, **22**, 116.
141. T. Vilkiner, D. Janasek and A. Manz, *Anal. Chem.*, 2004, **76**, 3373.
142. A. Rios, A. Escarpa and B. Simonet-Suau, *Miniaturization of Analytical Systems: Principles, Designs and Applications*, John Wiley & Sons, Chichester, 2009.
143. S. Han, J. H. Chob, I. H. Choa, E. H. Paek, H. B. Ohc, B. S. Kim, C. Ryu, K. H. Lee, Y. K. Kim and S. H. Paek, *Anal. Chim. Acta*, 2007, **587**, 1.
144. D. Hoegger, P. Morier, C. Vollet, D. Heini, F. Reymond and J. S. Rossier, *Anal. Bioanal. Chem.*, 2007, **387**, 267.
145. A. Rios, A. Escarpa, M. C. Gonzalez and A. G. Crevillen, *TrAC, Trends Anal. Chem.*, 2006, **25**, 467.

CHAPTER 6
Green Analytical Chemistry Through Flow Analysis

FÁBIO R.P. ROCHA AND BOAVENTURA F. REIS

Centro de Energia Nuclear na Agricultura, Universidade de São Paulo, Avenida Centenário 303, Piracicaba 13400-970, SP, Brazil

6.1 The Scope of Flow Systems in Chemical Analysis and Green Analytical Chemistry

Flow analysis encompasses a widespread group of analytical techniques that are extensively employed for automation in routine and research labs. The development of flow-based techniques has brought a new dimension to analytical chemistry, allowing the measurements to be carried out faster and with minimum intervention of the analyst. As a consequence, flow-based procedures are characterized by high sample throughputs, improved precision and reduced risks of exposure of the analyst to toxic substances. As measurements are usually carried out in closed systems, risks of sample contamination by external sources are also minimized. Manipulation of unstable reagents and products, reproducible measurements of on-line generated suspensions and kinetic discrimination have been effectively implemented in different flow configurations. Despite several examples of flow systems for simultaneous or sequential determinations, most of the applications are devoted to the quantification of a single analyte in a large number of samples, exploiting the above-mentioned advantages.

The potential to develop environmentally friendly analytical procedures is also inherent to flow-based methodologies. This can be exemplified by the capability to decrease the reagent consumption, reversible use of immobilized reagents, waste recycling and in-line waste detoxification. The main challenge is to achieve green analytical chemistry (GAC) without affecting the reliability of the analytical results or increasing the operational costs.[1]

New approaches in flow analysis have been pointed out as milestones in the development of GAC,[2] including flow injection analysis, sequential injection analysis, multicommutation in flow analysis and lab-on-valve. Potentially greener strategies often adopted in flow-based systems, such as solid-phase spectrophotometry, micro total analytical systems and single drop microextractions have also been highlighted. Strategies in flow systems towards GAC have been revised previously, including the definition of an order of priority for the development of cleaner procedures and discussion of successful examples of greener applications in flow systems.[3] A parallel can be traced between the evolution of flow analysis and GAC based on the replacement of toxic reagents and minimization of amounts and toxicity of wastes. Recent examples of successful greener approaches in flow analysis have also been discussed, complementing the applications considered previously.[3]

6.2 Brief Description of Flow Systems

6.2.1 Segmented Flow Analysis

Segmented flow analysis (SFA) was proposed by Leonard T. Skeggs in the 1950s, aiming to mechanize the routine work in clinical laboratories.[4] These systems were successfully commercialized as Technicom Autoanalyzers®. As schematized in Figure 6.1, in SFA, sample and reagents are continuously aspirated to the manifold, segmentation by air bubbles being exploited to minimize both axial sample dispersion and carryover effects. This approach is then suitable for implementing the long sample residence times often required in the analytical procedures used in clinical analysis. Measurements are thus usually carried out close to the steady-state condition, which improves sensitivity, although the sampling rate is rarely higher than 60 measurements per hour. Washing is carried out periodically by an inert solution aspirated through the analytical path.

The use of these systems was widespread some decades ago, but they were later replaced by unsegmented flow systems.

6.2.2 Flow Injection Analysis

In flow injection analysis (FIA), as shown in Figure 6.2, samples are injected in a carrier stream and processed under unsegmented flow. Concentration gradients are thus formed in the sample zone as a result of the sample dispersion in the carrier. The extent of the dispersion depends on physicochemical solution properties, such as viscosity, and system parameters, such as sample volume, dimensions and geometry of the reactor and flow rates. As a result of the concentration gradients, transient signals are obtained, quantification usually

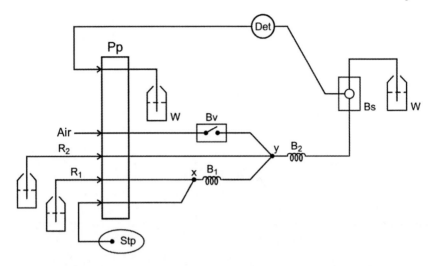

Figure 6.1 Diagram of a segmented flow analysis (SFA) system. B_1 and B_2, reaction coils; Bs, air bubble separation device; Bv, electromechanical switch to insert air bubbles; Det, detector; Pp, peristaltic pump; R_1 and R_2, reagents or washing solutions; Stp, sampling device; W, waste; x and y, confluence points.

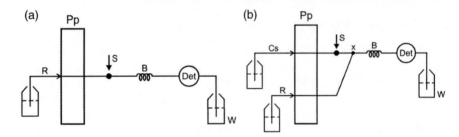

Figure 6.2 Diagram of flow injection systems: (a) single-line manifold; (b) confluent streams. B, reaction coil; Cs, carrier solution; Det, detector; Pp, peristaltic pump; R, reagent solution; S, sample injection port; W, waste vessel; x, confluence point. The arrows indicate the pumping direction.

being based on peak heights. In addition, gradients can be exploited for on-line dilutions, implementation of the standard additions method, titrations, *etc.*

In FIA, samples are submitted to highly reproducible processing conditions (*e.g.* dilutions, reagent additions and matrix separation) and times, making measurements feasible without achieving the steady state (residence times are usually less than 30 s). This opens up the possibility of exploiting kinetic aspects to improve selectivity, for simultaneous determination or to overcome matrix effects, for example.

6.2.3 Sequential Injection Analysis

Sequential injection analysis (SIA) is a robust alternative to FIA, which allows implementation of different flow methodologies with the same manifold

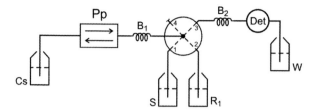

Figure 6.3 Diagram of a sequential injection analysis (SIA) system. B_1, holding coil; B_2, reaction coil; Cs, carrier fluid; Det, detector; Pp, pump; R_1, reagent solution; S, sample; W, waste. Numbers 1–4 represents ports of the selection valve.

(Figure 6.3).[5] This process exploits a computer-controlled multiport selection valve to sequentially collect sample and reagent aliquots required for the analytical determination. Solutions are aspirated towards a holding coil by a peristaltic pump or more usually a syringe pump. Mixing occurs by dispersion at the interfaces, and strategies such as flow reversal and the use of mixing chambers are usually exploited to increase the zone overlap. This is necessary mainly when the analytical procedure requires two or more reagent solutions and also to avoid artefacts caused by unsuitable zone overlap.[6] Generally, system operation can be designed to achieve an analytical performance comparable to that attained by the usual flow injection systems.

6.2.4 Monosegmented Flow Analysis

Monosegmented flow analysis (MSFA) matches the simplicity and high sampling rate achieved by FIA with the low sample dispersion achieved by SFA. In the scheme shown in Figure 6.4, the sample aliquot is inserted in the carrier stream sandwiched by two air bubbles in order to contain axial dispersion. The monosegment can then be submitted to long residence times required for relatively slow reactions without impairing the washing times and sampling rate.[7] MSFA also shows potential to minimize reagent volumes. This characteristic has been exploited by the simultaneous injection of sample and reagents[8] or by using optoswitches to locate the air bubbles, aiming to constrain the reagent addition to the sample zone.[9] MSFA was successfully applied to liquid–liquid extractions in single-phase[10] and two-phase[11] systems and also for the analysis of gaseous species.[12]

6.2.5 Multicommutation Approach

Flow systems based on the multicommutation approach (MCFA)[13] are designed with discrete computer-controlled commuting devices, such as solenoid valves (see the scheme in Figure 6.5). Each commutator can be independently controlled and the flow manifold can be reconfigured by software. This approach greatly

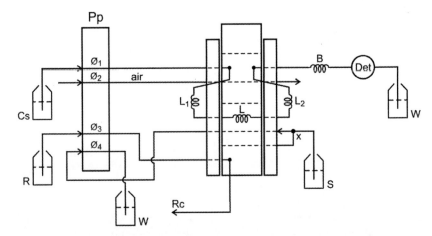

Figure 6.4 Diagram of a monosegmented flow analysis (MSFA) system. B, reaction coil; Cs, carrier fluid; Det, detector; L, sample loop; L_1 and L_2, loops for air load; $\emptyset_i$ = flow rates; Pp, peristaltic pump; R, reagent solution; Rc, reagent recirculation; S, sample; W, waste; x, confluence point. The three rectangles represent the sliding-bar commutator.

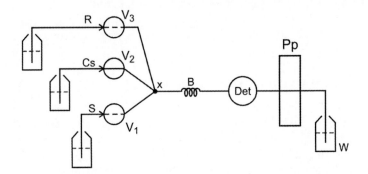

Figure 6.5 Diagram of a multicommuted flow system. B, reaction coil; Cs, carrier solution; Det, detector; Pp, peristaltic pump; R, reagent; S, sample; Vi, solenoid valves; W, waste vessel; x, confluence point. The continuous and dashed lines in the symbols of valves indicate the solutions pathway when the valves are switched off or on, respectively. The arrows indicate the pumping direction.

increases system versatility, because 2^n states of commutation can be established for n active devices.

In MCFA, each solution can be handled by an independent commutator and reagents are intermittently introduced into the flow system. Sample and reagent volumes are defined by the flow rates and the switching times of the corresponding commutators. It is then possible to introduce solutions simultaneously (merging zones approach) or alternately (binary sampling approach). The last strategy makes it possible to change the volumetric fraction of sample

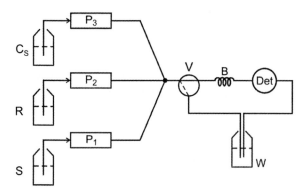

Figure 6.6 Diagram of a multipumping flow system. B, reaction coil; Cs, carrier solution; Det, detector; P_1, P_2, P_3, solenoid micropumps; R, reagent; S, sample; V, three-way solenoid valve; W, waste vessel.

and reagents by software, which is useful for optimization of the procedures and to implement volumetric (automatic) titrations.

6.2.6 Multipumping and Multisyringe Flow Systems

Solenoid micropumps, which deliver microlitre solution volumes in a reproducible way, are employed to design multipumping flow systems (Figure 6.6).[14] Generally, each device handles a solution, in a similar way to the MCFA systems. However, the micropumps can actuate as both commutators and propulsion devices. The pulsed flow inherent to these devices favours mixing and provides enhanced radial mass transport. Increased efficiency of heat transference in systems that exploits convective heating has also been verified.[15] Other advantages include portability and low energy requirements, making the systems suitable for *in situ* measurements. Reagent addition in multipumping flow systems is inherently intermittent, resulting in the ability to minimize reagent consumption and waste generation.

Multisyringe flow injection analysis[16] combines features of SIA, such as robustness, versatility and ability to reduce reagent consumption, with those of FIA (*i.e.* improved sampling rates and better analytical performance). Syringe pumps of diverse capacities (*e.g.* 0.5–25 ml) are used for solution handling. The pumps are coupled to a multiport valve or a set of solenoid valves and usually connected to the same step motor. By the synchronized actuation of the valves and the step motor, solutions can be directed to the analytical path or delivered back to their original reservoirs.

6.3 Evolution of System Design and Reduction of Waste Generation

The introduction and development of flow analysis was stimulated by the high demand for analysis and the need to obtain reliable analytical results in short time periods. High sample throughput was a key aspect in the initial development stage and this was achieved at costs of high reagent flow rates. Large effluent volumes were thus generated, which in some applications were even higher than in the corresponding batch procedures. In the first SFA systems (Figure 6.1), for example, about 10–20 ml of waste was generated per determination. New flow configurations have been proposed, aiming to achieve better analytical performance and implementation of more complex assays. This development came with minimization of both reagent consumption and waste generation. In recent flow configurations, good analytical performance can be achieved by consuming only micro amounts of reagents.

High sample throughput was also emphasized in the first stage of development of FIA. In the single-line manifolds (Figure 6.2a), the reagent stream also acted as sample carrier and flow rates as high as 8 ml min^{-1} were usual. Despite their inherent simplicity, single-line systems present drawbacks such as unnecessary waste of reagents and difficulties in mixing sample and reagents, which occurs only by dispersion at the interfaces. Double peaks can then be generated in conditions of limited sample dispersion, due to the lack of reagent in the central part of the sample zone. Mixing difficulties were circumvented in flow systems with confluent streams (Figure 6.2b), in which reagent solutions are equally distributed in the whole sample zone, providing a constant volumetric fraction. The amount of the reagent available to the chemical reaction can then be defined by the solution concentration and the flow rates. As an inert solution is used as carrier, reagent flow rate can be reduced without impairing the sampling rate, reducing the reagent consumption in relation to the single-line configuration. However, reagents are consumed even when a sample is not processed, contributing to increase reagent consumption and waste generation.

The merging zones[17] and intermittent flow[18] approaches (Figure 6.7) are also alternatives to minimize reagent consumption in flow injection systems. In the former, sample and reagent aliquots are simultaneously introduced into independent carrier streams, merging at the confluence point. As the volumetric fraction of the reagent is maintained, development of the chemical reaction and the analytical features are not affected. The potential of this approach was first demonstrated by spectrophotometric phosphate determination by the molybdenum blue method, in which the consumption of ascorbic acid was reduced to 9% of that required in the flow system with confluent streams. The merging zones approach was further exploited for simultaneous determinations[19] and implementation of the standard addition method.[20] The intermittent flow approach was formerly implemented by using a sliding-bar commutator or two independently controlled peristaltic pumps. Modern systems exploit computer-controlled valves for this purpose. The introduction of the reagent solution by confluence can be synchronized with the sample injection, the solution being recycled in the sampling stage. Reagent consumption is then reduced, similarly to the merging zones approach. As an example, the reagent consumption per determination in the spectrophotometric determination of calcium was reduced from 40 µg with continuous reagent addition to 0.27 µg with the intermittent flow strategy.[21] This approach was also explored to reduce the washing times[18]

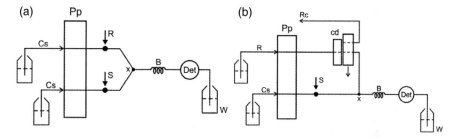

Figure 6.7 Diagram of flow injection systems with (a) merging zones and (b) intermittent flow. B, reaction coil; Cd, valve to direct the reagent stream towards either the confluence point (x) or to the recycling vessel (Rc); Cs, carrier solutions; Det, detector; Pp, peristaltic pump; R, reagent solution; S, sample injection port; W, waste vessel; x, confluence point. The arrows indicate the pumping direction.

and to remove reaction products adsorbed in the analytical path, aiming to minimize memory effects and baseline drifts.[22]

The potential to reduce reagent consumption was also presented by flow systems exploiting reagent injection in a sample flowing stream,[23] previously known as reverse flow injection analysis (r-FIA). For example, when this strategy was adopted for the sequential determination of analytes in water, the reagent consumption was reduced up to 240-fold and 4000-fold in comparison to flow systems with continuous reagent addition and batch procedures, respectively.[23]

The intermittent addition of reagents is a common characteristic of SIA, MCFA, multipumping and multisyringe flow systems, and contributes to the requirement for reagent saving. In all these approaches, the reagent volumes inserted into the analytical path are limited to those necessary for the reaction to develop (a few microlitres). In some circumstances, mixing occurs only by dispersion at the interfaces and increasing reagent volumes is not effective in increasing the availability of the chemical in the sample zone. This drawback can be circumvented by the binary sampling approach or by exploiting the pulsed flow inherent to the multipumping process, contributing to improved efficiency in using the chemical reagents. On the other hand, sample residence time is easily increased by the stopped-flow approach and, consequently, shorter reaction coils can be employed, requiring low sample and reagent volumes to attain similar dispersion conditions.

The efficiency of different flow approaches to minimize reagent consumption can be demonstrated by comparing the spectrophotometric procedures for determination of phosphate,[7,24–27] chloride,[25,28–31] nitrite[18,28,30,32,33] and total phenols[34–38] in Figure 6.8. Flow analysis has also contributed to avoid the drawbacks of other classical procedures, such as batch liquid–liquid extraction (LLE). In flow-based systems, the operation time, solvent amounts and risks to the analyst in LLE are minimized. Several flow approaches have been proposed for automation of this process, including FIA,[39,40] MSFA,[11] microextraction in MSFA,[41] MCFA,[42] and liquid–liquid extraction in a single 1.3 µl microdrop.[43]

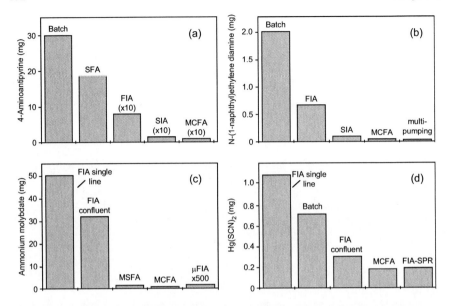

Figure 6.8 Reagent consumption per determination in spectrophotometric procedures for: (a) total phenols,[34–38] (b) nitrite,[18,28,30,32,33] (c) phosphate,[7,24–27] and (d) chloride.[25,28–31] Numbers in parentheses indicate the effluent volume per determination.

Table 6.1 Minimization of the amounts of organic solvents in liquid–liquid extraction in flow systems.

Flow approach	Solvent volume (μl)	Extraction	Reference
FIA	1200	Caffeine in $CHCl_3$	39
MCFA	225	Pb/ditizone complex with CCl_4	42
MSFA	100	Cd/PAN complex $CHCl_3$	11
FIA	30	Thiamine with $CHCl_3$	40
FIA–falling drop	1.3	Sodium dodecyl sulfate in a $CHCl_3$ microdrop	43
MSFA-microextraction	0.254	Caffeine in $CHCl_3$	41

The tendency to reduce the volume of organic solvent is shown in the applications listed in Table 6.1. For comparison, batch LLE usually consumes at least 10 ml of organic solvent.

6.4 Contributions of Flow-Based Procedures to Green Analytical Chemistry

6.4.1 Replacement of Hazardous Chemicals

Reagentless analytical procedures are scarce, but a significant contribution to GAC is minimization of waste toxicity, by avoiding the use of hazardous chemicals. Flow systems have been exploited to achieve this goal, as demonstrated by applications listed in Table 6.2.

Reliable determination of nitrate can be carried out in a flow system by coupling a minicolumn containing copperized cadmium filings for nitrate reduction to nitrite. The reduced form can then be quantified after a diazo-coupling (Griess) reaction.[44] The waste volume is considerably lower than that generated in the batch procedure, but residues of heavy metals and carcinogenic amines are still generated. With the aim of developing a greener procedure, the cadmium filings were replaced by a photochemical reduction step[45] or by enzymatic reduction with nitrate reductase obtained from corn leaves.[46] Reduction efficiency higher than 80% can be attained with these strategies under suitable working conditions. The use of the Griess reaction was avoided when the photochemically

Table 6.2 GAC in flow methodologies by the replacement of hazardous chemicals.

Analyte	Replaced reagents	Remarks	Reference
Ammonium	Hg(II), phenol, nitroprusside	Conductimetric detection after separation from the sample matrix by gas diffusion	52
Cyclamate	Carcinogenic amines or Pb(II)	Replacement of toxic reagents by nitrite/iodide and waste minimization with a multipumping flow system	73
Total phenols	4-aminoantipyrine and ferricyanide	Natural source of polyphenol oxidase (sweet potato root) and spectrophotometric detection of the formed o-quinones	54
Ethanol	$K_2Cr_2O_7/H_2SO_4$	Reagentless procedure based on detection by the solution density after analyte separation by pervaporation	59
Ethanol	$K_2Cr_2O_7/H_2SO_4$	Reagentless procedure based on photometric detection of falling drops	60
Hypochlorite	DPD	Direct detection at UV measurements before and after decomposition of the analyte to improve selectivity	58
Manganese	Organic solvents	Cloud point extraction with Triton X-114	81
Nitrate	Cd, NED, sulphanilamide	Direct UV measurements after separation of interfering species in a ion-exchange mini-column	48
Paraquat	Dithionite	Replacement by dehydroascorbic acid	74
Urea	Urease, phenol, nitroprusside	Natural source of urease (jackbean) and conductimetric detection of ammonium	53

DPD, N,N-diethyl-p-phenylenediamine; NED, N-(1-naphthyl)ethylenediamine.

generated nitrite was detected by biamperometry after reaction with iodide in acid medium.[47] Other environmentally friendly flow-based procedures exploited direct UV measurements, avoiding the use of toxic reagents in the derivatization step. Selectivity was improved by coupling an anion-exchange minicolumn for separation of the analyte from interfering species (*e.g.* humic substances) usually present in natural waters.[48] A diluted perchloric acid eluent solution was the only reagent employed, its consumption being equivalent to 18 µl of the concentrated acid per determination.

Hazardous chemicals (mercury(II) or phenol and sodium nitroprusside) were extensively used for ammonium determination by Nessler[49] or Berthelot[50] reactions. Waste toxicity was diminished by using salicylic acid as the phenolic compound in the Berthelot reaction,[51] achieving similar analytical performance. More recent work has exploited analyte volatilization as NH_3 for separation from the sample matrix, usually by gas diffusion through a PTFE membrane. This process can be reproducibly carried out in flow systems and the low transport efficiency can be compensated by a higher sample volume, different flow-rates of the donor and acceptor streams, increase of the back-pressure, *etc.* Selective detection at the acceptor stream can be performed by spectrophotometry with an acid–base indicator or by conductimetry, which does not require additional reagents. With the last alternative, the alkaline solution used to volatilize ammonia is the only required reagent[52] and a totally clean method is achieved by on-line neutralization of the donor stream after measurements. Similar strategies can be used for greener determination of other volatile analytes or species which can be volatilized in a suitable medium (*e.g.* carbonate, amines, cyanide, sulfide and sulfite).

Plant tissues or vegetable extracts are greener enzymatic sources with the potential to replace hazardous chemicals, as demonstrated by the determination of urea[53] and total phenols.[54] In the former application, a minicolumn filled with pieces of jack-bean (*Canavalia* sp.) was the source of *urease* used for up to 1000 determinations of urea in serum.[53] Conductimetric detection after gas diffusion through a PTFE membrane was explored to determine the ammonium ions originating from the enzymatic hydrolysis. A clean method for the spectrophotometric determination of total phenols in wastewater was based on oxidation to *o*-quinones by dissolved oxygen. A crude extract of sweet potato (*Ipomoea* sp.) root prepared in phosphate buffer was the source of *polyphenol oxidase*, which acted as the reaction biocatalyst.[54] This vegetable extract was also employed in clean flow procedures for determination of sulfite in wines[55] and L-dopa and carbidopa in pharmaceutical formulations.[56] More recently, an extract of guava (*Psidium* sp.) leaf was proposed for quantification of iron because it contains naturally occurring reducing and complexing species.[57]

Chemical derivatization of the analyte was avoided in a green procedure for the determination of hypochloride in bleaching products.[58] The procedure exploited the radiation absorption by the analyte at 292 nm, the analytical signal being based on the difference of the absorbances measured before and after reaction of the samples with solid cobalt oxide placed in a minicolumn. The solid reagent catalyses decomposition of hypochlorite to chloride and oxygen, thus allowing the selective determination of the analyte. The procedure requires only 20 mg of the reusable catalyst and a diluted NaOH solution.

A reagentless procedure was developed for ethanol determination in beverages, exploiting density measurements at the acceptor liquid after analyte separation by pervaporation.[59] The procedure yields results in agreement with the reference method, based on the oxidation of the alcohol with potassium dichromate and presents a sampling rate of 15 samples per hour. Analogous procedures could be applied for determination of other volatile analytes in high concentrations. Another reagentless procedure for determination of ethanol in wines exploits the effect of the alcohol on solution surface tension.[60] A FIA system reproducibly generated falling drops of sample between the photometric devices (an infrared LED and a phototransistor). The power of the radiation that strikes the photodetector is closely dependent on the drop size. As this parameter is a function of the ethanol concentration, analyte determination from 1 to 30% was feasible, with a sampling rate of 50 determinations per hour and relative standard deviation of 2.5%.

6.4.2 Reuse of Chemicals

Recycling and reuse of the waste generated in analytical procedures is interesting for GAC, as well as contributing to cost reduction. However, it is usually not easy to implement because of the complexity of the wastes. Some successful examples are presented in Table 6.3. On-line solvent recycling by distillation after the analytical measurement was efficiently adopted in procedures based on LLE. The toxic organic solvents were used for sample dissolution and as carriers for the simultaneous determination of propyphenazone and caffeine[61] or ketoptofen[62] in pharmaceutical preparations by Fourier transform infrared spectrometry. The flow systems generated about 1 ml min^{-1} of $CHCl_3$ or CCl_4, but this drawback was overcome by using a closed-loop configuration, which incorporates distillation and cooling units. In addition to this reuse of a toxic chemical, risks of the analyst's exposure to carcinogenic solvents were minimized.

The closed-loop configuration was also adopted for reagent reuse in the spectrophotometric determination of lead in gasoline,[63] based on complex formation with arsenazo(III). A minicolumn filled with a cation-exchange resin was placed after the flow cell in order to retain the heavy metal for further treatment. The released chromogenic reagent was returned to its vial for reuse. Reproducible analytical curves and no baseline drift were observed when 50 ml of the same recirculating reagent solution was used for 7 days. Other examples of reuse of chemicals in flow-based procedures are reagents immobilized on a solid phase (*e.g.* enzymes and complexing agents), as discussed in the next section.

6.4.3 Minimization of Reagent Consumption and Waste Generation

The amount and toxicity of the wastes are as important as any other analytical feature in the development of a new procedure or selection of a previously

Table 6.3 Some examples of the reuse of chemicals in flow-based systems.

Analyte	Sample	Reagents	Remarks	Reference
Caffeine[a]	Pharmaceuticals	$CHCl_3$	On-line solvent recycling by distillation	61
Iron[b]	Natural waters	Acetate buffer and ascorbic acid	Reversible retention of the analyte in C_{18}-TAN and measurements by solid-phase spectrophotometry	68
Ketoptofen[a]	Pharmaceuticals	CCl_4	On-line solvent recycling by distillation	62
Lead[a]	Gasoline	Arsenazo(III)	Release of the reagent by retention of the analyte in an ion-exchange minicolumn	63
Propyphenazone[a]	Pharmaceuticals	$CHCl_3$	On-line solvent recycling by distillation	61
Zinc[b]	Natural waters	Hexamine buffer	Reversible retention of the analyte in C_{18}-TAN and measurements by solid-phase spectrophotometry	67

[a]FIA system with closed loop configuration.
[b]FIA.

reported one. Minimization of the amounts of reagent consumed in the analytical procedure reduces both operational costs and toxicity of wastes. On the other hand, the effluent volume defines the amount of waste to be treated and the corresponding costs. Waste minimization can be considered the more general alternative to achieve GAC, because reliable analytical procedures can be implemented by consuming low amounts of chemicals exploring different strategies. Successful examples are presented in Table 6.4 and discussed in the following sections.

6.4.3.1 Immobilized Reagents

The use of immobilized reagents is a useful alternative for the development of environmentally friendly analytical procedures. Minicolumns filled with solid-phase reagents are usually employed for derivatization of the analyte or for on-line generation of unstable reagents.[64] Reagent immobilization often simplifies the manifolds, improves the radial mass transference and provides an excess of the chemical at the solid–liquid interface. Greener procedures can be developed because only stoichiometric amounts of the solid are consumed in the reaction. However, the strategy for reagent immobilization needs to be carefully selected in order to minimize risks of lixiviation and increase of back pressure.

Reagent immobilization contributes to the development of greener procedures even when toxic chemicals are employed. As an example, a minicolumn filled with a few milligrams of PbO_2 immobilized on polyester resin was employed for the indirect determination of dypirone by flame atomic absorption spectrometry.[65] The method was based on the measurements of the lead(II) ions released when the drug was oxidized by PbO_2. As more than 7000 determinations were carried out with the same minicolumn, the amount of the heavy metal consumed per determination was insignificant. Immobilization of $Hg(SCN)_2$ in epoxy resin was exploited to develop a greener procedure for the spectrophotometric determination of chloride, reducing in $c.1500$-fold the amount of Hg generated per determination in comparison to a single-line flow-injection manifold.[31] Another clean alternative for chloride determination exploited a minicolumn filled with silver chloranilate.[66] Formation of AgCl due to the reaction with the analyte released the absorbing chloranilate ions which were quantified by spectrophotometry in a 100 cm optical path liquid-core waveguide flow cell. This resulted in a 75-fold increase in sensitivity and reduction of waste generation to $c.100$ ng chloranilate per determination.

Analyte retention on a solid support can be exploited for direct measurements in the solid phase, aiming to improve sensitivity, selectivity or both. The reversible retention of the analyte is feasible in some applications and the reuse of the reagent yields greener analytical procedures. The process is suitably carried out in flow systems, the solid support being deposited in the measurement cell for detection by spectrophotometry (transmittance or reflectance modes), luminescence or infrared spectrometry. This approach was adopted to develop more environmentally friendly analytical procedures for zinc determination in pharmaceutical preparations[67] and iron in natural waters.[68] The solid support was C18-bonded silica modified with the azo-reagent 1-(2-thiazolylazo)-2-naphthol, and the analyte retention occurred as a result of the formation of a coloured complex. Elution of the analyte was carried out with a few microlitres of a diluted acid solution without removing the ligand from the solid support, which was used for at least 100 measurements. This results in reagent consumption of less than 1 µg per determination.

6.4.3.2 Multicommuted Flow Systems

Multicommutation shows a great potential for the development of environmentally friendly analytical procedures, because reagents can be added only at the instants and in the amounts strictly required for sample processing. This aspect has been emphasized in papers on MCFA, including comparison with the reagent consumption in other flow approaches. This advantage can be achieved without hindering the analytical performance, as demonstrated by

Table 6.4 Minimization of waste generation in flow-based systems.

Analyte	Sample	Flow configuration	Reagents	Remarks	Reference
Anionic surfactants	Natural waters	MCFA	Methylene blue and $CHCl_3$	Consumption of chloroform reduced by 98%	71
Anionic surfactants	Natural waters	FIA-falling drop	Methylene blue and $CHCl_3$	1.3 µl $CHCl_3$ per determination	43
Benzene	Gasoline	MCFA	Dilution in hexane	Consumption of hexane 3-fold lower than in a flow system with continuous pumping	72
Carbamate pesticides	—	µFIA	Trimethylaniline	< 300 nL toluene per determination	86
Carbaryl	—	MCFA and SIA	PAP/periodate	5 µg PAP (MCFA) and 11 µg PAP (SIA) per determination	69
Carbaryl	Natural waters		PAP/periodate	1.9 µg PAP and 5.7 µg KIO_4 per determination/waste mineralization by persulfate/UV	75
Chloride	Natural waters	FIA	$Hg(SCN)_2$ immobilized on epoxy resin	1500-fold reduction of Hg amount in the waste	31
Chloride	Natural waters	FIA	Silver cloranilate	c.100 ng chloranilate per determination	66
Chloride	Natural waters	MSFIA	$Hg(SCN)_2/Fe(III)$	3400-fold lower reagent consumption versus a single line FIA	80

Analyte	Sample	Technique	Reagents	Remarks	Ref.
Chloride	Natural waters	µFIA	$Hg(SCN)_2/Fe(III)$	Reagent consumption 300-fold lower than in the batch procedure and 20 ml of waste after 8-h working period	84
Cyanide	Natural waters	MCFA	OPA and glycine	5.6 µg OPA^2 and 6.7 µg glycine per determination	70
Cyclamate	Table sweeteners	Multi-pumping	$NaNO_2$, KI, H_3PO_4	3 mg KI, 1.3 µg $NaNO_2$ and 125 µmol H_3PO_4 per determination	73
Dypirone	Pharmaceutical	FIA	PbO_2 immobilized on polyester resin	>7000 determinations with the same minicolumn	65
Iron	Natural waters	µFIA	Nitroso-R salt	50 µl of effluent per determination	85
Nitrite		Multi-pumping	NED, sulfanil amide and diluted acid	55-fold lower consumption of NED in comparison to the batch procedure and in-line waste degradation	33
Paraquat	Natural waters	Multi-pumping	Dehidroascorbic acid	12-fold reduction of the reagent amount/2.0 ml of waste per determination	74
Phosphate	Natural waters	SIA-lab-on-valve	Molibdate/ascorbic acid	10 µl of reagent and 250 µl of waste per determination	83
Total phenols	Natural waters	MCFA	4-amino antipyrine and ferricyanide	200-fold lower reagent consumption by a MCFA system and increase of sensitivity by LPS avoiding LLE	34

LLE, liquid–liquid extraction; LPS, long pathlength spectrophotometry; NED, N,N-(1-naphthyl)ethylenediamine; OPA, o-phthalaldehyde; PAP, p-aminophenol.

comparing flow procedures for determination of the pesticide carbaryl.[69] The consumption of the most toxic reagent (*p*-aminophenol) per determination was reduced from 0.14 mg in a confluent flow injection system to 5 µg with MCFA. On the other hand, a SIA system consumed 11 µg of *p*-aminophenol to achieve a sensitivity threefold lower.

MCFA was exploited to develop a greener spectrophotometric procedure for determination of total phenols based on reaction with 4-aminoantipyrine in oxidizing alkaline medium,[34] resulting in reagent consumption 200-fold lower than in the batch procedure. Long pathlength spectrophotometry improved the sensitivity 80-fold, thus avoiding the need for a liquid–liquid preconcentration, which would consume $c.50$ ml $CHCl_3$ and generate $c.600$ ml of waste per determination.

The reference batch method for determination of acid-dissociable cyanide in natural waters is based on the reaction with barbituric acid and pyridine with detection by spectrophotometry. Replacement of hazardous chemicals and minimization of reagent amounts were exploited to develop an environmentally friendly procedure.[70] The MCFA system was designed for the independent handling of *o*-phthalaldehyde (OPA) and glycine, the isoindole derivative formed in the reaction being detected by fluorescence. Reagent consumption (5.6 µg of OPA and 6.7 µg of glycine per determination) was 230-fold lower than in a flow-based procedure with continuous reagent addition.

Also in the context of replacing a toxic reagent by a more environmentally friendly alternative or minimization of its consumption, the use of chloroform was reduced by 98% in the spectrophotometric determination of anionic surfactants in water.[71] MCFA allowed the implementation of sample dilutions, external calibration and the standard additions method in the determination of benzene in gasoline by FT-IR.[72] The consumption of the organic solvent (hexane) was threefold lower than in a flow system with continuous pumping, without affecting the analytical performance.

6.4.3.3 Multipumping Flow Systems

The potential of flow systems constructed with solenoid micropumps to minimize reagent consumption was demonstrated in the development of a greener procedure for nitrite determination in natural waters.[33] Two analytical methods were compared: the Griess reaction and formation of triiodide from nitrite and iodide in acid medium, a more environmentally friendly alternative. The Griess method was selected because of its better sensitivity and selectivity. The toxicity of the waste was minimized by using a multipumping flow system, which reduced by 55-fold and 20-fold the amount of the most toxic reagent (*N,N*-diethyl-*p*-phenylenediamine, NED) in comparison to the batch procedure and flow injection with continuous reagent addition, respectively. In addition, the residue was mineralized on-line by the photo-Fenton reaction, yielding a colourless solution and reducing the total organic carbon content by 87%. In another work, waste minimization and replacement of hazardous chemicals (e.g. Pb(NO$_3$)$_2$ and NED used in previous studies) were adopted to develop a greener procedure for determination of cyclamate in artificial sweeteners.[73] The method was based on the reaction of the analyte with nitrite in acid medium, the excess of reagent being determined by reaction with iodide. The procedure consumes only 3 mg KI, 1.3 µg NaNO$_2$ and 125 µmol H$_3$PO$_4$ per determination and generates a waste volume of 2.0 ml. Excess of the most toxic substance (nitrite ions) was completely decomposed in acid medium and the waste generated contained only triiodide and iodide ions.

Multipumping flow systems were investigated for the development of environmentally friendly analytical procedures for the determination of the pesticides paraquat[74] and carbaryl.[75] Replacement of sodium dithionite by dehidroascorbic acid and 12-fold reduction of the reagent amounts were the strategies adopted in the paraquat assay, the corresponding effluent volume being 2.0 ml per determination. Sensitivity was increased by using a 10 cm optical path flow cell in order to avoid the preconcentration step used in previous works, which would result in additional waste generation. Carbaryl was determined in a multipumping flow system coupled to a 100 cm optical path flow cell based on a liquid-core waveguide, reducing both the reagent consumption (1.9 µg of *p*-aminophenol and 5.7 µg of potassium metaperiodate) and the effluent volume (2.6 ml per determination). After waste degradation by UV irradiation in a medium of potassium persulfate, the total organic carbon was 94% lower and the residue was non-toxic for *Vibrio fischeri* bacteria. Cloud point extraction was adopted in the clean-up step, replacing the LEE recommended in EPA method 8318, which consumes 90 ml of methylene chloride per sample.[76]

The use of organic solvents was avoided in a procedure proposed for determination of anionic and cationic surfactants.[77] This was attained by employing a LED-based photometer and a flow system with solenoid mini-pumps for fluid propulsion.

6.4.3.4 Sequential Injection Analysis and Multisyringe Flow Injection Systems

A MSFIA system coupled to a selection valve was proposed for phosphorus fractionation in soils and sediments with detection by the molybdenum blue method.[78] The solid sample was directly deposited in a minicolumn and the sequential extraction of labile phosphorus, phosphorus species bound to iron and aluminium, and calcium-bound phosphorus was performed on-line with 1.0 mol L^{-1} NH$_4$Cl, 0.1 mol L^{-1} NaOH and 0.5 mol L^{-1} HCl according to the Hieltjes–Lijklema method. The proposed flow assembly has several advantages over fractionation analysis in the batch mode, such as reduction of the analysis time (from days to hours), minimization of energy expenditure and decrease of reagent consumption by *c.*92%.

A flow system with two syringe pumps coupled to a cold vapour atomic fluorescence detector was used for mercury speciation, after in-line sample clean-up and preconcentration.[79] Speciation was based on separation of the inorganic form as a tetrachloro complex of organic mercury on an anion-exchange membrane. Recoveries of the analyte spiked to the samples were close to 90% and results for a fish muscle certified reference material agreed with the certified value. The system reduced reagent consumption and the membrane separation replaces the classic LLE (with toluene and dichloromethane), thereby reducing waste amount and toxicity.

MSFIA was also hyphenated to a liquid core waveguide for determination of chloride in waters.[80] This classic analytical methodology exploiting reaction with $Hg(SCN)_2$ was improved by reducing reagent consumption by 3,400-fold and 600-fold in comparison to that required in a single-line flow system or with the reagent immobilized in epoxy resin,[31] respectively.

Cloud point extraction (CPE) is a green alternative for separation and preconcentration, in which toxic organic solvents are suitably replaced by nontoxic surfactants. This strategy can be carried out in flow-based systems, mechanizing laborious operations (*e.g.* heating, phase separation and removal of the surfactant-rich phase). CPE was implemented in a SIA system for manganese preconcentration before measurements by FAAS.[81] The analyte was separated by CPE with Triton X-114 after complexation with 4-(5-bromo-2-thiazolylazo)orcinol. The surfactant-rich phase was retained in a minicolumn filled with cotton in the preconcentration step, elution being carried out with a diluted acid solution. The enrichment factor was estimated as 14, attaining a detection limit of 0.5 μg L^{-1}.

6.4.3.5 Miniaturized Flow Systems

Another useful approach to GAC is the reduction of the manifold scale, with the aim of implementing analytical procedures with minimized reagent amounts (flow rates in the order of nL to μl min^{-1}).[82] Microfluidics and microelectronics can be exploited to integrate propulsion, mixing and detection units in a single microflow injection system (μFIA), which is commonly designed for a specific application. A more versatile alternative is the lab-on-valve approach, which exploits a microdevice integrating injection ports, microchannels and flow-cell coupled to conventional apparatus, including pumps and detection systems. Solutions can then be handled by conventional flow approaches, such as SIA. When the lab-on-valve approach was adopted for the spectrophotometric determination of phosphate, for example, reagent consumption and effluent generation were reduced to 10 μl and 250 μl per determination, respectively.[83]

The development of inexpensive microfluidic devices with polymeric materials has been the focus of recent work. As an example, a microflow system (total volume 7.0 μl) integrating the microchannels to a LED-based photometer was constructed by deep UV lithography on urethane–acrylate polymer.[84] When the device was used for determination of chloride, the consumption of $Hg(SCN)_2$ was *c.*300-fold lower than in the batch procedure, generating *c.*20 ml of effluent after an 8 hour working day. A microflow analyser was used for the determination of iron in water samples, based on formation of the complex with nitroso-R salt.[85] The miniaturized system drastically reduced waste generation, which was less than 2.0 ml h^{-1}, corresponding to 40 determinations.

More complex sample processing was involved in the determination of carbaryl pesticide in a glass microchip, in which pesticide hydrolysis, diazotization and extraction of the product in toluene was implemented. The enrichment factor was estimated as 50 with a toluene volume less than 300 nL per determination. The detection limit was estimated as 70 nmol L^{-1} with detection by thermal lens spectrometry.[86]

Another ingenious strategy for waste minimization is the analytical use of falling drops.[43] Liquid drops can be reproducibly formed at the end of tubes, enabling sample processing to take place on a removable reaction surface. The potential of this approach to GAC was demonstrated by the determination of sodium dodecyl sulfate by means of ion-pair formation with methylene blue and LLE in a 1.3 µl CHCl$_3$ microdrop. The analytical signal was measured directly in the drop surface with a LED-based photometer achieving a detection limit of 50 µg L^{-1} and coefficient of variation of 5%.

6.4.4 Waste Treatment

In addition to the strategies for minimizing the amount and toxicity of waste previously described, an additional step for on-line waste treatment can be efficiently coupled to flow systems. The general strategy is to introduce a suitable reagent after the flow cell in order to destroy or passivate the toxic species in the effluent. Detoxification of wastes can be achieved by photochemical, chemical, thermal or microbiological degradation processes. Photochemical degradation has been the most common alternative, resulting in effective degradation of dangerous organic compounds in time intervals compatible with the residence time in flow systems. Generally, the treatment is carried out in the presence of semiconductors such as TiO_2, which under UV irradiation can generate electron–hole pairs and catalyse the photoassisted degradation. In this sense, several flow-injection cleaner procedures have been developed by including the detoxification of wastes generated in the determination of carbamate pesticides[87,88] and resorcinol,[89] using a TiO_2 slurry and UV irradiation. As the semiconductor catalyst can be filtered and reused, the treatment step did not produce additional wastes. A photo-Fenton process [33] and treatment with persulfate under UV irradiation[75] were also efficiently used for degradation of wastes produced in flow-based procedures, as previously discussed.

Other efficient procedures for waste treatment, such as chemical or physical adsorption, precipitation and coprecipitation, ozonization, and microbiological as well as thermal processes have the potential to be carried out in flow systems.[90] If higher residence times are necessary, the waste of the flow system can be directed to a batch waste treatment unit. The combined effect of the minimization of reagent consumption and on-line treatment is very attractive in developing greener analytical procedures. A detailed discussion of strategies for on-line decontamination of analytical wastes is presented in Chapter 10 of this book.

6.5 Future Trends in Automation

A glance at this chapter shows a clear trend towards the use of electronic and informatics resources to design flow systems. Another trend has been integration of the basic units, including propulsion and detection devices and downscaling of the whole set-up. Control, data acquisition and processing tend to be performed by microcontroller devices, incorporated in a single chip,[87] providing facilities to control external devices, high resolution analogue/digital converters and data saving. This advantage has been achieved by designing the flow manifold with solenoid minipumps for fluid propulsion and LED-based photometers as detector.[91] This trend also holds in miniaturization, by designing microfabricated devices coupling all the stages of sample processing with waste generation in the order of nanolitres per determination, resulting in insignificant impact on the environment.

References

1. M. de la Guardia, *TrAC, Trends Anal. Chem.*, 2010, **29**, 577.
2. S. Armenta, S. Garrigues and M. de la Guardia, *TrAC, Trends Anal. Chem.*, 2008, **27**, 497.
3. F. R. P. Rocha, J. A. Nobrega and O. Fatibello, *Green Chem.*, 2001, **3**, 216.
4. L. T. Skeggs Jr, *Clinical Chemistry*, 2000, **46**, 1425.
5. J. Ruzicka and G. D. Marshall, *Anal. Chim. Acta*, 1990, **237**, 329.
6. E. A. G. Zagatto, F. R. P. Rocha, P. B. Martelli and B. F. Reis, *Pure Appl. Chem.*, 2001, **73**, 45.
7. C. Pasquini and W. A. Oliveira, *Anal. Chem.*, 1985, **57**, 2575.
8. V. O. Brito and I. M. Raimundo Jr, *Anal. Chim. Acta*, 1998, **371**, 317.
9. I. M. Raimundo Jr and C. Pasquini, *Analyst*, 1997, **122**, 1039.
10. I. Facchin, J. W. Martins, P. G. P. Zamora and C. Pasquini, *Anal. Chim. Acta*, 1994, **285**, 287.
11. I. Facchin and C. Pasquini, *Anal. Chim. Acta*, 1995, **308**, 231.
12. M. D. H. da Silva and C. Pasquini, *Anal. Chim. Acta*, 1997, **349**, 377.
13. F. R. P. Rocha, B. F. Reis, E. A. G. Zagatto, J. L. F. C. Lima, R. A. S. Lapa and J. L. M. Santos, *Anal. Chim. Acta*, 2002, **468**, 119.
14. J. L. F. C. Lima, J. L. M. Santos, A. C. B. Dias, M. F. T. Ribeiro and E. A. G. Zagatto, *Talanta*, 2004, **64**, 1091.
15. P. R. Fortes, M. A. Feres, M. K. Sasaki, E. R. Alves, E. A. G. Zagatto, J. A. V. Prior, J. L. M. Santos and J. L. F. C. Lima, *Talanta*, 2009, **79**, 978.
16. M. Miró, V. Cerdà and J. M. Estela, *TrAC, Trends Anal. Chem.*, 2002, **21**, 199.
17. H. Bergamin Filho, E. A. G. Zagatto, F. J. Krug and B. F. Reis, *Anal. Chim. Acta*, 1978, **101**, 17.
18. E. A. G. Zagatto, A. O. Jacintho, J. Mortatti and H. Bergamin-Filho, *Anal. Chim. Acta*, 1980, **120**, 399.
19. E. A. G. Zagatto, A. O. Jacintho, L. C. R. Pessenda, F. J. Krug, B. F. Reis and H. Bergamin-Filho, *Anal. Chim. Acta*, 1981, **125**, 37.

20. M. F. Giné, B. F. Reis, E. A. G. Zagatto, F. J. Krug and A. O. Jacintho, *Anal. Chim. Acta*, 1983, **155**, 131.
21. F. R. P. Rocha, P. B. Martelli, R. M. Frizzarin and B. F. Reis, *Anal. Chim. Acta*, 1998, **366**, 45.
22. F. J. Krug, E. A. G. Zagatto, B. F. Reis, O. Bahia-Filho, O. Jacintho and S. S. Jørgensen, *Anal. Chim. Acta*, 1983, **145**, 179.
23. P. B. Martelli, F. R. P. Rocha, R. C. P. Gorga and B. F. Reis, *J. Braz. Chem. Soc.*, 2002, **13**, 642.
24. J. Ruzicka and J. W. B. Stewart, *Anal. Chim. Acta*, 1975, **79**, 79.
25. E. H. Hansen and J. Ruzicka, *Anal. Chim. Acta*, 1976, **87**, 353.
26. R. N. Fernandes and B. F. Reis, *Talanta*, 2002, **58**, 729.
27. G. N. Doku and S. J. Haswell, *Anal. Chim. Acta*, 1999, **382**, 1.
28. A. D. Eaton, L. S. Clesceri and A. E. Greenberg, *Standard Methods for the Examination of Water and Wastewater*, American Public Health Association, Washington DC, 1995.
29. F. J. Krug, L. C. R. Pessenda, E. A. G. Zagatto, A. O. Jacintho and B. F. Reis, *Anal. Chim. Acta*, 1981, **130**, 409.
30. F. R. P. Rocha, P. B. Martelli and B. F. Reis, *Anal. Chim. Acta*, 2001, **438**, 11.
31. C. R. Silva, H. J. Vieira, L. S. Canaes, J. A. Nóbrega and O. Fatibello Filho, *Talanta*, 2005, **65**, 965.
32. J. F. van Staden and T. A. van der Merwe, *Microchim. Acta*, 1998, **129**, 33.
33. W. R. Melchert, C. M. C. Infante and F. R. P. Rocha, *Microchem. J.*, 2007, **85**, 209.
34. K. O. Lupetti, F. R. P. Rocha and O. Fatibello-Filho, *Talanta*, 2004, **62**, 463.
35. B. Ettinger, C. C. Kuchiioft and H. J. Lixhli, *Anal. Chem.*, 1951, **23**, 1783.
36. W. Frenzel, J. O. Frenzel and J. Möller, *Anal. Chim. Acta*, 1992, **261**, 253.
37. P. D. Goulden, P. Brooksbank and M. B. Day, *Anal. Chem.*, 1973, **45**, 2430.
38. R. A. S. Lapa, J. L. F. C. Lima and I. V. O. S. Pinto, *Analusis*, 2000, **28**, 295.
39. B. Karlberg and S. Thelander, *Anal. Chim. Acta*, 1978, **98**, 1.
40. A. Alonso, M. J. Almendral, M. J. Porras and Y. Curto, *J. Pharm. Biomed. Anal.*, 2006, **42**, 171.
41. K. Carlsson and B. Karlberg, *Anal. Chim. Acta*, 2000, **415**, 1.
42. A. L. D. Comitre and B. F. Reis, *Talanta*, 2005, **65**, 846.
43. H. Liu and P. K. Dasgupta, *Anal. Chem.*, 1996, **68**, 1817.
44. M. F. Giné, H. Bergamin Filho, E. A. G. Zagatto and B. F. Reis, *Anal. Chim. Acta*, 1980, **114**, 191.
45. S. Motomizu and M. Sanada, *Anal. Chim. Acta*, 1995, **308**, 406.
46. C. J. Patton, A. E. Fischer, W. H. Campbell and E. R. Campbell, *Environ. Sci. Technol.*, 2002, **36**, 729.
47. I. G. Torró, J. V. G. Mateo and J. M. Calatayud, *Anal. Chim. Acta*, 1998, **366**, 241.
48. W. R. Melchert and F. R. P. Rocha, *Talanta*, 2005, **65**, 461.

49. F. J. Krug, J. Ruzicka and E. H. Hansen, *Analyst*, 1979, **104**, 47.
50. F. J. Krug, B. F. Reis, M. F. Giné, J. R. Ferreira, A. O. Jacintho and E. A. G. Zagatto, *Anal. Chim. Acta*, 1983, **151**, 39.
51. J. A. Nóbrega, A. A. Mozetto, R. M. Alberici and J. L. Guimarães, *J. Braz. Chem. Soc.*, 1995, **6**, 327.
52. C. Pasquini and L. C. Faria, *Anal. Chim. Acta*, 1987, **193**, 19.
53. L. C. Faria, C. Pasquini and G. Oliveira-Neto, *Analyst*, 1991, **116**, 357.
54. I. C. Vieira and O. Fatibello-Filho, *Anal. Chim. Acta*, 1998, **366**, 111.
55. O. Fatibello-Filho and I. C. Vieira, *Anal. Chim. Acta*, 1997, **354**, 51.
56. O. Fatibello-Filho and I. C. Vieira, *Analyst*, 1997, **122**, 345.
57. T. Settheeworrarit, S. K. Hartwell, S. Lapanatnoppakhun, J. Jakmunee, G. D. Christian and K. Grudpan, *Talanta*, 2005, **68**, 262.
58. J. G. March and B. M. Simonet, *Talanta*, 2007, **73**, 232.
59. J. González-Rodríguez, P. Pérez-Juan and M. D. Luque de Castro, *Talanta*, 2003, **59**, 691.
60. S. S. Borges, R. M. Frizzarin and B. F. Reis, *Anal. Bioanal. Chem.*, 2006, **385**, 197.
61. Z. Bouhsain, S. Garrigues and M. de la Guardia, *Analyst*, 1997, **122**, 441.
62. M. J. Sánchez-Dasi, S. Garrigues, M. L. Cervera and M. de la Guardia, *Anal. Chim. Acta*, 1998, **361**, 253.
63. M. Zenki, K. Minamisawa and T. Yokoyama, *Talanta*, 2005, **68**, 281.
64. J. M. Calatayud and J. V. G. Mateo, *TrAC, Trends Anal. Chem.*, 1993, **12**, 428.
65. L. L. Zamora and J. M. Calatayud, *Talanta*, 1993, **40**, 1067.
66. V. G. Bonifácio, L. C. Figueiredo-Filho, L. H. Marcolino Jr. and O. Fatibello-Filho, *Talanta*, 2007, **72**, 663.
67. L. S. G. Teixeira, F. R. P. Rocha, M. Korn, B. F. Reis, S. L. C. Ferreira and A. C. S. Costa, *Anal. Chim. Acta*, 1999, **383**, 309.
68. L. S. G. Teixeira and F. R. P. Rocha, *Talanta*, 2007, **71**, 1507.
69. B. F. Reis, A. Morales-Rubio and M. de la Guardia, *Anal. Chim. Acta*, 1999, **392**, 265.
70. C. M. C. Infante, J. C. Masini and F. R. P. Rocha, *Anal. Bioanal. Chem.*, 2008, **391**, 2931.
71. E. Rodenas-Torralba, B. F. Reis, A. Morales-Rubio and M. de la Guardia, *Talanta*, 2005, **66**, 591.
72. E. Ródenas-Torralba, J. Ventura-Gayete, A. Morales-Rubio, S. Garrigues and M. de la Guardia, *Anal. Chim. Acta*, 2004, **512**, 215.
73. F. R. P. Rocha, E. Ródenas-Torralba, A. Morales-Rubio and M. de la Guardia, *Anal. Chim. Acta*, 2005, **547**, 204.
74. C. M. C. Infante, A. Morales-Rubio, M. de la Guardia and F. R. P. Rocha, *Talanta*, 2008, **75**, 1376.
75. W. R. Melchert and F. R. P. Rocha, *Talanta*, 2010, **81**, 327.
76. http://www.epa.gov/sam/pdfs/EPA-8318a.pdf, accessed in November, 2009.
77. A. F. Lavorante, A. Morales-Rubio, M. de la Guardia and B. F. Reis, *Anal. Chim. Acta*, 2007, **600**, 58.

78. J. Buanuam, M. Miró, E. H. Hansen, J. Shiowatana, J. M. Estela and V. Cerdà, *Talanta*, 2007, **71**, 1710.
79. A. M. Serra, J. M. Estela and V. Cerdà, *Talanta*, 2009, **78**, 790.
80. F. Maya, J. M. Estela and V. Cerdà, *Anal. Bioanal. Chem.*, 2009, **394**, 1577.
81. V. A. Lemos, P. X. Baliza, A. L. Carvalho, R. V. Oliveira, L. S. G. Teixeira and M. A. B., *Talanta*, 2008, **77**, 388.
82. S. J. Haswell, *Analyst*, 1997, **122**, 1R.
83. J. Ruzicka, *Analyst*, 2000, **125**, 1053.
84. A. Fonseca, I. M. Raimundo Jr, J. J. R. Rohwedder and L. O. S. Ferreira, *Anal. Chim. Acta*, 2007, **603**, 159.
85. S. Kruanetr, S. Liawruangrath and N. Youngvises, *Talanta*, 2007, **73**, 46.
86. A. Smirnova, K. Mawatari, A. Hibara, M. A. Proskurnin and T. Kitamori, *Anal. Chim. Acta*, 2006, **558**, 69.
87. M. de la Guardia, K. D. Khalaf, V. Carbonell and A. Morales-Rubio, *Anal. Chim. Acta*, 1995, **308**, 462.
88. M. J. Escuriola, A. Morales-Rubio and M. de la Guardia, *Anal. Chim. Acta*, 1999, **390**, 147.
89. M. de la Guardia, K. D. Khalaf, B. A. Hasan, A. Morales-Rubio and V. Carbonell, *Analyst*, 1995, **120**, 231.
90. G. Lunn and E. B. Sansone, *Destruction of Hazardous Chemicals in the Laboratory*, 2nd edn, John Wiley & Sons, New York, 1994.
91. E. Rodenas-Torralba, F. R. P. Rocha, B. F. Reis, A. Morales-Rubio and M. de la Guardia, *J. Autom. Meth. Manag. Chem.*, 2006, article ID 20384.

CHAPTER 7
Green Analytical Separation Methods

MIHKEL KALJURAND AND MIHKEL KOEL

Institute of Chemistry, Faculty of Science, Tallinn University of Technology, Ehitajate tee 5, 19086, Tallinn, Estonia

7.1 Why Green Separation Methods Are Needed in Analytical Chemistry

Green chromatography or *green electrophoresis* may seem like odd terms at first sight. The chemical industry is more concerned about environmental protection and environmental friendliness than analytical chemistry in general and separation science in particular. However, a significant number of analytical methods relate to the preparation, treatment and separation of individual components of samples (by some estimates up to 80% of analysis time[1]) and most of the chemicals and solvents involved in analysis are consumed in this step. For one run, the amounts may be minimal. But consider, for example, the amount of acetonitrile (ACN) consumed per year by an individual high-performance liquid chromatograph (HPLC) in an average laboratory, assuming that 2000 runs is a typical instrument load. If we assume that for a single run the test and rinse times equal 40 min + 20 min with a flow rate of $2\,\text{ml min}^{-1}$, the eluent consumption per year would be $60 \times 0.002 \times 2000 = 240\,\text{L}$. If more than one chromatograph is operating, then the ACN consumption calculations transform an academic exercise into an economic and environmental problem.

ACN is one of the most important and popular solvents in analytical chemistry. It has minimal chemical reactivity, low acidity and a reasonably low

RSC Green Chemistry No. 13
Challenges in Green Analytical Chemistry
Edited by Miguel de la Guardia and Salvador Garrigues
© Royal Society of Chemistry 2011
Published by the Royal Society of Chemistry, www.rsc.org

boiling point. Its miscibility with water, wide range of achievable polarities with water mixtures, low viscosity compared to other organic solvents (resulting in low pressure drop, even with water binary systems), and low ultraviolet cut-off (down to 192 nm) make it ideal for reversed-phase HPLC applications. The unique properties of ACN make it the solvent of choice in separations of pharmaceuticals. These physical and chemical properties, as well as its former abundance, led to the validation of many HPLC methods using ACN as an eluent. In addition to the uses described above, ACN is also a highly functional solvent for sample preparation techniques for dilution and protein precipitation in biological fluids, and as an eluent for solid-phase extraction. ACN is used in synthetic biochemistry for solid-phase oligonucleotide–DNA synthesis, for peptide synthesis, and in drug manufacturing.

However, ACN is toxic to humans and slightly persistent in water, with a half-life of between 2 and 20 days. It also has acute and chronic toxicity to aquatic life. Reducing the use of ACN in the laboratory is therefore sensible from a green perspective. Academic research related to green chromatography has provided many ACN-free separation protocols and methodologies that might not seem as efficient or effective as ACN-based solutions, but they do accomplish the task. Analysts are neither instantly nor universally convinced of the benefits of green chromatography. Analytical chemists have always lagged behind in the application of green chemistry principles, because their first priority has always been analytical performance.

ACN is a by-product of the automobile parts industry. In 2009, precipitated by the downturn in the global economy, this industry decreased production substantially, with knock-on effects on the cost and availability of ACN.[2] Until this crisis, no one seriously thought about replacing acetonitrile as an HPLC eluent. The labs and corporations who are involved in preparative- and process-scale chromatographic purification and use ACN as eluent have been hit particularly hard. The traditional business model, in which economic and environmental concerns are presumed to be mutually exclusive, does not support green ideas, but now, in light of the ACN crisis, corporations are considering green chromatographic solutions seriously.

In the following sections, we describe HPLC techniques as well as solutions provided by capillary electrophoresis (CE) and microfluidics to make analytical separation science more environmentally benign. We will show that separation science can meet the requirements of green chemistry with regard to the reduction of the use of solvents and other reagents and lowering energy consumption by increasing the speed of analysis, and through portability and miniaturization of equipment.

7.2 Green Chromatography

7.2.1 Gas-Phase Separations

Gas chromatography (GC) is a relatively green separation method because its eluents are usually helium and hydrogen, gases that are harmless to the atmosphere. However, GC sample preparation methods should be revisited in the light of green chemistry principles. Eliminating or minimizing the amount of solvent used in sample preparation techniques before the final chromatographic analysis is highly recommended. Various approaches to the implementation of green chemistry principles in GC so-called direct chromatographic analysis are pre-eminent[3] because they permit the determination of analytes in a sample without any pretreatment or sample preparation. Concerns about the disposal of toxic solvents have given rise to cleaner extraction methods, among which are those commonly described as solventless sample preparation techniques,[4] as opposed to the customary liquid–liquid or liquid–solid extractions which use an organic solvent (often followed by clean-up and preconcentration steps).

There are two truly solventless approaches in gas phase separation: gas extraction and membrane-based techniques. Supercritical fluid extraction is an example of a solventless extraction method when carbon dioxide, which is a gas at normal conditions, is used, and if the extraction unit is directly connected to a gas chromatograph.

The U.S. Environmental Protection Agency (EPA) recommends another solventless extraction method: dynamic gas extraction, *i.e.* purging inert gas through the solution under investigation and trapping extracted analytes on a suitable sorbent (*e.g.* Tenax). This is the widely accepted method for routine analysis of volatile organic compounds in water.

Static headspace microextraction is also becoming very popular, mainly because it does not require such sophisticated instruments as the purge technique. Gas extraction is the most widely used method for isolating volatile pollutants from different matrices, primarily because it is a proecological (solvent-free) means of isolation and enrichment. Gas extraction provides the required sensitivity (up to the parts per thousand level) and can be automated by combining it with GC. Thermal desorption of pollutants collected on a sorbent is a standard method for the measurement of volatile organic compounds in workplace or environmental air.[5] The advantages of thermal desorption over conventional solvent extraction include an improvement in detection limits (by three orders of magnitude), no chromatographic interference from solvents or solvent impurities, enhanced sample throughput, and lower cost per analysis. Another advantage is that thermal desorption is a straightforward gas extraction process. In light of these considerations, thermal desorption meets all the requirements of green chemistry for chromatographic analysis.

Membrane-based separation can be considered a solventless method when the analytes are volatile and the accepting media is a gas phase. In this technique, the separation of analytes is performed by a membrane: a stream of gas flushes the external side of a membrane and the gas flow from the internal side of the membrane is used to dispense the analytes into the inlet of the gas chromatograph. Another option is the collection of analytes from a stream of gas or liquid flushing the external side of a membrane, on the internal trap side of the membrane, and after the collection cycle, desorption of the analytes for GC analysis.

In addition to sample preparation, fast GC is gaining attention for greening gas phase separations. Fast GC reduces overall analysis time leading to significant savings in time and energy.[6,7] Fast chromatography can be especially attractive for laboratories where many routine samples are analysed every day. It can also be advantageous in situations where quick results are needed. However, increasing the speed of analysis requires modification of commercially available instrumentation. Faster separation can be achieved by decreasing the inner diameter of the capillary columns, reducing the thermal mass of the column thermostat for fast temperature programming, applying shorter columns, or working in turbulent flow either by using a vacuum outlet operation or working above optimal carrier gas velocities.

7.2.2 Liquid Phase Separations

7.2.2.1 Green HPLC Methods

Liquid chromatography is one of the biggest consumers of organic solvents in analytical chemistry; greening this method could therefore start with reducing the use of solvents. The ACN crisis has spawned lively discussion in the chromatographic community about how to deal with this situation. A consensus seems to have been reached that reducing the use of toxic solvents via miniaturization, column temperature programming and the use of solvents less toxic than ACN could provide the solution. Mayors has recently discussed these options at length.[8] He has identified the following opportunities:

- *Shorter columns with the same internal diameter:* Because column length, L, is proportional to analysis time t_r, but resolution is proportional to the square root of the number of plates, N (*i.e.* the square root of t_r) then reducing the column length (and keeping the flow rate constant) will decrease proportionally the amount of solvent required. Consequently, less resolution is sacrificed than one might expect and sufficient separation may be achieved. However, for gradient elution, the gradient time must be reduced proportionally. Of course, a change in conditions could have a detrimental effect on resolution, but this could easily be tested.
- *Shorter columns with smaller particles:* A proportional reduction in particle size can compensate for the loss of resolution due to decreased column length. Working with shorter columns implies a loss of plates because $L \sim N$, but since $N \sim 1/d_p$ (where d_p is the average particle diameter), the number of plates is increased and the overall resolution should be maintained. For example, the amount of solvent can be decreased fivefold and approximately the same resolution maintained by replacing 25 cm columns with 5 cm columns and reducing the particle diameter from 5 μm to 1.8 μm while keeping the flow rate and column diameter constant.[8] Because the minimum of the van Deemter curve shifts to the higher flow rate, the separation time can be further shortened. The new submicrometre-sized particles are particularly attractive in this situation

because separation times and solvent usage can be reduced up to one order of magnitude compared with longer columns packed with larger particles.
- *Reducing the internal column diameter:* An easy way to reduce solvent consumption is to change the internal diameter of the column (with or without decreasing the column length). The decreasing factor is $(d_l/d_n)^2$ where d_l and d_n are diameters of wider and narrower bore columns. Decreasing the column diameter while simultaneously reducing flow rate (to maintain constant linear velocity) allows the separation time to remain the same, and the new chromatogram should look very similar to the original. In order to avoid extracolumn chromatographic band-broadening, the injection volume, the tubing volumes, and detector flow cell volume might need to be reduced. If the internal diameter of the column is reduced to less than 1 mm, the chromatograph must be completely redesigned. Most commercial chromatographs intended to operate with capillary columns achieve extremely low extracolumn band-broadening contribution to zone by implementing smaller flow cells that are designed to function well at flow rates below 100 μl min^{-1}. Nano-LC instruments that operate with few hundred nanolitres per minute are now available commercially; these are being applied to proteomics, where samples are mass-limited and mass spectrometry is used for detection. These types of systems yield the ultimate solvent savings. With chip-based systems, a litre of acetonitrile could last a month or more!
- *Using other organic modifiers in reversed-phase HPLC (RP-HPLC):* The most popular replacement for ACN is methanol. Other solvents that have been considered are tetrahydrofuran, isopropanol, ethanol, and n-propanol. These solvents are available in high purity for HPLC applications. Methanol has some disadvantages for HPLC use compared with ACN due to its higher viscosity (which influences the column pressure drop) and higher UV cut-off (210 nm compared to 192 nm for ACN). It is also possible that not only the elution order of the analytes could change, but that peaks that were well resolved with the ACN/water solvent system could become unresolved if CH_3OH/water were used as the mobile phase. Because methanol is a weaker solvent than ACN, the percentage must be increased to produce similar retention times. Another way to reduce the use of ACN might be to employ a different washing solvent to remove unwanted compounds from the HPLC column.
- *Changing the type or amount of the stationary phase:* Using a less hydrophobic stationary phase or the same stationary phase with lower surface coverage decreases the retention of organic analytes. Substituting a C8 or even a C4 phase for a C18 phase can reduce the amount of organic modifier in RP-HPLC and still achieve the same retention time. Some suppliers provide RP-HPLC columns with different amount of the same stationary phase.
- In *hydrophilic interaction chromatography (HILIC)*, separation is achieved by partitioning between a water-enriched layer on the surface of a polar stationary phase and a mobile phase that contains a high

percentage of organic solvent (mostly ACN).[9] HILIC is used for the separation of highly polar ionizable solutes. In recent research on the use of HILIC, dos Santos Pereira *et al.* demonstrated that the HILIC mechanism could be reversed.[10] The features of the method they named 'per aqueous LC' were illustrated in the analysis of catecholamines, nucleobases, acids, and amino acids. In keeping with the principles of green chromatography, the more environmentally friendly ethanol was used.

- *Changing the pH of eluent for ionizable compounds:* Lowering or raising the pH below/above the pK_a of ionizable compounds might elute them from the column more rapidly than using an organic solvent, thereby saving it. The authors are unaware of any examples where this method has been used as a green alternative to common HPLC with the aim of reducing organic solvent consumption.

- *Using increased column temperature:* Elevated-temperature HPLC has developed rapidly in recent years and, as an important mode of green chromatography, it deserves a more thorough treatment. Chromatographers frequently overlook temperature as an optimization parameter in HPLC, but it can play an important role in selectivity because nearly all of the physical parameters of liquid chromatographic separation depend on it. As the temperature of a two-phase water–organic system increases, retention decreases. This fact can be exploited to reduce the amount of organic solvent in the eluent and still maintain the previous retention time, thus conserving solvent. Moreover, using only pure water as the eluent and increasing the temperature from 25 °C to 250 °C decreases properties such as polarity, surface tension, and viscosity, and results in solvation behaviour that resembles that of an organic solvent like methanol or ACN. Consequently, instead of using a water–acetonitrile gradient, one can use temperature programming as in GC. With the introduction of new phases capable of operating up to 200 °C and higher, the use of pure water as a mobile phase produces a truly 'green' analytical technique. For this reason, temperature has recently attracted the attention of chromatographers. An extremely attractive feature of temperature as a separation parameter is the fact that it can be set instrumentally. It is much easier to adjust temperature, simply by turning a knob on an instrument, than to adjust eluent composition or the buffer pH during the method development stage, which requires wet chemistry manipulations. Still, most HPLC separations are currently performed at temperatures ranging from 18–30 °C, *i.e.* at room temperature. If the column temperature is higher, the term 'high-temperature HPLC' is frequently used. The authors prefer to use the term 'elevated temperature', because many applications of interest to chromatographers take place at temperatures in the 40–200 °C range, which are certainly elevated with respect to ambient temperature but would not normally be considered high. When planning to use elevated-temperature chromatography, one must keep in mind an important prerequisite: the stability of the analytes at those temperatures. Obviously, this is a vital consideration for biologically significant analytes.

Reviews summarizing the development and applications of elevated-temperature HPLC have been published recently.[11–17] In 2001 a special issue of the *Journal of Separation Science* was dedicated to the role of temperature in LC.[18] To illustrate the effect of temperature, (Figure 7.1) shows a separation of a mixture of 10 triazine and 10 phenylurea pesticides.[19] A temperature gradient was combined with a solvent gradient to enhance the selectivity because it is impossible to reach the complete resolution of all pairs of peaks under isothermal conditions; however, the use of a moderate temperature gradient made it possible to achieve higher resolution between solute pairs.

Even more sophisticated use of temperature can be made. The relative retention of two solutes is sensitive to changes in the conformation of the stationary phase as the temperature is varied. Thus, temperature-dependent stationary phases can be synthesized. For example, retention can be modulated by a temperature-responsive polymer with reversible hydrophilic–hydrophobic conformation.[20,21] Such temperature-responsive stationary phases also contribute to green chromatography. The temperature-responsive properties of the coupled phase were demonstrated using only water as a mobile phase: an

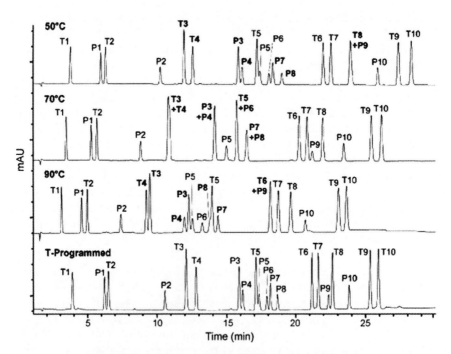

Figure 7.1 Separation of a mixture of 10 triazine and 10 phenylurea pesticides on a Zorbax StableBond-C18 column (150 mm × 4.6 mm I.D., 1.8 μm d_p). Flow rate 1 ml min^{-1}; gradient: water/ACN 80:20 to 45/55 in 30 min. Temperature programme (lower chromatogram): initial 40 °C hold 0–1 min, ramp 40–60 °C at 1.3 °C min^{-1}, and ramp 60–90 °C at 10 °C min^{-1}. (Reproduced from Ref. 19, with kind permission of Wiley-VCH).

increase in retention was observed with rising temperature. Kikuchi and Okano describe several applications using surfaces and interfaces modified with stimuli-responsive polymers for stimuli-responsive surface property alteration and posit their application in the separation sciences for affinity separation of proteins.[22] Particular attention is directed to the temperature-responsive polymer poly(N-isopropylacrylamide) (PIPAAm) and its derivatives as surface modifiers for green chromatography, in which only an aqueous mobile phase was used for separating bioactive compounds. This polymer exhibits hydrophilic properties below 32 °C and becomes hydrophobic above that temperature. Several effects of bioactive compounds on separation were investigated and discussed, including the effects of temperature-responsive hydrophilic–hydrophobic changes, copolymer composition, and graft polymer molecular architecture. Compounds covering a wide polarity range including phenones, alkylbenzenes, phenols, alkylated benzoic acids, anilines, sulfonamides and carbamates were analysed, and the retention, peak shapes and plate counts were compared under identical conditions. For retained solutes, an increase in retention as a function of temperature was observed between 25 and 55 °C, and it was higher for analytes containing a longer hydrophobic chain. Compounds with similar hydrophobic chains but additional polar functions showed increased retention and improved peak shapes, suggesting that a mixed mode interaction mechanism also exists at temperatures above the transition temperature of the polymer.

7.2.2.1.1 Supercritical Fluid Chromatography. An important direction in greening separation methods is to seek replacements for existing organic solvents, as discussed above. A second approach is to use common gases and solvents under different conditions of pressure and temperature. It was clearly demonstrated above that the use of solvents above their normal temperature and pressure conditions will continue to expand the range of analytical methods and should be seen as part of an overall spectrum of solubility, polarity and volatility properties of solvents and mobile phases.

Every compound has a critical point on the temperature and pressure scale at which the difference between its gaseous and liquid states disappears and the compound is in a supercritical state. Supercritical fluids have served to link gases and liquids, providing a continuum of mobile phase properties to the analyst. The possibility of making seamless changes using supercritical fluids is very attractive for researchers. Slight changes in temperature and/or pressure around the critical point of supercritical fluids cause significant changes in density and other physical properties that make it possible to tune the solubility and other parameters of the solvent. With supercritical fluids, a greater range of solvent properties can be achieved with a single solvent, through careful manipulation of temperature and pressure. Therefore, using solvents in their supercritical condition expands their spectrum of solubility, polarity and volatility. The ability to fine-tune the properties of the solvent medium allows it to replace specific solvents in a variety of different processes, or to create new methods for processing (analysing) samples. The same solvent can be used for different applications and procedures. However, although the multitude of supercritical fluid parameters lends flexibility to the method, the lack of fundamental knowledge about how these parameters affect the process makes straightforward method development difficult, and the technique remains largely empirical. Nevertheless, using supercritical fluids, especially carbon dioxide, instead of organic solvents for chromatography is becoming more popular. Supercritical fluids can be used as mobile phases in chromatography because of their properties. They can act as substance carriers like the mobile phases in GC, and also dissolve these substances like solvents in liquid chromatography (HPLC).[23]

The supercritical fluid chromatography (SFC) technique has been in existence for several decades and has recently been effectively applied to chiral separations, especially for preparative work. SFC significantly reduces organic solvent usage and waste by using carbon dioxide as the mobile phase; it has therefore been labelled a 'green' technology. Commercial instrumentation is available, including accessories that can convert a conventional HPLC system to one that is SFC-capable. Supercritical carbon dioxide exhibits liquid-like densities and can be pumped like a liquid as long as the system is kept at the proper pressure and temperature to maintain the supercritical state. Supercritical carbon dioxide has very low viscosity and is highly diffuse, enabling fast and highly efficient separations, often exceeding those of HPLC. SFC could be considered a subset of normal-phase HPLC because carbon dioxide is a non-polar eluent, even at high density. To increase its polarity, the supercritical fluid must be doped with an organic solvent such as methanol, but usually at less than 30%. Columns used in SFC are similar to those used in normal-phase HPLC, such as cyano and amino, but SFC-specific stationary phases such as 2-ethyl pyridine have been developed that provide even better selectivity. Although traditionally regarded as a technique only for non-polar compounds, SFC has been applied to cationic, anionic, and, as mentioned previously, chiral compounds, as well as to proteins and drugs.

During the development of packed column SFC, researchers realized that the characteristics of chromatographic separation are present irrespective of whether the fluid is defined as a liquid, a dense gas, or a supercritical fluid. In some instances, the initial pressure used in SFC is actually below the critical pressure. The differences between SFC, subcritical fluid chromatography, enhanced fluidity chromatography and HPLC have been overstated in the past. When the outlet pressure is elevated and the pressure and temperature are controlled for the mobile phase, the resulting techniques are similar and the behaviours of conventional GC and HPLC are completely and seamlessly bridged.[24] Each type of chromatography represents part of a continuum of increasing mobile phase solvating power coupled with increasing mobile phase viscosity and decreasing mobile phase diffusivity. In principle, with a supercritical carbon dioxide carrier it is possible to perform GC, SFC, and HPLC in the same chromatograph (Figure 7.2).

SFC can be interfaced with most detection systems including mass spectrometers, with the only qualification being the requirements of a high-pressure

Green Analytical Separation Methods 177

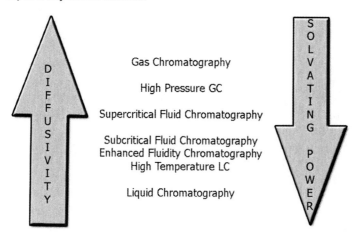

Figure 7.2 The chromatography continuum.

flow cell or back-pressure regulator and flow-splitter at the exit of the column. Of course, an existing HPLC method would have to be completely redesigned; SFC is therefore a disruptive technology for most laboratories. Nevertheless, SFC could be a strong competitor with HPLC for certain applications, and the reduction of organic solvent used in SFC is significant.

7.2.2.1.2 Consequences of the Acetonitrile Shortage. As mentioned in the introduction, the ACN crisis is a good reason for laboratories to move to 'greener' solvent systems or to miniaturized liquid phase separation methods such as CE or microchips. Laboratories that must comply with regulations (*e.g.* those in the pharmaceutical industry) are deeply concerned about the ACN shortage, as changing a method requires validation. The U.S. Food and Drug Administration (FDA) has reacted to the ACN shortage by issuing a warning that appropriate method validation procedures and relevant good manufacturing practices (GMP) must be followed.[25]

An important component of laboratory procedure is solvent recycling and purification for reuse. Pure ACN is rarely used as a mobile phase in RP-HPLC. The solvent is generally used in a binary system with water or buffered water; the column effluent therefore contains a mixture of solvents. Some ingenious solutions can be devised, such as collecting solvent when there are no chromatographic zones eluting from the column.[8] This can be performed by an intelligent system controlled by a detector signal that directs the solvent to the collection bottle by means of a switching valve at the end of the chromatographic column. ACN/water azeotrope can also be distilled as a purified binary mixture containing 14% water, which could be collected and used as material for eluent preparation.

7.2.2.1.3 The Greening of Chromatography Is Not Black and White. It is obvious that many of the suggestions outlined above could reduce ACN usage and make HPLC greener. With the exception of SFC, none of the techniques requires the purchase of new instruments, although the use of high-temperature ovens for HPLC might pose a technical problem for laboratories performing routine work. However, the more difficult problem for those laboratories would be changing approved methods because, as stated above, regulatory agencies have strict directives against altering column dimensions and experimental conditions without a full revalidation.

One should not take a simplistic view that green chromatography is merely a matter of using aqueous rather than organic eluents. The whole process should be considered when evaluating the greenness of a particular form of chromatography. The most promising green separation method—elevated-temperature chromatography—consumes more energy. At the time this book was being written (early in 2010) the energy aspects of green chromatography were mostly being ignored. A remarkable exception is the work of Vorst et al.[26] who performed an *exergetic* life cycle analysis of a chromatographic separation of enantiomers of a racemic mixture of phenylacetic acid derivatives in order to compare preparative HPLC with preparative SFC. The exergy of a system is the maximum work that is required to bring the system into equilibrium in a process.[27] Their conclusions are as follows: if one considers instrumentation alone, the exergy consumption related to the preparative HPLC technique is about 25% higher than for preparative SFC because of its inherently higher use of organic solvents. Considering the whole undertaking (plant, lab, *etc.*), one must take into account exergy calculations for the physical boundaries of the production site and the resources crossing those boundaries. Resources have to be purchased, including the cooling and industrial water, the cooling and heating medium, and steam. Storing the product entails costs. From this perspective, preparative SFC is more favourable because it consumes about 30% less resources than preparative HPLC as quantified in exergy. However, an analysis of the cumulative exergy extracted from the environment to deliver the mass and energy to the plant and a chromatograph 'boundary' via the industrial network reveals that preparative SFC requires about 34% more resources than preparative HPLC. The conclusion of their work[26] is astonishing: the most sustainable process in terms of integral resource consumption is preparative HPLC (using ACN as the eluent modifier). The authors reason that the requirement for electricity for heating and cooling and the production of liquid carbon dioxide argues against the use of preparative SFC.

Although the study[26] is limited to the preparative separation of a particular sample, its findings could be far-reaching. Another apparently green chromatography technique, elevated-temperature HPLC, also avoids the use of harmful solvents by using temperature (*i.e.* electric power) to achieve its purposes. Therefore, the actual greenness of elevated-temperature chromatography remains unproven until a thorough exergy life cycle analysis has been performed. The exergy calculations should also be applied to recycling and disposal processes on a case-by-case basis. Moreover, solvents that are considered innately green (such as ionic liquids) may not be completely green. Even 100% aqueous waste effluent cannot be flushed into municipal sewers without cleaning, which consumes energy and has an environmental impact. When energy consumption is taken into account, completely green chromatography is probably impossible, in the sense that it is not sustainable as the term is defined in green chemistry.[28] Even GC, the greenest mode of chromatography, consumes purified carrier gases, the extraction of which from the atmosphere results in a measurable carbon footprint.

At present, green chromatography is being developed by a small group of proponents. Opposition to its application on a wider scale has been based not on chromatographers' ignorance of the principles of green chemistry, but rather on the fact that green chromatography cannot solve some important separation problems faced by the chemical industry and regulatory agencies. If green HPLC makes further progress and becomes widely accepted, its exergy calculations will transform an academic exercise into organizational policy issues. One can envision conflicting scenarios in which green chromatography is environmentally acceptable and economically attractive in local laboratories and institutions, but not sufficiently benign for the environment and society as a whole.* Cynically speaking, green laboratory practices might involve the transportation of wastes to regions where regulations are not as stringent as they are where the pollution is being produced.

Finally, two lesser-known chromatographic preparative techniques also deserve mention because they qualify as green separation methods. These are *steady-state recycle chromatography*[29] and *simulated moving-bed chromatography*.[30] Steady-state recycle chromatography is a discontinuous, single-column separation technique that involves recycling unresolved fractions back into the column. It combines high throughput and low solvent consumption. Simulated moving-bed chromatography uses the counter-current flow of the stationary and mobile phases in continuous mode. Continuous chromatography provides two significant benefits: higher throughput (due to smaller runs and columns) and up to a hundredfold reduction in solvent consumption.

7.2.2.2 Capillary Electrophoresis

7.2.2.2.1 Capillary Electrophoresis as an Unrecognized Green Alternative to HPLC. One liquid-based separation method that can offer substantial competition to liquid chromatography is CE, a family of separation methods

---

*To view green chromatography from a wider green philosophical perspective it is interesting to consider the work of Karl Rebane, an Estonian physicist. He writes that history indicates that the species and societies that act more quickly and consume more high-quality energy and materials are the 'winners': in other words, those that cause more pollution and promote the faster growth of entropy. This could be a reason why protection of the environment is inherently difficult and why it is almost impossible to significantly reduce human consumption of energy and materials in a competitive world. To escape this inevitably fatal evolutionary outcome, fundamentally different thinking is needed—thinking that makes the survival of humankind the foremost value. A completely green process could be one in which pollution is transported away from Earth as infrared radiation (K.K. Rebane, 'Energy, entropy, environment: why is protection of the environment objectively difficult?', *Ecol. Econ.*, 1995, **13**(2), 89–92.)

that use narrow-bore fused-silica capillaries to separate mixtures of large and small molecules. Sample introduction is accomplished by immersing the end of the capillary into a sample vial and applying pressure or vacuum or voltage. Depending on the types of capillary and electrolyte used, CE technology can be divided into several separation techniques. CE typically uses high electric fields to separate analytes, which may be based on their charge, size, or hydrophobicity. CE is a powerful separation technique that involves minimal solvent consumption. The typical CE column is a capillary with an internal diameter of 50 μm and a length of 50 cm having a volume of about 5 μl. The analysis runtime is typically 10 min and electroosmosis, the eluent driving force in CE, has a rate of sub-μl per minute. This means that eluent consumption during a CE run is almost non-existent, with a buffer/sample volume of approximately 100 μl. ACN and other harmful solvents are rarely used as eluents in CE because most of the separations are conducted in ordinary aqueous buffer systems (see Figure 7.3).

CE was very actively studied at the beginning of the 1990s, but its development has stagnated in recent years. Although more than 2000 papers in fundamental studies in CE were published each year during the last decade, acceptance of CE by the industry has been slow. It has struggled to replace HPLC for the analysis of conventional small-molecule pharmaceuticals. The lack of familiarity with CE can lead to methods being developed that are not robust, perform poorly and give CE a negative reputation. Analytical chemists have learned that CE is highly sensitive to parameter changes and that it is not a

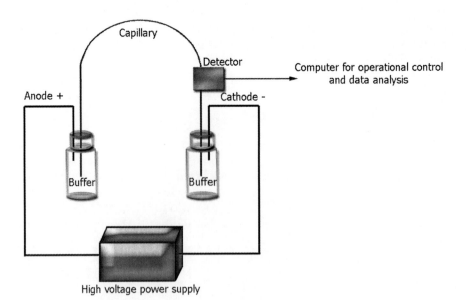

Figure 7.3 CE equipment. For sampling, the buffer reservoir is replaced by a sample reservoir.

very reproducible technique, but modern instrumentation has eliminated most of the early variability issues. As robust interfaces with mass spectrometers are developed, the obstacle of CE not being compatible with MS will soon be overcome.[31]

Surprisingly, CE has received little recognition as a genuinely green separation method. The authors of this chapter have pointed out that its greenness is due to the microscopic volumes of solvent it consumes.[32,*] Moreover, the solvents used are usually harmless aqueous buffers. As the ACN shortage continues to influence the field of analytical separations, CE is gathering increased attention as confirmed in the report from one panel discussion.[33] Suppliers of CE equipment have responded to the need for improved robustness and reliability. They have made improvements in recently upgraded equipment. There is also a trend towards well-controlled and validated chemistry/capillary kits to improve performance, for example, in inorganic anions and metal ion determinations. The emergence of more powerful, automated instruments makes this technique more accessible than ever. There are some signs that the future of CE is bright. It is expected[33] that the use of CE for small inorganic anion and metal ion analyses will replace ion-exchange chromatography. Chiral analysis by CE is well established. CE is used for screening and characterization (determination of pK_a, solubility, etc.) of compounds. Dedicated equipment and related kits/capillaries will be applied to specific protein characterization/assays. CE is far superior to either SDS-PAGE or isoelectric focusing gels. However, CE should not replace HPLC: it is a technique that is complementary/orthogonal to HPLC. Proponents of CE believe that outdated thinking among the part of separation scientists is the biggest obstacle to acceptance of the technique. The method still needs the support of instrument manufacturers (to build better and more robust instrumentation), consumables vendors (to continue to devise kits and reagents), and scientists themselves (to develop novel methodologies and applications).[33] Without this type of investment, the technique cannot expand, but without expansion, no one wants to invest. Wider acceptance of green chemistry by analysts may help to break this impasse.

7.2.2.2.2 Capillary Electrophoresis as a Method of Choice for Portable Instruments.
Portable instruments are commonly believed to be more economical than their stationary counterparts. Whether or not this is true, portability is an important and obvious feature of green analytical instruments.

---

*Although green philosophical arguments in favour of the benefits to humanity may not influence a chromatographer, the political-economic situation in a particular country may persuade scientists to accept green solutions. The authors have personal experience in this regard. The rise of CE at the beginning of the 1990s coincided with the political changes in eastern Europe that caused a dramatic reduction in funding for fundamental research. In this situation it was almost impossible to use HPLC because of the lack of supplies and the requirement for large amounts of solvents and spare parts. On the other hand, it was relatively easy to assemble CE instruments from old colour television sets (which contained a high-voltage power supply) and modify the cells of discarded optical HPLC detectors, and thereby carry on research at a reasonably competitive level.

The consumption of resources (either power or chemical) and the generation of waste are limited in the case of portable instruments. Moreover, they are designed to be taken to an analysis site, *i.e.* a point of care (POC). A POC can be any place, such as a hospital, home or crime scene. The portability of an instrument is generally understood to be its potential to work where a sample can be collected and analysed *in situ*. In fact, many field analyses could be accomplished without traditional sampling. Field analytical chemistry (FAC) is a growing trend that promises to liberate the analyst from tedious and inconvenient sample manipulations. Analytical methods typically involve sample preparation; a time- and labour-consuming collection step usually needs to take place before samples can be transported to a laboratory for analysis. Sample collection can often be problematic. For instance, when analysing polluted soil, samples have to be collected in many locations according to a time-consuming and laborious sampling plan to avoid missing the 'hot spot'. Samples must usually be treated with specific reagents and stored in containers under certain conditions to maintain their integrity before analysis. Furthermore, in many cases, there are difficulties in sampling, such as at hazardous polluted sites or with precious cultural relics and archaeological objects. FAC can not only eliminate the need for sample transportation but also greatly shorten the analysis time or even provide real-time results, thereby providing rapid warning and accurate feedback. Thus, the logic and nature of FAC are such that analysis is done with little or no sample collection and preparation. Transportation of the sample to a laboratory is eliminated. He *et al.* summarize the green characteristics of FAC instruments.[34] In order to perform field analyses, the ideal analytical instruments should meet several requirements: they should (1) have a fast response time to be able to acquire the necessary information on a real-time or near real-time basis; (2) be capable of *in situ*/at-site rather than just on-site[*] analysis and need little or no sample preparation; (3) be portable for field use with a minimum requirement for energy (battery-powered is desirable), consumables (gases/solvents), or clean space for handling samples; and (4) perform a cost-effective analysis. However, Turl and Wood are more specific with regard to the characteristics an FAC instrument requires in order to be useful at the POC.[35] A step out of the lab into the field requires a giant leap from technology to capability because the instrument, although small and light enough to be portable and consuming very little power, must:

- perform adequately
- work in a harsh environment
- have a minimal number of failures and be easy to maintain

---

[*]He *et al.*[35] provide a definition of the terms '*in situ*' and 'on site'. In this chapter, 'on-site analysis' is understood to be a common analysis procedure that involves sample collection/preparation using a field-portable instrument. '*In situ*' analysis leaves the sample site virtually undisturbed. '*In situ*' analysis could be done with an X-ray spectrometer but it is difficult to imagine how a chromatographic analysis of art or soil samples would be possible.

- ensure the security of classified data, and
- be easy to operate.

In addition,

- an appropriate sampling method must be developed
- operator health and safety risks must be minimized, and
- operator training must be simple.

It is difficult to imagine a portable gas or liquid chromatograph meeting all these conditions, although it might be technically possible (portable gas chromatographs are used successfully in space applications). Nevertheless, chromatography is not well suited to portability because eluent is required for analysis. There are, however, publications on field-portable gas[36,37] and liquid chromatographs.[38] On the other hand, CE is a promising technology for field instruments.

Portable CE can be constructed easily because the power consumption of the electrophoresis process in a capillary is small, so small, high-voltage power supplies can be used. Optical detection is not well suited for portable instruments; however, the emergence of small light-emitting diodes (LEDs) will soon change that situation. Da Silva[39] and Zemann[40] have developed a useful detection device: a contactless conductivity detector (CCD) that measures the conductance of a small cap between tubular electrodes laid on the separation capillary. The device is intrinsically small in size, and most analytes can be detected with a CCD. Several groups have developed new portable CE instrument designs. Hauser's group[41] has developed and optimized a portable CE instrument with CCD for the sensitive field measurement of ionic compounds in environmental samples. It is battery-powered, and the high-voltage modules are capable of delivering up to 15 kV at either polarity for more than 1 day. Inorganic cations and anions, including heavy metal ions and arsenates, can be determined with detection limits of approximately 0.2 to 1 mM. The instrument was field-tested in a remote region of Tasmania. Nitrite and ammonium were determined on-site at concentrations as low as 10 ppb in the presence of other common inorganic ions at concentrations two to three orders of magnitude higher. In another publication, Haddad's group demonstrated the use of CE for the detection of explosives in the environment.[42] Instead of CCD, Haddad's team used indirect photometric detection. They proved that it is possible to analyse blast residues at a crime scene, where they can be sampled simply by wiping hard surfaces with a wet cloth, rather than by transporting the residues back to the laboratory. They found that they could separate and detect the 12 cations at concentrations as low as $0.11\,\mathrm{mg\,L^{-1}}$ and separate and detect the 15 anions at concentrations as low as $0.24\,\mathrm{mg\,L^{-1}}$. In both cases, the analyses took less than 10 minutes. However, they found that CCD performed better than indirect photometric detection.[43]

Seiman et al.[44] attempted to develop a robust sampling procedure for on-site analysis. In this project, the CE analyser consisted of two pieces of capillary that were separated by a narrow gap (30 µm). To introduce the sample, a plastic syringe was inserted into a socket connected to the gap, and the background electrolyte (BGE) in the cross-section of the sampler was flushed out by the sample stream injected by the syringe. Then the sample between the capillaries was carried into the separation channel by electroosmotic flow (EOF), and BGE filled the junction between the two capillaries as soon as high voltage was applied. By this method, the manipulation of buffer vials is reduced. The method developed for this instrument has an LOD of 4–8 µM for phosphonic acids and 0.3–0.5 µM for cations, and an RSD (internal standard) of 8%.

The complexity of construction can be reduced even further. A possible design for what is conceivably the simplest portable instrument is demonstrated in Figure 7.4. An analysis compartment with two buffer vials and a CCD sensor is located in the front part of the instrument and a touch-screen computer on the top. A typical electropherogram of the phosphonic acids recorded with this instrument is shown in Figure 7.5.

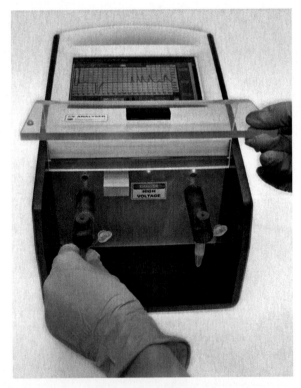

Figure 7.4 Portable CE instrument controlled via a touch-screen computer (screen is on the top of the instrument). The electronics and high-voltage power supply are located below the computer. In front is the CE compartment with two sample/buffer vials and CCD bloc. High voltage is applied to vials via electrodes located inside the two plastic insulators (black rods). (Photograph of the working instrument from the authors' laboratory.)

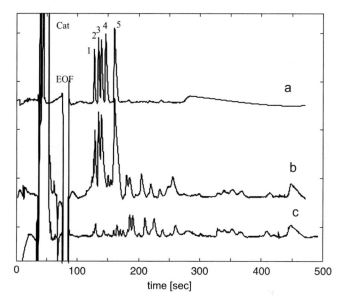

Figure 7.5 An example of on-site analysis of phosphonic acids by a portable CE instrument. Procedure: a water solution of 75 µM of phosphonic acids sprayed on the floor. After 24 h the polluted area was wiped with wetted Millibore filter paper and the wipe was extracted with MilliQ water. (a) Electropherogram of the standards. (b) Electropherogram of the extract: peaks: 1, pinacolylmethylphosphonic acid; 2, 1-butylphosphonic acid; 3, propylphosphonic acid; 4, ethylphosphonic acid; 5, methylphosphonic acid; Cat, peak of cations, EOF, marker of electro-osmotic flow. (c). Blank. (From the authors' laboratory.)

As we will see in the following sections, electrophoresis is a key technology for micronizing analytical separation methods even further by making use of an advanced concept based on lab-on-a-chip platforms. It is believed that this will open the way to many inexpensive POC medical diagnostic devices. Many reports on portable CE analysers based on microfluidics platforms have been published in recent years (see references 45 and 46 for examples) CE-based microfluidic devices are described in the next section.

7.3 Miniaturization of Separation Methods

Analytical chemistry deals with acquiring and processing information about chemicals in our environment. It should consume no more resources than are needed to obtain information about the sample. Contemporary analytical chemistry must consume fewer resources, and the most acute problem is the usage of toxic compounds and solvents. This is the driving force behind the miniaturization of analytical chemistry. As we saw when discussing 'black and white' issues in green chromatography (see section 7.2.2.1.3), the greenness of elevated-temperature chromatography remains ambiguous because the consumption of energy is not taken into account. Miniaturization of various analytical methods is on the cutting edge of research at the moment because it can provide the solution to energy economy as well. It is not surprising that many miniaturized analytical methods and technologies were inspired by developments in information technology (such as computer chips), where the miniaturization process has been under way for many decades. Frequently miniaturized analytical instruments copy the architecture of computer components. Miniaturization, known as 'microfluidics', has influenced separation science in general and electrophoresis in particular. Microfluidics handles volumes of fluid of the order of nanolitres and picolitres.

7.3.1 Continuous-Flow Microfluidics

Continuous-flow microfluidics deals with the precise control and manipulation of fluids that are geometrically constrained to small (typically submillimetre) channels. 'Micro' in this context means the following features: small volumes (nl, pl, fl), small size, low energy consumption, and other effects of the micro scale on fluids. The flow of liquid is actuated either by external pressure sources (external pumps or integrated micropumps) or by electrokinetic mechanisms. Microfluidic devices are well suited for many simple biochemical applications and for chemical separations. Continuous-flow operation is the mainstream approach to microfluidics.

Methods of fabricating microfluidic devices have been inspired by photolithography, a technology for manufacturing computer chips. Initially, most systems were made of silicon, but because of demand for such features as specific optical characteristics, bio- or chemical compatibility, lower production costs and faster prototyping, various new substrates such as glass, ceramics and metal etching, deposition and bonding have been proposed. Soft lithography, used for fabricating microfluidic channels in polydimentylsiloxane (PDMS) is especially popular. Typically, a 'negative' template is etched on silicon or glass, which is then coated with 'Slygard', a commercially available product, and cured to make a PDMS polymeric substrate for the device.[47] The channels are covered with a glass or polymer plate after curing, and detectors and other supporting pieces of equipment such as electrodes for carrying voltage to the microchip are attached. Microfluidics has increasingly more in common with lithography-based microsystem technology, nanotechnology and precision engineering.

One disadvantage of continuous-flow microfluidic systems is that they are less suitable for tasks requiring complicated fluid manipulations or a high degree of flexibility. The fluid flow at any location in the channels is dependent on the properties of the entire system. Permanently etched microstructures also lead to limited reconfigurability and poor fault tolerance.

7.3.2 Droplet and Digital Microfluidics

Droplets are used and manipulated in continuous microfluidics as a separate liquid phase differentiated from the carrier by chemical composition. This approach is referred to as 'droplet' microfluidics. Novel alternatives to the closed-channel continuous-flow systems described above are open structures, in which discrete, independently controllable droplets are manipulated on a planar substrate. Using an analogy to digital microelectronics, this approach is referred as *'digital' microfluidics* (DMF), and was pioneered at Duke University.[48] It should not be confused with *droplet microfluidics*. In droplet microfluidics, individual droplets are held in channels and the droplets are not controlled independently. DMF uses discrete unit-volume droplets, and one unit of fluid is moved over one unit of distance. This facilitates the use of microfluidic biochip design approaches such as cell-based assays. Because each droplet can be controlled independently, these systems can be reconfigured (reprogrammed) to change their functionality during the concurrent execution of a set of bioassays.

In droplet microfluidics, two of the most common methods of forming droplet streams in channels are T-channel geometry and flow-focusing.[49] In T-channel geometry, a perpendicular flow of continuous oil phase meets the inlet of the aqueous phase, where droplets are generated. Separating and analysing the content of droplets is highly desirable; CE is particularly suited for separating and analysing ultra-small-volume droplets because of its high separation efficiency, speed, and sensitivity. The content of droplets can be analysed by directing them fluidically to the CE separation channel via sophisticated microchannel manifolds.[50,51] The flow-focusing method relies on three parallel flows. The aqueous phase is located between two oil-phase flows. These three flows converge on an orifice to break the aqueous flow into droplets. In this way, eluted CE zones can be converted into individual droplets, thus preventing their dilution and loss of the separated components, and facilitating their downstream manipulation and analysis. Edgar *et al.*[52] performed the integration of CE with droplet generation driven by electro-osmotic flow, which enabled the compartmentalization of molecular components separated by CE into a series of droplets (see Figure 7.6). The droplet-confined bands can be docked for further study.

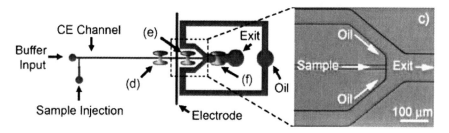

Figure 7.6 Schematic representation of the fluidic design used to integrate CE output with droplet compartmentalization. (c) Droplet-formation region shown in detail. (d), (e) and (f) are the locations of detection spots. Oil channels 50 × 50 mm; exit channel 50 × 100 mm; CE channel 10 × 10 mm. (Reproduced from Ref. 52, with kind permission from John Wiley & Sons.)

DMF devices are not yet very common, so we will devote a few words to their preparation. One common actuation method for digital microfluidics is the electrowetting-on-dielectric (EWOD) phenomenon. (Other techniques for droplet manipulation using surface acoustic waves, optoelectrowetting, *etc.* have also been demonstrated recently). Briefly, EWOD is the phenomenon by which an electric field changes the contact angle and thus the wetting behaviour of a polarizable and/or conductive liquid droplet in contact with a hydrophobic, insulated electrode. The insulating film is frequently made of two layers of Paralyene C and Teflon-AF. The application of voltage to a series of adjacent electrodes (which can be turned on or off) can be used to manipulate droplets because the semispherical shape of the droplet becomes asymmetrical and the internal tension of the droplet actuates its movement.[53] This effect is illustrated in Figure 7.7. Droplets can be sandwiched between two parallel plates to prevent evaporation, or an open format can be used without an upper glass plate, enabling easier access to the droplets for further study.

DMF is an attractive platform for biological applications, which often require the use of expensive or rare reagents and small amounts of samples.

DMF has been applied for solving various biological problems such as enzymatic and immunoassays, proteomics and DNA analysis. Applications such as cell-sorting and cell-based assays, PCR and pyrosequencing have been reported. These and several others are described by Miller and Wheeler.[54]

The number and variety of analyses being performed on chip has increased with the need to perform multiple sample manipulations. As in the case of droplet microfluidics, components that produce a signal of interest must be isolated in order to be detected. Although DMF could be a good platform for miniaturizing separation techniques, mass separation methods such as CE are not currently an established part of the digital microfluidic toolkit and the integration of separation methods seems to present a significant challenge. There have been very few attempts to perform molecular separation on a digital microfluidic platform. In one example, Shah and Kim achieved high-purity separation using EWOD-based droplet microfluidics by introducing a 'fluidic conduit' between a sample droplet and a buffer droplet. The long, narrow

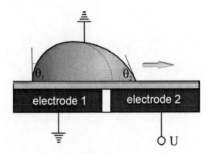

Figure 7.7 Droplet deformation due to the application of electric potential to one of the electrodes in an open set-up. Movement of the droplet results from a change in its geometrical shape.

fluidic path minimizes the diffusion and fluidic mixing of the two droplets, which eliminates non-specific mass transport but provides a channel between them for actively transporting particles (thus allowing specific transport). The effectiveness of the technique was demonstrated by eliminating approximately 97% of non-magnetic beads in one purification step, while maintaining high collection efficiency (>99%) of the magnetic beads.[55] A recent publication by Abdelgawad et al.[56] described a hybrid microfluidic device in which a sample was delivered to the separation chip channel by a DMF device. In another study, a common CE device with a CCD was interfaced to DMF platforms.[57] In this research, to pursue 'the spirit of low-cost digital microfluidics' advanced by Wheeler,[58] it was demonstrated that the actuation of droplets can be achieved using an electrode system prepared from the copper substrate of a common printed circuit coated only with food wrap (without the hydrophobic layer). The DMF sample injection was performed by transporting sample and buffer droplets in succession under the end of the CE capillary inlet, immersing the capillary in the sample/buffer droplet, and performing CE separation by applying a high voltage between the (grounded) buffer droplet and the CE outlet reservoir. Using a DMF sampler, CE separation of a mixture of vitamins was achieved.[57]

7.3.3 World-to-Chip Interfacing and the Quest for a 'Killer' Application in Microfluidics

The connection between the components of a microfluidic device and the macro-environment of the world is often referred to as a macro-to-micro interface,[59] interconnect,[60–63] or world-to-chip interface.[64–68] This is a problem that jeopardizes the greenness of microfluidic devices. The difficulty results from the fact that samples and reagents are typically transferred in quantities of microlitres to millilitres (or even litres), but microfluidic devices deal only with nanolitres or picolitres of samples or reagents because their reaction chambers and channels typically have micro scale dimensions. This challenge can be overlooked in research environments such as academic laboratories, but it erodes the founding pillar of green chemistry: minimal use of organic solvents. The world-to-chip problem cannot be ignored in routine applications. A good illustration is provided by the photographs in works, picked at random by a Google search on the keyword 'microfluidics': a tiny microchip is surrounded by bulky instruments and a bundle of tubing needed for the fluid supply (pumps), actuation (by gas) and detection.[69,70] In the photograph, (Figure 7.8) there are dozens of connecting tubes, indicating the number of actuating (perhaps syringe) pumps that are required. Supporting equipment is not usually shown in illustrations of microfluidic devices. Although microchips can perform many sophisticated operations, the analytical process can only be considered to have been truly microminiaturized if the instrumentation supporting unit operations in the chip is also miniaturized by many orders of magnitude.

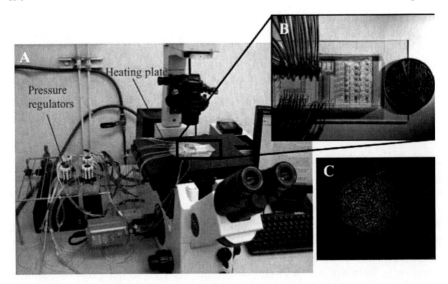

Figure 7.8 Connecting microfluidic devices to the world: an image of an actual microfluidic device for single hESC mRNA extraction. (A) The system includes a microscope, a computer to control air pressure with pressure regulators, and a heating stage to heat the microfluidic chip to necessary temperature. Insert (B) A typical microfluidic chip. (C) Merged image of fluorescent and light microscopes. (Reproduced from Ref. 70, with kind permission from The Royal Society of Chemistry.)

The so-called 'world-to-chip' interface problem has plagued microfluidics since its inception. Microfluidics must integrate all components of the system on the same chip to ensure the portability and minimum energy consumption required by the principles of green analytical chemistry. Pumps, valves, mixers, *etc.* must be miniaturized in order to achieve an integrated system, which forces a choice between active methods—efficient, but requiring energy sources and difficult to miniaturize—and the passive methods provided by non-instrumental microfluidics—easier to integrate, but less efficient. This is a huge challenge, especially in biotechnology, where the volume of the targets of study located in the macroscopic environment must be reduced to accommodate the microscopic environment of the microfluidic device. Finally, the huge surface/volume ratio of miniaturized systems could modify the physical behaviour of the system, giving rise to new problems, such as adherence of target molecules to the solid walls, or the effect of capillary forces that may prevent the fluid from entering the microchannels.

According to some researchers, the dilemma of the world-to-chip interface is one of the bottlenecks in the development of μ-TAS.[70] It is critical for high-throughput applications where manual manipulation is not economic and a macro-to-micro interface must be developed. In a review article,[59] solutions that have appeared in the literature are presented and discussed. The solutions described above—droplet and digital microfluidics for sample processing and injection, continuous-flow microfluidics or CE for molecular separation and CCD and LED for detection—could well be combined into a portable instrument consuming little energy and material. The main breakthrough could well come from such integration.

The lab-on-a-chip has frequently been designed for a very specific application. Academic institutions and corporations have developed a plethora of lab-on-a-chip devices for different applications, but what is really needed is a universal application that could trigger widespread use of microchips in biomedicine. There is a general consensus that miniaturization should have advanced more quickly than it has. A few years ago, many analytical chemists were enthusiastic about chip-based analyses, but now one can sense disappointment and disillusionment. If there is no widely used application, then the development of microfluidic devices is not economically profitable. This phenomenon is known as the absence of a 'killer application'. This term '... is commonly used to describe a product which has such highly desirable properties that it generates very large revenues with attractive margins in a comparatively short amount of time. In addition to this purely economic description it also helps to promote the underlying technology, thus helping typically "disruptive technologies" (a technology that enables products which dramatically change markets due to their (often unexpected) performance and which are not achievable by simple linear extrapolation of existing products or technologies'.[71] Examples of killer applications are digital photography and large flat-panel television screens. Let us hope that a killer application will soon be found, that the world-to-chip problem will be solved as quickly as possible, and that obstacles to developing the ultimate green analysis method will be surmounted. A hype cycle model introduced in 1995 can be used to describe the stages of development of microfluidics. According to Mukhopadhyay,[72] the model has five stages: a technology trigger, a peak of inflated expectations, a trough of disillusionment, a slope of enlightenment, and finally, a plateau of productivity. He believes that microfluidics is now on the slope of enlightenment. Many microfluidics proof-of-concept have been advanced, but the gap is large between an academic proof-of-concept and the needs of industry. Only 1% of proofs-of-concept become commercialized.[72] Most innovative technologies take longer than anticipated to create large markets. So there are reasons to be hopeful and positive about the future of microfluidics.

7.3.4 Non-Instrumental Microfluidic Devices

As we saw above, the typical set-up for a microfluidic experiment consists of a small custom-made chip that is surrounded by desktop-sized analysis instruments and power supplies. The microminiaturization of the main analysis process is only part of the story, and to make analytical methods environmentally friendly, one must solve this world-to-chip interface problem. One way to miniaturize supporting instruments is to replace complex elements in analysers with passive components that operate without external power by manipulating fluids using gravity, air pressure, or simple manual actions.

The need to develop such simple and possibly disposable devices is motivated primarily by the need for simple POC tests in developing countries, where non-instrumental analytical devices could be put to good use in medical diagnostics. In a recent report on the top 10 biotechnologies for improving health in developing countries, 'modified molecular technologies for affordable, simple diagnosis of infectious diseases' were ranked as the number one priority.[73] Such devices can also find application in developed countries where, although most medical diagnostics are performed in centralized, well-equipped hospital laboratories, home tests have a place as well. Two such applications are home glucose and pregnancy testing, and detection by first responders of natural bioemergencies or bioterrorism.

For these applications, paper is the best prospective material. It can be used as much more than a material for writing, printing and packaging. It has potential as an inexpensive, biodegradable, renewable, flexible polymer substrate for designing lab-on-a-chip prototypes. Paper-based three-dimensional microfluidic devices can be constructed with capabilities that are difficult to achieve using conventional open-channel microsystems made from glass or polymers. In particular, paper-based devices wick fluids and can distribute microlitre volumes of samples from single inlets to arrays of detection points. This capability opens the way to carrying out a variety of new analytical protocols simply and inexpensively on a piece of paper without external pumps. Much of this line of study has been pursued by Whitesides and his colleagues at Harvard University. They have achieved the ultimate operational robustness and cost reduction with microfluidic devices to date. They demonstrated that three-dimensional microfluidic devices could be made from stacked layers of ordinary paper and sticky tape.[74] Because of paper's wicking ability, the devices do not require external pumps to drive the liquids through. To define the pathways of the liquids in such a paper-based microfluidic device, the team impregnated each layer of paper with a common photoresist, and patterned them with UV light. With channels thus established on a sheet of paper, layers of paper were alternated with layers of double-sided tape; holes cut in the tape connected the channels in adjacent layers of paper. The complex routing that can be achieved can wick liquid horizontally and vertically to an array of 1024 detection zones underneath. With reagents or antibodies placed in detection zones prior to assembly, such devices could provide highly parallel, independent assays. The results could be transmitted to a central hospital using camera phones.[75] Wax printing simplifies the fabrication of paper-based microfluidics even further. The fabrication of paper-based microfluidic devices in nitrocellulose membrane for protein immobilization was reported in.[76] The fabrication process, which can be finished within 10 minutes, consists mainly of printing and baking.

The success of this approach depends on the absorption of test fluids into hydrophilic areas of porous paper and the use of capillary forces for fluid actuation. The products of reactions occurring inside such labs-on-paper (LOP) cannot easily be extracted for further biochemical analysis. If wicking is undesirable, one approach would be to develop an LOP device capable of storage, transfer, mixing and sampling of liquid drops by making the surface of the paper superhydrophobic (by oxygen plasma etching and fluorocarbon film deposition) and marking it with high surface energy ink patterns (lines and dots) using simple software similar to that used for word processing.[77] Surface energy and gravitational force can then be used to manipulate and transfer drops, thus eliminating the need for an external power source.

Paper-based microfluidic devices are especially appropriate for use in distributed healthcare in the developing world and in environmental monitoring and water analysis.[78] An example of a non-instrumented LOP is the immunochromatographic strip (ICS) assay used for pregnancy tests. Urine, saliva, serum, plasma, or whole blood can be used as specimens. To perform the test, a sample is placed on a pad at one end of the strip. The signal reagent is solubilized and binds to the antigen or antibody in the sample and moves through the membrane by capillary action. If a specific analyte is present, the signal reagent binds to it. The complex proceeds to the second antibody or antigen, and is immobilized at a test line on the strip, which captures the complex. If the test is positive, a pink/purple line develops. Moving further along the strip, the signal reagent encounters a second set of antibodies and forms an antibody–analyte–antibody sandwich matrix with a visible (control) signal. Because ICS relies on inexpensive reagents and components, the cost is less than 1.5€ to the end user in many cases. ICS strips require little or no sample processing, and they do not require an external instrument.

It is obvious that components of non-instrumental, microfluidics-based disposable diagnostic devices must function on simple physicochemical phenomena like capillary action, evaporation, endo/exothermic reaction, gravity, and laminar flow in microchannels. Some of the energy would be supplied by the analyst because a power supply would likely not be available. In their review, Weigl *et al.* give many examples of components designed to perform unit operations:[79]

- Liquid transport (pumps) can be achieved by wicking and capillary action, gravity or finger-operated bellows fabricated from polydimethylsiloxane (PDMS).
- Mixing reagents and samples in microfluidic structures is challenging because of the laminar flow of liquids in microchannels. Spiral microchannels, expansion vortices, channel obstacles, lamination splits and recombining have been designed for passive methods such as transverse mixing wells.
- Flow switching can be implemented by means of microvalves, but the complexity of manufacturing has largely limited the application of this method.
- Molecular separation can be conducted by diffusion. An H-filter is a device based on the parallel laminar flow of two or more miscible streams in contact with each other. Because of the laminarity of the flow, the streams do not mix, but chemicals can diffuse from one stream to the other. Smaller molecules diffuse faster than larger ones and this

- phenomenon can be exploited to extract targeted components from one stream to another.
- A sample can be concentrated by evaporation.
- Localized heating and cooling can be effected by positioning the endothermic (*e.g.* evaporation of acetone) and exothermic (*e.g.* dissolution of concentrated sulfuric acid in water) processes in microreaction chambers near the reactant flow interface.[80]
- Detection in non-instrumental microfluidics relies solely on the physical senses and therefore can only be based on colorimetry.

In summary, the eventual goal of this line of research is to develop methods that do not require electronics, such as a piece of paper that can detect disease markers or pathogens, in the way that litmus paper detects pH. Sample preparation and, if needed, even target amplification, are possible without the use of instrumentation, but is difficult to imagine how very low signal intensities could be observed without electronic signal amplification. In other words, detection will become a bottleneck for POC design. This may limit the applicability of this type of assay where extremely high sensitivity is required. Therefore, the incorporation of simple instruments in POC devices could be tolerated. The successful integration of paper-based microfluidics and electrochemical detection has been demonstrated.[81] Whitesides *et al.* developed an inexpensive handheld colorimeter that generates quantitative data in a point-of-care analytical system.[82] The aluminium-cased colorimeter has a tricolour LED for illuminating coloured spots on paper, and a manifold to hold the paper device in proper alignment with the LED for measurements. The LED light is modulated and detected by a narrow-bandpass detector; measurements therefore can be made in any lighting conditions. Whitesides speculates that there is ample scope for ingenious chemistry, because many colorimetric tests are potentially relevant to this type of analytical problem. Developing new dyes that have more pronounced colour changes and are more stable would be very useful for POC diagnostics. Another example of an extremely simple non-instrumental microfluidics device is a chip fabricated by Grudpan *et al.*[83] They made a simple chip by drilling channels in a piece of acrylic plastic. The chip was tilted to actuate sample and reagent flows by gravity. Different uses of the chip were demonstrated by reactions involving colour detection. To eliminate the need for instruments such as spectrophotometers, detection was done visually, based on the migration time of the reaction zone, using a simple stopwatch.

Consumer electronics-based analytics could support non-instrumental microfluidics. Filippini *et al.* demonstrated how a combination of a computer monitor and an inexpensive webcam could be used as a spectrophotometer.[84] This computer screen photo-assisted technique (CSPT) is based on the fact that a computer screen can easily be programmed to display millions of colours that combine three narrow-band emission profiles. The light emitted from a computer screen is not monochromatic, but a combination of three polychromatic primary colours that excite the human perception of red, green and blue. Personal computers are inexpensive and widely available. In this approach, the computer monitor acts as a light source and the camera as a detector. It is possible to envision the application of this type of system to non-instrumental microfluidics devices with colorimetric output. No reports of such applications are available, so we do not know what the detection limits of such devices would be, but the premise seems reasonable.

7.4 Challenges in Miniaturization of Separation Methods

Greening separation science is possible, but it may involve increasing the overall energy consumption of the process. Even if green analytical laboratory practices are locally viable, global green analytical chemistry might be regarded with suspicion for this reason, as we saw in section 7.2.2.1.3. One way to reduce energy consumption is to miniaturize the entire analytical process. Because analytical chemistry is an information science, it need not consume more resources than are necessary to support the process of analysis. Therefore, the greening of analytical chemistry via miniaturization seems to be possible in principle, and the development of green instrumentation depends primarily on analysts' creativity. Research and development of 'smart', miniaturized point-of-care instrumentation would be highly advantageous. The most acute issue seems to be solving the world-to-chip problem. An emerging new science—non-instrumental microfluidics—looks very promising in this respect because it is a genuinely green separation method. Still, much research in this field remains to be done. It might not be too farfetched to say that because the small number of assays would limit the possibilities of a class of methods based only on the colour perception of the operator, then gradient elution RP-HPLC-ESI-Q-TOF-MS/MS would be the only option for solving an analytical problem.

References

1. T. Hyötyläinen, *J. Chromatogr. A*, 2007, **1153**, 14.
2. A. Tullo, *Chem. Eng. News.*, 2008, **86**, 27.
3. W. Wardencki and J. Namieśnik, *Polish J. Environ. Stud.*, 2002, **11**, 185.
4. J. Namiesnik and W. Wardencki, *J. High Resolut. Chromatogr.*, 2000, **23**, 297.
5. M. Harper, *J. Chromatogr. A*, 2000, **885**, 129.
6. R. Sacks, H. Smith and M. Nowak, *Anal. Chem.*, 1998, **70**, 29A.
7. C. A. Cramers, H.-G. Janssen, M. M. Van Deurse and P. A. Leclerc, *J. Chromatogr. A*, 1999, **856**, 315.
8. R. E. Majors, The continuing acetonitrile shortage: how to combat it or live with it, *LCGC North Am.*, 1 June 2009.
9. A. J. Alpert, *J. Chromatogr.*, 1990, **499**, 177.
10. A. dos Santos Pereira, F. David, G. Vanhoenacker and P. Sandra, *J. Sep. Sci.*, 2009, **32**, 2001.

11. T. Greibrokk and T. Andersen, *J. Chromatogr. A*, 2003, **1000**, 743.
12. P. Jandera, L. Blomberg and E. Lundanes, *J. Sep. Sci.*, 2004, **27**, 1402.
13. B. A. Jones, *J. Liq. Chromatogr. Relat. Technol.*, 2004, **27**, 1331.
14. C. R. Zhu, D. M. Goodall and S. A. C. Wren, *LCGC Eur.*, 2004, **17**, 530.
15. C. R. Zhu, D. M. Goodall and S. A. C. Wren, *LCGC N. Am.*, 2005, **23**, 54.
16. K. Hartonen and M.-L. Riekkola, *Trends Anal. Chem.*, 2008, **27**, 1.
17. T. Teutenberg, *Anal. Chim. Acta*, 2009, **643**, 1.
18. *J. Sep. Sci.* 2001, **24**, special issue.
19. G. Vanhoenacker and P. Sandra, *J. Sep. Sci.*, 2006, **29**, 1822.
20. J. Kobayashi, A. Kikuchi, K. Sakai and T. Okano, *Anal. Chem.*, 2001, **73**, 2027.
21. H. Kanazawa, *J. Sep. Sci.*, 2007, **30**, 1646.
22. A. Kikuchi and T. Okano, *Progr. Polymer Sci.*, 2002, **27**, 165.
23. E. Klesper, A. H. Corwin and D. A. Turner, *J. Org. Chem.*, 1962, **27**, 700.
24. T. L. Chester and J. D. Pinkston, *Anal. Chem.*, 2004, **76**, 4606.
25. FDA, *Acetonitrile Shortages: Recommendations for Reporting Changes in Analytical Procedures.* http://www.fda.gov/downloads/AboutFDA/CentersOffices/CDER/UCM171776.pdf.
26. G. van der Vorst, H. van Langenhove, F. de Paep, W. Aelterman, J. Dingenen and J. Dewulf, *Green Chem.*, 2009, **11**, 1007.
27. P. Perrot, *A to Z of Thermodynamics*, Oxford University Press, New York, 1998.
28. P. T. Anastas and J. C. Warner, *Green Chemistry: Theory and Practice*, Oxford University Press, New York, 1998.
29. J. W. Lee and P. C. Wankat, *Ind. Eng. Chem. Res.*, 2008, **47**, 9601.
30. J. W. Lee and P. C. Wankat, *Ind. Eng. Chem. Res.*, 2009, **48**, 7724.
31. E. J. Maxwell and D. D. Y. Chen, *Anal. Chim. Acta*, 2008, **627**, 25.
32. M. Koel and M. Kaljurand, *Pure Appl. Chem.*, 2006, **78**, 1993.
33. Technology forum: capillary electrophoresis, *e-Separations Solutions*, 12 August 2009.
34. Y. He, L. Tang, X. Wu and X. Hou, *Appl. Spectrosc. Rev.*, 2007, **42**, 119.
35. D. E. P. Turl and D. R. W. Wood, *Analyst*, 2008, **133**, 558.
36. J. A. Contreras, J. A. Murray, S. E. Tolley, J. L. Oliphant, H. D. Tolley, S. A. Lammert, E. D. Lee, D. W. Later and M. L. Lee, *J. Am. Soc. Mass Spect.*, 2008, **19**, 1425.
37. H. Q. Lin, Q. Ye, C. H. Deng and X. M. Zhang, *J. Chromatogr. A*, 2008, **1198**, 34.
38. M. A. Nelson, A. Gates, M. Dodlinger and D. S. Hage, *Anal. Chem.*, 2004, **76**, 805.
39. J. A. F. da Silva and C. L. do Lago, *Anal. Chem.*, 1998, **70**, 4339.
40. A. J. Zemann, E. Schnell, D. Volgger and G. K. Bonn, *Anal. Chem.*, 1998, **70**, 563.
41. P. Kuban, H. T. A. Nguyen, M. Macka, P. R. Haddad and P. C. Hauser, *Electroanalysis*, 2007, **19**, 2059.
42. J. P. Hutchinson, C. J. Evenhuis, C. Johns, A. A. Kazarian, M. C. Breadmore, M. Macka, E. F. Hilder, R. M. Guijt, G. W. Dicinoski and P. R. Haddad, *Anal. Chem.*, 2007, **79**, 7005.

43. J. P. Hutchinson, C. Johns, M. C. Breadmore, E. F. Hilder, R. M. Guijt, C. Lennard, G. Dicinoski and P. R. Haddad, *Electrophoresis*, 2008, **29**, 4593.
44. A. Seiman, J. Martin, M. Vaher and M. Kaljurand, *Electrophoresis*, 2009, **30**, 507.
45. G. V. Kaigala, M. Behnam, C. Bliss, M. Khorasani, S. Ho, J. N. McMullin, D. G. Elliott and C. J. Backhouse, *IET Nanobiotechnol.*, 2009, **3**, 1.
46. G. V. Kaigala, V. N. Hoang, A. Stickel, J. Lauzon, D. Manage, L. M. Pilarski and C. J. Backhouse, *Analyst*, 2008, **133**, 331.
47. Y. Xia and G. M. Whitesides, *Angew. Chem. Int. Ed.*, 1998, **37**, 551.
48. M. G. Pollack, R. B. Fair and A. D. Shenderov, *Appl. Phys. Lett.*, 2000, **77**, 1725.
49. D. T. Chiu, R. M. Lorenz and G. D. M. Jeffries, *Anal. Chem.*, 2009, **81**, 5111.
50. J. S. Edgar, C. P. Pabbati, R. M. Lorenz, M. Y. He, G. S. Fiorini and D. T. Chiu, *Anal. Chem.*, 2006, **78**, 6948.
51. G. T. Roman, M. Wang, K. N. Shultz, C. Jennings and R. T. Kennedy, *Anal. Chem.*, 2008, **80**, 8231.
52. J. S. Edgar, G. Milne, Y. Q. Zhao, C. P. Pabbati, D. S. W. Lim and D. T. Chiu, *Angew. Chem. Int. Ed.*, 2009, **48**, 2719.
53. J. Berthier, *Microdrops and Digital Microfluidics*, William Andrew, Norwich, NY, 2008.
54. E. M. Miller and A. R. Wheeler, *Anal. Bioanal. Chem.*, 2009, **393**, 419.
55. G. J. Shah and C.-J. Kim, *Lab Chip*, 2009, **9**, 2402.
56. M. Abdelgawad, M. W. L. Watson and A. R. Wheeler, *Lab Chip*, 2009, **9**, 1046.
57. J. Gorbatsova, M. Jaanus and M. Kaljurand, *Anal. Chem.*, 2009, **81**, 8590.
58. M. Abdelgawad and A. R. Wheeler, *Microfluid. Nanofluid.*, 2008, **4**, 349.
59. G. Jesson, G. Kylberg and P. Andersson, in *Micro Total Analysis Systems 2003*, ed. M. A. Northrup, K. F. Jensen and D. J. Harrison, pp. 155–158.
60. V. Nittis, R. Fortt, C. H. Legge and A. J. deMello, *Lab Chip.*, 2001, **1**, 128.
61. A. Puntambekar and C. H. Ahn, *J. Micromech. Microeng.*, 2002, **12**, 35.
62. C. Gonzalez, S. D. Collins and R. L. Smith, *Sens. Actuators B*, 1998, **49**, 40.
63. H. Chen, D. Acharya, A. Gajraj and J.-C. Melners, *Anal. Chem.*, 2003, **75**, 5287.
64. J. M. Ramsey, *Nat. Biotechnol.*, 1999, **17**, 1061.
65. S. Attiya, A. Jemere, T. Tang, G. Fitzpatrick, K. Seiler, N. Chiem and D. J. Harrison, *Electrophoresis*, 2001, **22**, 318.
66. N. H. Bings, C. Wang, C. D. Skinner, C. L. Colyer, P. Thibault and D. J. Harrison, *Anal. Chem.*, 1999, **71**, 3292.
67. Z. Yang and R. Maeda, *Electrophoresis*, 2002, **23**, 3474.
68. J. Liu, C. Hansen and R. Q. Stephen, *Anal. Chem.*, 2003, **75**, 4718.
69. Y. Wang, W. Y. Lin, K. Liu, R. J. Lin, M. Selke, H. C. Kolb, N. Zhang, X.-Z. Zhao, M. E. Phelps, C. K. F. Shen, K. F. Faull and H. R. Tseng, *Lab. Chip*, 2009, **9**, 2281.

70. F. Zhong, Y. Chen, J. S. Marcus, A. L. Scherer, S. R. Quake, C. R. Taylor and L. P. Weiner, *Lab. Chip*, 2008, **8**, 68.
71. H. Becker, *Lab. Chip*, 2009, **9**, 2119.
72. R. Mukhopadhyay, *Anal. Chem.*, 2009, **81**, 4169.
73. A. S. Daar, H. Thorsteinsdottir, D. K. Martin, A. C. Smith, S. Nast and P. A. Singer, *Nat. Genet.*, 2002, **32**, 229.
74. A. W. Martinez, S. T. Phillips and G. M. Whitesides, *Proc. Natl. Acad. Sci. U. S. A.*, 2008, **105**, 19606.
75. A. W. Martinez, S. T. Phillips, E. Carrilho, S. W. Thomas, H. Sindi and G. M. Whitesides, *Anal. Chem.*, 2008, **80**, 3699.
76. Y. Lu, W. Shi, J. Qin and B. Lin, *Anal. Chem.*, 2010, **82**, 329.
77. B. Balu, A. D. Berry, D. W. Hess and V. Breedveld, *Lab. Chip*, 2009, **9**, 3066.
78. A. W. Martinez, S. T. Phillips, G. M. Whitesides and E. Carrilho, *Anal. Chem.*, 2010, **82**, 3.
79. B. Weigl, G. Domingo, P. LaBarre and J. Gerlach, *Lab. Chip*, 2008, **8**, 1999.
80. R. M. Guijt, A. Dodge, G. W. van Dedem, N. F. de Rooij and E. Verpoorte, *Lab. Chip*, 2003, **3**, 1.
81. W. Dungchai, O. Chailapakul and C. S. Henry, *Anal. Chem.*, 2009, **81**, 5821.
82. A. K. Ellerbee, S. T. Phillips, A. C. Siegel, K. A. Mirica, A. W. Martinez, P. Striehl, N. Jain, M. Prentiss and G. M. Whitesides, *Anal. Chem.*, 2009, **81**, 8447.
83. K. Grudpan, S. Lapanantnoppakhun, S. Kradtap Hartwell, K. Watla-iad, W. Wongwilai, W. Siriangkhawut, W. Jangbai, W. Kumutanat, P. Nuntaboon and S. Tontrong, *Talanta*, 2009, **79**, 990.
84. D. Filippini, S. Svensson and I. Lundström, *Chem. Commun.*, 2003, **2**, 240.

CHAPTER 8
Green Electroanalysis

LUCAS HERNÁNDEZ,[1] JOSÉ M. PINGARRÓN[2] AND PALOMA YÁÑEZ-SEDEÑO[2]

[1] Departamento de Química Analítica y Análisis Instrumental, Facultad de Ciencias-Módulo C-XIII, Universidad Autónoma de Madrid, C/. Francisco Tomás y Valiente, 7, Campus de Cantoblanco, 28049 Madrid, Spain;
[2] Departamento de Química Analítica, Facultad de Ciencias Químicas, Universidad Complutense, 28040 Madrid, Spain

8.1 The Role of Electroanalytical Chemistry in Green Chemistry

Electroanalytical chemistry fulfils many of the requirements that characterize green chemistry, owing to its capacity for obtaining measurements in real time with minimal impact on the environment.[1] In various important ways modern electroanalysis is in accordance with the principles of green analytical chemistry, such as the use of electrodes prepared with non-toxic materials, the reduction of sample size, or the application of miniaturized devices and microfluidics. These aspects directly affect the application of electroanalytical techniques and the development of electrochemical sensors and biosensors. In general, electrochemical detection is known to be very sensitive, so that a small amount of sample is needed and a small amount of waste is generated.[2] Furthermore, many electroanalytical procedures can be used without any need for sample pretreatment or derivatization, so reagents are not needed and alternative solvents such as ionic liquids can be used instead common organic solvents. New monitoring strategies, including continuous and on-line flow injection procedures, are also relevant to green methodologies and are intimately related to electrochemical detection. Also, the synthesis and application of new materials to be used as electrode modifiers, including polymers, composites, alloys, biomaterials, nanomaterials and synthetic receptors, following environmentally friendly procedures, are interesting trends in this field.

All these aspects will be considered briefly in the following sections, with some illustrative examples from the recent literature to emphasize the fundamental principles.

8.2 Green Stripping Voltammetric Methods for Trace Analysis of Metal and Organic Pollutants

Stripping analysis can be considered as a key tool to improve one of the most important areas of research and development in the field of green chemistry, as is the detection, measurement and monitoring of chemicals in the environment.[1] For several decades, mercury electrodes were the most popular electrochemical transducers for the stripping voltammetric analysis of trace metals because of the high sensitivity, reproducibility, and surface renewability that they offered.[3] However, because of the toxicity of mercury, alternative electrode materials have been investigated. The relatively recently introduced bismuth electrodes offer a very attractive alternative. They have demonstrated to display well-defined, undistorted and highly reproducible responses, favourable resolution of neighbouring peaks, high hydrogen evolution, and good signal-to-background ratios, characteristics all comparable to those of mercury electrodes. This attractive stripping behaviour of bismuth electrodes can be attributed to the ability of bismuth to form homogeneous multicomponent alloys with heavy metals. Bismuth film electrodes are most commonly prepared by plating bismuth ions on to a carbon substrate. However, in some cases, this method is not applicable for the *in situ* and on-line trace metal measurement because it introduces additional bismuth ions.

8.2.1 Determination of Trace Metal Ions with Bismuth Film Electrodes

An illustrative example of application of bismuth film electrodes is the anodic square-wave stripping voltammetric determination of Tl(I) trace levels in non-deoxygenated solutions. A rotating-disc bismuth film electrode plated *in situ* was used for this purpose. A 10.8 nM detection limit was achieved, and the relative standard deviation (RSD) (n = 15) for 0.1 µM Tl(I) was 0.2%, for a 120 s accumulation time.[4] Bismuth film microelectrodes (BiFMEs) were also employed for the adsorptive cathodic stripping voltammetric (AdCSV) determination of trace Co(II). A comparison of the stripping performance between bismuth and mercury film microelectrodes revealed a distinct practical advantage of the BiFME.[5] Using square-wave voltammetric detection, a detection limit of 70 ng L^{-1} was achieved as well as excellent reproducibility

Green Electroanalysis

with 2.4% RSD at the 1 μg L^{-1} level (n = 10) with only 2 min preconcentration time in the presence of dissolved oxygen. In addition, the usefulness of the *ex situ* prepared BiFME for both anodic stripping voltammetry of Cd(II) and Pb(II) and AdCSV of Co(II) and Ni(II) was demonstrated. Furthermore, a screen-printed microband electrode prepared by *ex situ* bismuth deposition was recently reported for the sensitive determination of cadmium in non-deaerated and unstirred solutions. A detection limit of 1.3 μg L^{-1} was obtained by using square-wave anodic stripping voltammetry and a deposition time of 120 s. The method was successfully applied to a non-treated river water sample.[6]

8.2.2 Determination of Organic Compounds with Bismuth Electrodes

The wide operational potential window of the bismuth electrode from −200 mV to −1400 mV (*vs* SCE) makes it very convenient for the sensitive adsorptive stripping voltammetric (AdSV) determination of organic compounds. For example, the drug daunomycin can be determined in the nM concentration range by accumulation at −0.65 V (*vs* SCE) for 2–10 min and further recording of the SW voltammogram between −0.65 and −0.25 V. The voltammetric signal for this compound at the bismuth electrode is higher and better shaped than that obtained at gold electrodes. Nevertheless, the electrode must be regenerated between two consecutive measurements by polishing on damp paper or by stirring in 96% ethanol for 5 min.[7] The pesticide methyl parathion was also determined using a bismuth film electrode prepared by *ex situ* bismuth deposition on to a glassy carbon electrode. In this case, the sufficiently wide negative potential window available made the bismuth film electrode suitable for application in cathodic electrochemical detection, SWV between −0.3 and −0.9 V *vs* SCE being used for this purpose. The electrode exhibited similar behaviour to that of a mercury electrode, or even better. The cathodic voltammetric response was proportional to the concentration of methyl parathion from 3.0 to 100 ng mL^{-1} with a detection limit (3 s$_b$) of 1.2 ng mL^{-1}. Due to the easy preparation and regeneration of bismuth film electrodes, together with their good reproducibility and stability, the combination of these electrodes and cathodic voltammetry opens up new opportunities for fast, simple, and sensitive analysis of organophosphorous compounds.[8]

8.2.3 Stripping Voltammetry at Other Modified Electrodes

Other environmentally friendly materials have also been used as electrode modifiers for the development of stripping methods for metal traces and organic compounds. This is the case for β-cyclodextrin (β-CD), a versatile oligosaccharide acting as a host molecule for a wide variety of possible guests. Different CD derivatives have been employed for this purpose. For example, a highly sensitive and reproducible lead sensor based on a CD-modified gold electrode was described. A self-assembled monolayer (SAM) of thiolated

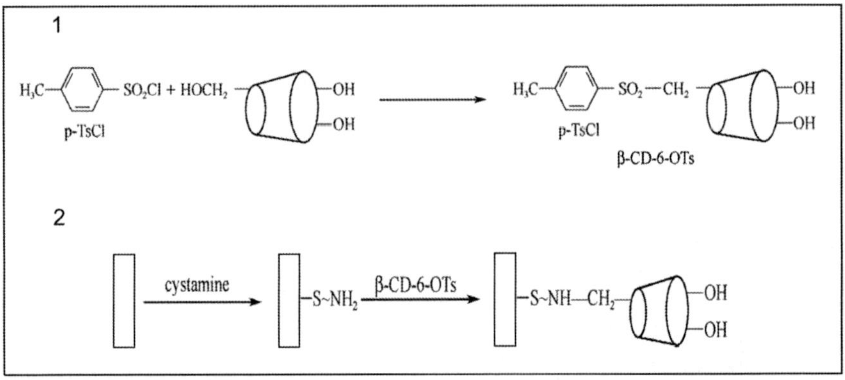

Figure 8.1 Synthesis reaction of β-CD-6-OTs (1) and fabrication of the β-CD-modified electrode (2). (Adapted from Y.Y. Li et al., Biomed. Environ. Sci., 2008, **21**, 479, with kind permission of Elsevier Ltd.)

β-cyclodextrin (6-(2-mercapto-ethylamino)-6-deoxy-β-cyclodextrin (MEA-β-CD)) was prepared on the gold electrode to construct a mercury-free sensor for Pb^{2+} with a linear response over the 1.7×10^{-8}–9.3×10^{-7} M concentration range and a detection limit of 7.1×10^{-9} M.[9] Similar configurations were also used for the determination of organic compounds. An example is the development of a sensor to detect azobenzene, a toxic industrial pollutant, by using a gold electrode modified with a 6-*O*-toluenesulfonyl-β-CD (β-CD-6-OTs) layer. The inclusion complex formed at the modified electrode allowed determination of azobenzene in solution by square-wave voltammetry with a detection limit of 1.0×10^{-10} M. An additional advantage is the reusability of the β-CD-modified electrode. When azobenzene on the β-CD layer is irradiated by UV light (360 nm) in the stirred solution, it is detached from the β-CD layer.[10] The sensor design, including the synthesis reaction of β-CD-6-OTs, is schematized in Figure 8.1.

8.3 Electrochemical Sensors as Tools for Green Analytical Chemistry

Among the 12 principles of green chemistry, principle no. 11 refers to the necessity for real-time analysis to prevent pollution.[1] As stated above, the detection of toxic substances in the environment, at the same time avoiding or minimizing the production of residues in the analytical process, is a general objective for green analytical chemistry. In 2007, Professor C.M.A. Brett[11] in his article 'Novel sensor devices and monitoring strategies for green and sustainable chemistry processes' considered the application of green chemistry principles to the development of sensors. In particular, chemical sensors are needed for real-time in-process monitoring and control in order to avoid the formation of hazardous substances. On-line monitoring also allows for continuous optimization of the efficient use of reagents, and permits determination of the composition of waste and effluents and their variation over time.

8.3.1 Electrochemical Detection in Flow Injection Analysis and Other Injection Techniques

Among other advantages, flow analysis permits the automation and miniaturization of analytical methods and also the use of more environmentally friendly procedures.[12] Some decades ago, *flow injection analysis* (FIA) was recognized as an efficient approach to improve the response time in analytical experiments. In this well-known technique, a small sample volume is injected into a carrier stream where it is diluted by dispersion in the course of an imposed flow, and may be processed on-line in this way by a detector. Numerous methods using FIA coupled to *electrochemical detection* (ED) were developed in these years because of their remarkable sensitivity, inherent miniaturization, independence of sample turbidity and low cost. In the FIA system, convection leads to an increase in sensitivity at the detector, a decrease in the detection limit and an increase in reproducibility.[13] These characteristics convert FIA methods into attractive tools able to provide green methodologies. For example, various ED-flow analysis procedures aimed to the determination of environmentally important organic compounds have been described in the literature. Among these, phenolic compounds constitute one of the most significant groups.

As well as natural phenolics and polyphenolic derivatives, other synthetic phenols widely employed in many industrial applications can be detected in water samples. Various FIA methods have been reported for the determination of total phenol concentrations in order to gain quantitative information about the degree of pollution from these compounds. An electrochemical flow injection method for the determination of phenols in wastewater was developed by coupling the oxidation of phenols at a platinum wire electrode with the reduction of MnO_4^- at another equivalent electrode to enable biamperometric detection with a difference on applied potential of 0 V.[14] A detection limit of 4.0×10^{-7} M was achieved using this methodology. In comparison with the 4-aminoantipyrine (4-AAP) standard method, this electrochemical alternative does not require the use of hazardous chemicals and, consequently, does not generate toxic wastes. Furthermore, it is simple, economic and rapid, with a sample throughput of $180\,h^{-1}$.

The *chemical oxygen demand* (COD) is an index widely used to assess the amount of oxygen required to oxidize totally the oxidizable material contained in a water sample. In the classical determination, the method involves the use and generation of large amounts of extremely hazardous wastes. ED is probably the most environmentally friendly alternative for COD measurement. The quantification of COD using FIA with amperometric detection can be performed directly, with no need for sample treatment, and the reported methods are characterized by short analysis times, simplicity, low environmental impact, a limited reagent consumption and easy automation. For example, an F-doped PbO_2 modified electrode was used for COD monitoring in wastewaters by measuring the corresponding anodic current response which was proportional to the COD value. Under the optimized experimental conditions, the linear range was 100–1200 mg L^{-1}.[15] Another electrochemical method involved the use of a $Ti/TiO_2/PbO_2$ electrode for the photoelectrosynergistic catalysis oxidation of organics. A FIA system was utilized, the proposed electrode functioning as an amperometric detector to determine COD by applying a detection potential of $+1.35$ V under UV illumination.[16] Furthermore, a simple, environmentally friendly, continuous flow method was also developed making use of a boron-doped diamond electrode (BDD) as the detector electrode. A range of linearity between 2 and 175 mg L^{-1} and a detection limit of 1 mg L^{-1} were obtained.[17]

Sequential injection analysis (SIA) is based on the formation of well-defined zones of samples and reagents, which interpenetrate as a result of combined axial and radial dispersion. Between the aspiration of small volumes of sample and reagents, the propulsion device is stopped; the flow direction is then reversed and the zones in the holding coil are directed to the detector. The reaction products are formed in well-defined areas of concentration gradient and the transitory signals generated provide reproducible analytical results.[18] The use of electrochemical detectors in SIA systems possesses some advantageous characteristics. The sensitivity of the methods is not dependent on the flow-through cell, and it is possible and relatively easy to carry out simultaneous determinations using arrays of electrochemical sensors. SIA is very suitable for implementing stripping technique applications such as the determination of heavy metals in different matrices by anodic stripping voltammetry. Furthermore, the well-known medium-exchange methodology can be easily adapted to SIA systems yielding selective determinations.

SIA methods profiting from the green advantages inherent to the use of bismuth electrodes as detectors have been developed. For example, a sequential injection lab-on-valve (LOV) system was reported for the determination of cadmium by anodic stripping voltammetry at a bismuth film electrode.[19] A miniaturized electrochemical flow cell was fabricated for the LOV system, using a Nafion-coated bismuth film electrode as the working electrode (Figure 8.2). The analyte was electrodeposited on the electrode surface from a bismuth solution and measured from the subsequent stripping scan. Under optimal conditions, the system responded linearly to cadmium concentration over the 2.0–100.0 µg L^{-1} range, with a detection limit of 0.88 µg L^{-1}. The methodology was applied to the analysis of trace cadmium in environmental water samples with good results. Another related approach involved coupling of a cost-effective *sequential injection monosegmented flow analysis* (SI-MSFA) system with anodic stripping voltammetry for the determination of Cd(II) and Pb(II). The bismuth film electrode was employed for accumulative preconcentration of the metals at a potential of -1.10 V *vs* Ag/AgCl for 90 s. The SI-MSFA device provided a convenient way for the preparation of a homogeneous solution zone containing both the sample in an acetate buffer electrolyte solution and Bi(III) solution for *in situ* plating of the bismuth film electrode, ready for ASV measurement at a

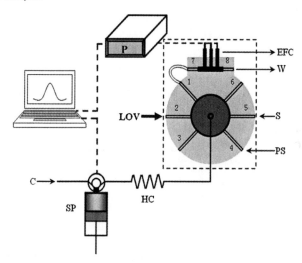

Figure 8.2 Configuration of the SI–LOV manifold used for the determination of cadmium by ASV. C, carrier; EFC, electrochemical flow cell; HC, holding coil; P, potentiostat; PS, plating solution; S, sample; SP, syringe pump; W, waste. (From Y. Wang et al., Talanta, 2010, **80**, 1959, with kind permission of Elsevier BV)

flow-through thin-layer electrochemical cell. Linear calibration graphs between 10 and 100 µg L^{-1} Cd(II) and Pb(II) were obtained with detection limits of 1.4 and 6.9 µg L^{-1}, respectively. The system was successfully applied for the analysis of water samples collected from a zinc mining draining pond.[20]

The usefulness of polymer-coated bismuth film electrodes was also assessed for the simultaneous on-line determination of Cd(II), Pb(II) and Zn(II) by square-wave anodic stripping voltammetry in the SIA mode.[21] The polymeric Nafion coating was initially plated on a glassy carbon electrode that formed part of the flow-through electrochemical cell. All the steps concerning the bismuth layer generation, analyte preconcentration, voltammetric measurement and electrode cleaning were performed on-line. A sample volume of 1.2 mL was used, and the limits of detection achieved were 2 µg L^{-1} for Cd(II) and Pb(II) and 6 µg L^{-1} for Zn(II).

Batch injection analysis (BIA) was developed in 1991.[22] It involves the injection of a small volume of analyte sample by an automatic micropipette directly above the sensing surface of the detector, which is immersed in a large volume of blank solution, and records a transient signal during the flow of sample over the detector surface. The main characteristics offered by this technique are short response time, essentially zero dispersion, but a limited possibility of sample on-line processing. Various examples of green analytical methods involving the use of BIA with ED have been described in the literature. For example, a glassy carbon electrode modified with a porphyrin film formed by the [Co(TPyP){Ru(bipy)$_2$Cl}$_4$](TFMS)$_5$(H$_2$O)] complex was applied to the analysis of sodium metabisulfite in pharmaceuticals.[23] The sensor was rapidly and easily prepared by drop-casting a microlitre volume of a diluted methanolic solution of the complex on to the electrode surface. The combination of the amperometric sensor with the BIA technique led to results that combine good repeatability of the current responses, wide linear dynamic range, high sensitivity and low limits of detection. The excellent reproducibility of the analytical signals obtained can be seen in Figure 8.3, which shows the current–time recordings obtained using this methodology. These results compared well with those reported by applying a polarographic method, with the important advantage of substituting a more environmentally friendly modified electrode in place of the toxic mercury electrode.

Another example of coupling of BIA with ED is the determination of the herbicide paraquat using square-wave voltammetry upon sample injection. The herbicide could be detected at $\mu g\,L^{-1}$ levels with small injection volumes ($<100\,\mu L$) and with a measurement time less than 2 s.[24] Batch injection was also used for stripping analysis. In BIA stripping voltammetry, the electrodeposition or adsorption occurs during injection, after which, a potential scan is applied to oxidize or reduce the accumulated species. For example, fibrinogen-coated bismuth films deposited on carbon paste electrodes (Fbg-BiFEs) were employed for the simultaneous determination of trace amounts of lead and cadmium using square-wave anodic stripping voltammetry combined with a batch injection analysis technique (BIA-SWASV). The BIA system provided the possibility of rapid and simple electroanalysis of microlitre samples. Its use in combination with Fbg-BiFEs allowed the determination of metals at trace levels in environmental samples, because the surfactant interference observed

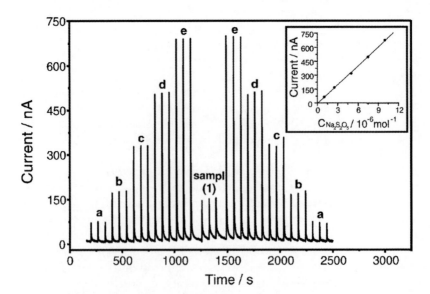

Figure 8.3 BIA results obtained with a glassy carbon electrode modified with a CoTRP film for the determination of sodium metabisulfite. (From M.S.M. Quintino et al., *Talanta*, 2006, **68**, 1281, with kind permission of Elsevier BV)

with stripping analysis at bismuth electrodes could be minimized as a consequence of the fibrinogen coating. The resulting limits of detection were 0.2 µg L^{-1} and 0.1 µg L^{-1} for Cd and Pb, respectively.[25]

8.3.2 Microsystems

The miniaturization of analytical systems has constituted a widely studied field of research in recent decades, with the ultimate objective of achieving a 'lab-on-a-chip' or micro total analysis systems (µ-TAS) in which miniaturization affects the overall analytical process.[26] An important characteristic of these systems, related to the principles of green chemistry, is the extremely low volumes of sample and reagents, usually of the order of nL or pL, that are needed. This means much smaller consumption of chemicals than in FIA, as well as the possibility of all analysis steps being performed using microfluidic arrangements built on the same device (pretreatment, separation and detection integrated). These systems have demonstrated to be robust, probably disposable, and requiring only a minimum of direct operator intervention, circumstances that provide advantages in relation to the cost and analytical throughput although they have some disadvantages regarding detection systems.[26] Because of its inherent miniaturization with no losses of sensitivity, high compatibility with microfabrication techniques, and independent responses from sample turbidity, ED constitutes an important and effective alternative. Different microsystem designs have been developed in which analytical configurations partially or totally integrate the electrochemical cell in the separation system.

Microfabrication and screen-printing are interesting techniques used for the fabrication of planar electrodes to be used as electrochemical sensors. For example, in the area of *in situ* determination of heavy metals, bismuth electrodes along with integrated Ag/AgCl reference and gold counter electrodes are a good candidate to overcome the limitations of previous designs. Furthermore, the incorporation of the on-chip electrochemical sensors with microfluidic components makes it possible to fabricate µ-TAS which provide platforms for chemical and biological analysis in a miniaturized format.[27] Polymer substrates such as cyclic olefin copolymer (COC) have also been extensively used for labs-on-a-chip instead of the traditional silicon and glass substrates because of their unique properties of biocompatibility, high optical transparency and very low cost.

A straightforward example of these strategies is the development of an environmentally friendly, disposable, heavy metal ion sensor with on-chip planar bismuth working electrode and microfluidic channels, which was used for the determination of Pb(II) and Cd(II). The sensor aimed for very low cost and mass production, small analyte consumption and waste generation, fast sensing time, and ease of use. Moreover, it was also suited for fast *in situ* environmental monitoring (Cd(II) measurements in soil porewater and groundwater), and on-line biological and clinical measurement (cell culture

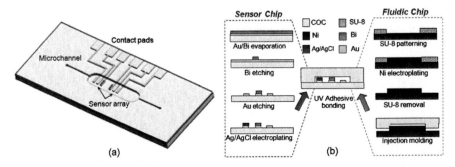

Figure 8.4 (a) Scheme of the disposable on-chip heavy metal sensor array using microfabricated bismuth electrodes with microfluidic channels. (b) Summary of the fabrication processes. (From Z. Zou et al., Sens. Actuators, B, 2008, **134**, 18, with kind permission of Elsevier S.A.)

media).[27] A scheme of the miniaturized sensor chip is shown in Figure 8.4. The detection and quantification of Pb(II) and Cd(II) were statically performed using anodic stripping voltammetry inside the microchannels, in the Pb(II) concentration range of 25–400 ppb ($R^2 = 0.991$) with a detection limit of 8 ppb for 60 s deposition, and in the Cd(II) concentration range of 28–280 ppb ($R^2 = 0.986$) with a detection limit of 9.3 ppb for 90 s deposition.

Capillary electrophoresis (CE) was one of the first separation techniques applied in μ-TAS systems. The use of microchips in CE reduces the time of analysis to microseconds, and high separation efficiencies are reached. It should be noted that the very important conflict between the high voltages employed for the separation and the detection potential is not applicable at the micro scale. Among all possible configurations, the end-channel one is the most appropriate because of its simplicity and facility for electrode replacement, although alignment of the electrode with the capillary output is crucial.[28] In a seminal work published in 2000, J. Wang's group described a miniaturized analytical system for separating and detecting toxic phenolic compounds, based on the coupling of a micromachined CE chip with a thick-film amperometric detector. The integrated microsystem offered a rapid (4 min) simultaneous measurement of seven priority chlorophenols as well as negligible waste production, then meeting the requirements of green analytical chemistry.[29] Subsequently, a microchip CE with a BDD electrochemical detector was applied to analyse aromatic amines.[30] The diamond electrode gave favourable analytical performance, including low noise levels, high peak resolution, enhanced sensitivity and improved resistance to electrode passivation.

The use of nanomaterials as electrode modifiers has demonstrably enhanced the capabilities of electrochemical sensing. Amperometric detection in CE making use of carbon nanotube (CNT)-modified electrodes allows this detection to be carried out at significantly lower operating potentials yielding substantially enhanced signal-to-noise ratios as well as a high resistance to surface fouling and hence enhanced reproducibility. The broad and significant catalytic activity exhibited by CNT-based CE detectors suggests great promise

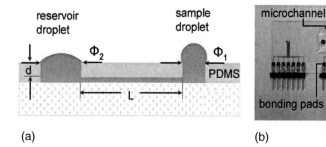

Figure 8.5 (a) Schematic cross-sectional view of the polydimethylsiloxane (PDMS)–glass hybrid microfluidic system used in passive pumping experiments. d, depth of the entry and revervoir ports; L, channel length; $\varnothing_1$ and $\varnothing_2$, diameters of the sample and reservoir ports, respectively. (b) Glass slide with three electrochemical cells patterned on to its surface and a microchannel in the middle of the slide above the central cell. (Adapted from I.J. Chen et al., Anal. Chem., 2009, **81**, 9955, with kind permission of American Chemical Society)

for a wide range of bioanalytical and environmental applications.[31] Illustrative examples are the simultaneous separation and sensitive detection of catecholamines and their O-methoxylated metabolites in mouse brain homogenate,[32] and the separation and detection of phenolic pollutants with limits of detection ranging between 9 nM for phenol and 24 nM for 2,4-dichlorophenol, using a CNT/PMMA composite electrode as amperometric detector in microchip CE.[33]

Recently, a lab-on-chip analysis system without external pump and valves and integrated with an in-line electrochemical detector was reported.[34] In this system, surface energy in small droplets was used to drive samples through microchannels. A so-called 'passive pumping device', in which the sample fluid is spontaneously driven through a solution-filled microchannel with liquid droplets on its entry (sample) and exit (reservoir) ports, was constructed. Figure 8.5 shows as the flow system is integrated with microfabricated electrochemical sensors inside the microfluidic channel. The whole device can be considered as a lab-on-a-chip flow injection manifold that does not require an external pump or injection valves, adequate for the analysis of microlitre volume samples.

8.4 Alternative Solvents

The search for alternative solvents is an important step on the way to greener methods. In this task, the main target should be not just the replacement, but introduction of additional advantages from different properties of these solvents to improve the selectivity, sensitivity, and reliability of analytical methods as well to reduce analysis time.

8.4.1 Ionic Liquids

Ionic liquids (IL) have already demonstrated their usefulness in electroanalytical chemistry for both their use as electrolytes and in the preparation of electrodes.[35] They exhibit high conductivity, large electrochemical windows, excellent thermal and chemical stability, and negligible evaporation. Regarding the fabrication of electrochemical sensors, a significant number of papers has appeared recently in the literature describing the improvements and advantages arising by using these compounds as an integral part of the electrode material. For instance, room-temperature ionic liquids (RTILs) were employed as an efficient binder in place of nonconductive organic binders for the preparation of carbon ionic liquid electrodes (CILEs), which were regarded as a new kind of chemically modified electrodes with the advantages of high conductivity, fast electron transfer rate and low overpotential for biomolecules.[36] IL-modified electrodes have been used for the detection of various electroactive species.[37,38] For example, electrochemical methods for the determination of the nucleoside guanosine, which is important in various biological processes, generally require relatively high potentials to accomplish the oxidation reaction, which limits the sensitivity and selectivity. An *N*-butylpyridinium hexafluorophosphate (BPPF6)-modified carbon paste electrode was proposed for this purpose.[36] Due to the π–π interaction of pyridinium ion with graphite and also to the layer of cationic pyridinium film in the electrode, this exhibited ion-exchange properties and adsorptive ability. A detection limit of 0.26 μM guanosine was achieved and the electrochemical sensor was applied with good results to the analysis of urine samples.

Another interesting configuration involves the electrodeposition of gold nanoparticles on a gel composed of an IL and multiwalled CNTs (MWCNTs). The resulting composite was used to construct a sensor for the determination of guanine and adenine, in which the components of the composite not only efficiently promoted the electron transfer between the analytes and electrode surface but also enhanced the accumulation efficiency.[39] Guanine and adenine exhibited well-separated and well-defined oxidation peaks with detection limits down to nanomole level. In addition, the composite film coated electrode exhibited good reproducibility, long-term stability and simplicity of operation, and it was employed for the simultaneous detection of guanine and adenine in various real samples. A green electrochemical methodology in a supporting electrolyte-free solution using a simple polyelectrolyte-functionalized ionic liquid (PFIL)-modified electrode demonstrated its applicability for amperometric detection in a flow system and HPLC.[40]

Regarding potentiometric sensors, ILs have been used as a membrane component of ion-selective electrodes showing enhanced responses towards hydrophilic anions, increasing the dielectric constant of the membrane.[41] They have also been used in the design of all-solid-state miniaturized reference electrodes suitable for flow-through analysis.[42]

On the other hand, the application of ILs as media for electrochemical synthesis is expected to grow steadily since they offer wider potential windows than some other certain solvents and microemulsions. For example, in polymer science, ILs have been used as polymerization media in several processes, as components of polymeric matrices (such as polymer gels), as templates for porous polymers and as novel electrolytes for electrochemical polymerizations.[43] The use of ILs as solvents in polymerization instead of classical organic solvents not only offers some general advantages such as low volatility and nonflammability, but also markedly affects polymerization rate and degree. ILs are also ideal supporting electrolytes for the electrochemical generation of conjugated polymers because of their excellent oxidative and reductive stability, which allows access to potentials that cannot be provided in the smaller electrochemical windows of molecular solvent/electrolyte systems. For example, polypyrrole electropolymerized in pure ILs was demonstrated to be more electroactive (when swollen) and had a smoother morphology than similar films grown in conventional molecular solvents/electrolyte systems.[44]

8.4.2 Supercritical Fluids

The use of supercritical fluids (SCFs) is often highlighted as an important strategy within green chemistry to replace volatile organic compounds (VOCs), and to enable new, clean technologies. However, SCFs have rarely been used in electrochemical experiments because of the low dielectric constant of these solvents, which severely limits the solubility of salts, and the resulting high resistivity of the supercritical solution.[45] Among SCFs, supercritical carbon dioxide is non-toxic, easy to purify, and relatively inert.[46] Electrochemical sensors have been developed to study the voltammetric behaviour of various compounds for their potential amenability to ED coupled to supercritical fluid separations. For example, a device that combined a working and a counter electrode connected to each other via a poly(ethylene oxide) film containing electrolyte forming a self-contained electrochemical cell was used to inspect the voltammetry of p-benzoquinone and anthracene in carbon dioxide-based fluids.[47] In a chromatographic application, a self-contained electrochemical cell consisting of a working and a quasi-reference electrode coated with a thin film of a conductive polymer was also used as a postcolumn detector for SFC-ED of ferrocene, anthracene, p-benzoquinone and hydroquinone mixtures after their elution with unmodified, acetonitrile-modified or methanol-modified carbon dioxide.[48]

In recent years, SCFs have become important tools for materials processing. The adverse effects of the residual solvents, from both processing and environmental standpoints, have led to alternative methods.[49] Supercritical carbon dioxide and water are extensively being used in the preparation of a great variety of nanomaterials. For example, SnO_2/MWCNT nanocomposites were prepared via oxidation of $SnCl_2$ in a supercritical carbon dioxide–methanol mixture containing MWCNTs. SnO_2 nanoparticles 3–5 nm in size were uniformly and tightly decorated on the MWCNTs. Among other characteristics, the SnO_2/MWCNTs composites exhibited extremely high efficiency for detecting H_2S.[50] Recently a simple and green chemistry approach based on organic-ligand assistance to achieve the shape control and self-assembly of ceria particles under supercritical conditions has been proposed.[51] A simple strategy for the synthesis of metal oxide nanocrystals in the organic-ligand-assisted SCF technique is shown in Figure 8.6.

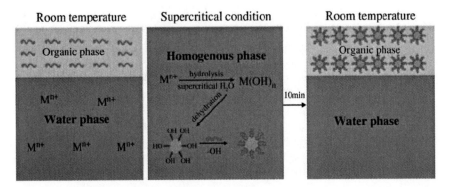

Figure 8.6 Strategy for the synthesis of metal oxide nanocrystals in the organic-ligand assisted SCF technique. (From K. Bryappa *et al.*, *Adv. Drug Del. Rev.*, 2008, **60**, 299, with kind permission of Elsevier BV)

Furthermore, CNTs can also be synthesized in SCFs.[52] Carbon monoxide was employed as the carbon source as it was successfully used as a gaseous reactant by the HiPCO (high pressure CO) process. The high concentration of supercritical carbon monoxide used in the reactor (working at 750 °C) results in a high yield of CNTs. The diameters of the MWCNTs synthesized ranged between 10 and 20 nm with lengths of several tens of micrometres.

The electrochemical polymerization of pyrrole in a supercritical carbon dioxide in water (C/W) emulsion in the presence of a surfactant was also reported. Black polypyrrole films were formed on platinum electrodes, whose conductivity was comparable to non-oriented polypyrrole prepared in conventional solvents.[53] Although supercritical carbon dioxide itself is a non-polar solvent and immiscible with water, the addition of a surfactant to the water allows electrochemical polymerization in the C/W emulsion.

8.5 New Electrode Materials

Investigation of new electrode materials offering enhanced electrochemical performance for applications in the development of sensors, energy storage, electrocatalysis, *etc.* has been a regular trend in recent decades. Green methods for the preparation of nanomaterials are of particular interest in this context.

8.5.1 Metal Nanoparticles

An interesting example is the straightforward and rapid approach proposed for the fabrication of gold nanostructured films. A gold electrode was first oxidized at a high potential of 10 V and β-D-glucose was chosen as a non-toxic reducing agent to reduce the gold oxide. This process produced a fractal gold nanofilm on top of the gold electrode in a completely green chemical way. The electrode so formed served as a sensitive enzyme-free sensor for the detection of glucose at a potential of $-0.15\,\text{V}$ with no interference from ascorbic acid, uric acid or acetylaminophenol.[54] Another example is the preparation of a gold nanoparticles–chitosan nanocomposite gel for the immobilization of K562 leukaemia cells. Gold nanoparticles were produced *in situ* in a solution of chitosan (a non-toxic natural polysaccharide). The methods for preparation of the gel and immobilization of cells were simple and environmentally friendly. The nanocomposite gel showed improved immobilization capacity for cells and good biocompatibility for preserving the activity of immobilized living cells. Further modification of a glassy carbon electrode with the gel allowed an irreversible voltammetric response to be obtained as well as an increased electron transfer resistance with a good correlation to the logarithmic value of concentration and a LOD of 7.71×10^2 cells mL^{-1}. The whole process did not introduce any environmental toxicity or biological hazards, according to the fundamental principles of green chemistry.[55] Using a related approach, a green method for the preparation of platinum nanoparticles by reduction of H_2PtCl_6, using nanocrystalline cellulose from cotton as the reducing agent, has also been recently described.[56]

8.5.2 Hybrid Nanocomposites

A facile and green one-step procedure that allows the aqueous synthesis of a nanocomposite composed of single-walled carbon nanotubes (SWNTs) and gold nanoparticles was reported.[57] In this procedure, chitosan served as a polymer for wrapping the SWNTs, allowing dispersion of SWNTs in aqueous solution. Furthermore, it also functioned as a reducing agent for gold cations and a stabilizing agent for the resulting gold nanoparticles (GNPs). The synthesized GNPs are shown by TEM to coat the side walls of CNTs. The further modification of a gold electrode with a nanocomposite film showed promotion of direct electron transfer for immobilized microperoxidase-11 (MP-11), which retained its bioelectrocatalytic activity for the reduction of oxygen. New nanocomposite films based on chitosan and bacterial cellulose nanofibrils were also prepared by a fully green procedure. The films were highly transparent, flexible and displayed better mechanical properties than the corresponding unfilled chitosan films. These renewable nanocomposite materials also had reasonable thermal stability and low O_2 permeability.[58]

8.5.3 Oxide Nanoparticles

Non-aqueous and surfactant-free sol–gel strategies represent versatile synthesis approaches to many classes of inorganic nanomaterials for electrochemical applications.[59] In comparison to solvothermal and hot injection methods, the development of synthesis strategies for the fabrication of tailor-made materials at lower temperatures and without the use of any surfactant permits the reduction of energy and raw material consumption as well as of waste, toxic and side products. For example, titanium dioxide nanoparticles in the 4–8 nm size range were easily prepared by non-aqueous routes using $TiCl_4$ and benzyl alcohol.[60] These nanoparticles exhibit enhanced electrochemical properties when arranged into mesoporous films.[61] Among such nanomaterials, Fe_3O_4 nanoparticles exhibit unique characteristics, such as half-ferromagnetism, large specific area, non-toxicity and biocompatibility, which paves the way for a variety of potential applications. Their synthesis has been implemented using various techniques including arc-discharge, solution-based non-hydrolytic, sol–gel, gas-phase chemical reaction and inverse microemulsion polymerization.[62] However, large-scale and low-cost synthesis of Fe_3O_4 nanoparticles is a challenge, especially if a good size and narrow size distribution is needed. Recently, a simple, low-cost and environmental friendly method based on arc-discharge submerging in deionized water has been described. The resulting Fe_3O_4 nanoparticles had a uniform spherical shape and their size could be controlled by adjusting the processing parameters. Since no vacuum system was used, the synthesizing process was greatly simplified. In addition, only cheap deionized water and industrial iron bar were used and no pollutant or harmful by-products were found in the synthesis process.

8.5.4 Polymers

Poly(3,4-ethylenedioxythiophene) (PEDOT) is a stable conducting polymer with a high electrical conductivity. However, its application for the preparation of electrochemical biosensors is limited because polymerization of the monomer EDOT is inhibited by the interaction of water molecules with thienyl cation radical, the polymerization intermediate.[63] However, the polymer can be prepared using an aqueous solution of the environmentally friendly amino acid-based surfactant sodium *N*-lauroylsarcosinate (SLS). This compound is a mild biosurfactant with good biocompatibility, low toxicity, good solubilization, and biodegradability. The moderate interactions between a neutral SLS-aqueous micellar solution and EDOT monomer (see Figure 8.7) led to the decreased onset oxidation potential of EDOT.[63]

8.5.5 Solid Amalgams

Solid amalgam electrodes (SAE) represent a suitable alternative to mercury electrodes because of their similar electrochemical performance and the non-toxicity of the amalgam material. This makes SAE compatible with the concept of green analytical chemistry. Among other advantages, their mechanical robustness allows their application in FIA or HPLC as electrochemical detectors.[64] For example, silver SAE were used for the voltammetric and amperometric determination of nitroquinolines, as a model of environmentally important reducible species.

8.6 Electrochemical Biosensors

So far there are few examples of the preparation of electrochemical biosensors invoking the principles of green analytical chemistry.

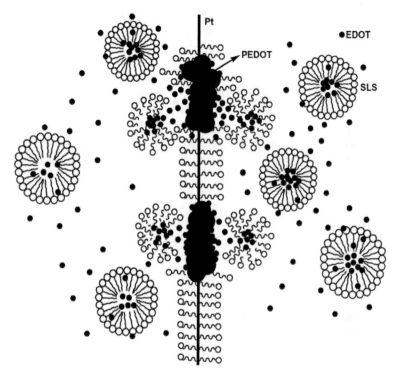

Figure 8.7 Polymerization of EDOT in water–SLS micellar solution on a platinum electrode. (From Y. Wen et al., J. Electroanal. Chem., 2009, **634**, 49, with kind permission of Elsevier Science)

8.6.1 Environmental Applications

As already mentioned, in general, the development of sensors for environmental monitoring is an aspect intimately related with the philosophy of green analytical chemistry. In particular, biosensors for environmental applications have been shown to provide improvements in sensitivity, selectivity and simplicity for the detection of specific compounds or compound classes such as pesticides, hazardous industrial chemicals, toxic metals, and pathogenic bacteria.[65]

Cholinesterase enzymes (AChE, BChE) and urease have been widely used in the design of electrochemical biosensors for the detection of organophosphate and carbamate pesticides.[66] A variety of analytical devices, based on the inhibition of cholinesterase have been proposed. For example, screen-printed electrodes incorporating the enzymes into the carbon-based ink were prepared and use in the determination of chlorfenvinphos and diazinon, pesticides used for the control of insect pests in wool. The BChE-based biosensor was shown to be more sensitive, with LOD 0.5 $\mu g\, g^{-1}$.[67] Screen-printed electrodes chemically modified with 7,7,8,8-tetracyanoquinodimethane (TCNQ) were demonstrated to be useful for the mediated ED of acetylcholinesterase activity.[68] The AChE biosensor allowed detection of carbaryl at a $10\,ng\,g^{-1}$ concentration, which corresponds to the limit value set by the European legislation.

DNA biosensors based on different modes of DNA recognition besides base-pairing hybridization events have been also employed in environmental monitoring, for example as screening devices for the rapid bioanalysis of environmental pollution. The binding of small molecules to DNA immobilized on disposable screen-printed electrodes was measured through the variation of the electrochemical signal of guanine. These biosensors were applied to the rapid detection of genotoxic compounds in soils from polluted areas with advantages with respect to the most common biological tests for ecological/environmental risk assessment.[69]

Ultrasound is considered as a green technology owing to its high efficiency, low instrumental requirements, significantly reduced process time compared with other conventional techniques, and economically viable performance.[70] Ultrasound irradiation has been used in the preparation of electrochemical biosensors to facilitate enzyme immobilization. This is the strategy followed in the so-called sonogel–carbon electrodes, which have been used mainly to detect pesticides in wastewaters. For example, a biosensor for detection of phenols, based on tyrosinase immobilization with alumina sol–gel on a sonogel–carbon transducer, was prepared using high-energy ultrasound applied directly to the precursors. The alumina sol–gel provided a microenvironment for retaining the native structure and activity of the entrapped enzyme and a very low mass transport barrier to the enzyme substrates.[71] Amperometric biosensors based on immobilization of acetylcholinesterase on a sonogel–carbon electrode for detection of organophosphorous compounds were also proposed. The enzyme was immobilized by simple entrapping in Al_2O_3 sol–gel matrix on the sonogel-carbon, or by adsorption and modification via a polymeric membrane and Nafion.[72]

The search for more environmentally acceptable and user-friendly systems for field determination has long been pursued.[73] Various green approaches to field nitrate analysis based on enzyme biosensors have been described. The inherent selectivity and catalytic ability of the enzyme nitrate reductase to convert the nitrate to nitrite can be exploited for the development of an electrochemical nitrate biosensor based on the scheme depicted in Figure 8.8. The enzyme catalyses the reduction of nitrate and the analytical signal comes from the electrochemical reduction of an appropriate mediator on the electrode surface. Various compounds, such as methylviologens[74] or microperoxidase,[75] were used as mediators for this electrocatalytic cycle.

8.6.2 Biosensors Using Ionic Liquids

Configurations based on the use of ILs can be considered as interesting alternatives within this area. ILs have be shown to be suitable media for enzyme catalysis. Enzymes in ILs exhibit high conversion rates and enantioselectivity and, furthermore, their stability has been proved to be increased over organic solvents. All these advantages, in combination with the green, designable properties, make ILs a potential material for enzyme immobilization and therefore for the development of a new generation of enzyme bioelectrodes.

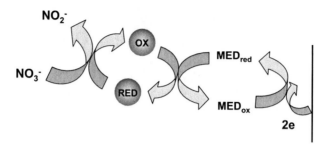

Figure 8.8 Scheme of the functioning of an electrochemical biosensor for nitrate based on nitrate reductase.

RTILs have been used as modifiers by incorporation within biocomposite matrices. For example, N-butylpyridinium hexafluorophosphate (BPPF6), sodium alginate (SA), and graphite were used to construct an electrochemical horseradish peroxidase biosensor for the determination of H_2O_2. The use of BPPF6 instead of silicone oil for the preparation of the composite electrode greatly enhanced the current and allowed long-term stability to be achieved. Entrapping the enzyme into a biocompatible SA–RTIL composite provided an effective means for realizing direct electrochemistry of a variety of enzymes and for fabricating novel biosensors.[76]

ILs were also used in combination with conducting polymers for the preparation of enzyme biosensors. The *in situ* one-step electropolymerization of monomers provided a green method for preparation of bioelectrode materials. For instance, a composite based on polyaniline–ionic liquid–carbon nanofibre (PANI–IL–CNF) was greenly prepared by *in situ* one-step electrochemical polymerization of aniline in the presence of IL and CNF. The fibrillar morphology of the composite was beneficial to the loading of enzymes such as tyrosinase, and thus improved the capacity for immobilization. The combination of the advantages of PANI, IL and CNF with this beneficial morphology for enzyme loading allowed a high sensitivity, wide linear range and excellent stability for the detection of phenols.[77]

Nanostructured materials, and in particular nanoparticles, have been used in recent years as support for biomolecule immobilization.[78] Nowadays, functionalization of nanoparticles is an important trend because of the synergistic effect of the functional groups and the amplification effect associated with the nanoscale dimension.[79] ILs, as nanoparticle modifiers, also play a significant role in this area. For example, enzyme biosensors using magnetic nanoparticle-supported ILs were developed. The approach involved covalent bonding of ILs–silane on magnetic silica nanoparticles and enzyme immobilization on to the magnetic supports via physical adsorption. Lipase was used as a model enzyme, and the functional ILs were employed not only for the formation and stabilization of magnetic nanoparticles, but also as media for lipase catalysis. ILs increase the enzyme activity and, furthermore, their structure can be adjusted to give a higher hydrophobicity that accelerates the enzyme reaction.[80]

RTILs have been also widely used in the preparation of DNA biosensors and for the investigation of redox proteins by direct electrochemistry. For example, active nano-interfaces were designed from gold nanoparticles embedded on 1-butyl-3-methylimidazolium tetrafluroborate ([bmim][BF$_4$]) IL for DNA immobilization. The modified film was applied to the evaluation of DNA damage by electrocatalytic activity toward HCHO oxidation.[81] DNA was also accumulated on the positively charged surface of CILE with BPPF6 as binder. Further myoglobin (Mb) immobilization onto the DNA film by electrostatic interaction formed an Mb/DNA/CILE electrode which exhibited direct electrochemistry of Mb that was reflected by a pair of well-defined, quasi-reversible cyclic voltammetric peaks.[82]

8.6.3 Natural Biopolymers

Recently, natural biopolymers have been widely used for the immobilization of biomolecules. They are characterized by their chemical inertness, versatility and damage resistance, among other characteristics. Chitin and its derivative chitosan are abundant biopolymer products found in the exoskeleton of crustaceans, in fungal cell wall and other biological materials. Their biodegradability, non-toxicity and biocompatibility and, especially, their susceptibility to chemical modification due to reactive amino and hydroxyl functional groups, make them excellent substrates for enzyme immobilization. For example, an environmental friendly biosensor for the determination of adrenaline was prepared using a corn (*Zea mays* L.) homogenate as a source of peroxidase immobilized in chitin chemically cross-linked with carbodiimide and glyoxal. In the presence of hydrogen peroxide, peroxidase catalyses the oxidation of adrenaline to the corresponding *o*-quinone, which is electrochemically reduced at a potential of -0.23 V *vs* Ag/AgCl.[83] Chitosan has been demonstrated to be a good option for keeping live microbial cells on the electrode surface under operational conditions. So, bacterial sensors based on the entrapment of cells in chitosan matrices have been developed and applied, for example, to the determination of different carbohydrates. The oxygen consumption due to the respiratory activity of the microorganisms in the presence of sugars provides the electrochemical signal from oxygen reduction at the electrode surface.[84]

8.6.4 Microsystems-Based Biosensors

The attractive green properties of microsystems have also been used for the development of biosensing assays. A pioneering example was the multienzyme on-chip assay reported by Wang's group in 2001 based on precolumn reactions of glucose oxidase and alcohol dehydrogenase to generate, respectively, hydrogen peroxide and NADH species that were separated on the basis of their different charge and detected at the same electrode.[85] Nowadays, analytical microdevices have wide applications in the preparation of electrochemical biosensors because of their low cost and the minimal consumption of energy, space, samples and reagents, as well as the small volume of wastes and versatility.[86] Recently, a portable and inexpensive amperometric biosensor for rapid salicylate determination in blood was constructed by using an electrochemical microcell fabricated by photolithography to make gold electrodes on a polyester film, and immobilization of the enzyme salicylate hydroxylase on to the working electrode.[86]

On-chip integration of immunological and enzymatic reactions, amperometric detection, and microchip operations makes it possible to perform immunoassays more rapidly, easily, and economically. An illustrative example is a microchip-based electrochemical immunoassay with a surface-functionalized poly-(dimethylsiloxane) (PDMS) channel. The microchip was fabricated by photolithography on a glass substrate, and consisted of a system of three electrodes assembled with the antibody-immobilized PDMS channel by plasma treatment. The internal surface of the PDMS channel was chemically modified with a silane monolayer containing poly(ethyleneglycol) groups, for effective immobilization of antibodies. The small channel dimensions (50 μm) enable a rapid immunoassay by reducing the mass transfer resistance between electrodes and immobilized antibodies.[87]

Protein chip based electrochemical sensors have become essential tools in bioanalysis. These consist of immobilized biomolecules (proteins, peptides, oligonucleotides) spatially arranged on planar substrates that interact with sample analytes for molecular recognition events.[88] Their use has several advantages over conventional methodologies: the very small analyte volumes that can be employed and easy automation, as well as rapidity and possibility of detection of many addressable elements in a single experiment. Protein chip based sensors can be also incorporated into microfluidic components for the development of lab-on-a-chip devices. Figure 8.9 shows a scheme of a low-cost disposable immunosensor array prepared from a screen- printed chip consisting

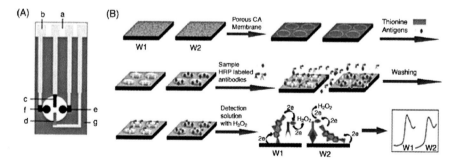

Figure 8.9 Schematic diagrams of (A) a four-electrode screen-printed carbon electrode (SPCE) system and (B) the preparation of an immunosensor array and the simultaneous multianalyte immunoassays (SMIAs) procedure. (a) Nylon sheet; (b) silver ink; (c) graphite auxiliary electrode; (d) Ag/AgCl reference electrode; (e) W1; (f) W2; (g) insulating dielectric. (Adapted from J. Wu et al., Biosens. Bioelectron., 2007, **23**, 114, with kind permission of Elsevier Science)

of two carbon electrodes (W1 and W2), with two different antigens and a redox mediator (thionine) immobilized by means of a cellulose acetate membrane.[89] Two simultaneous competitive immunoreactions using immobilized HRP-labelled antibodies enable fast and simultaneous determination of the antigens.

8.7 Future Trends in Green Electroanalysis

Modern electroanalytical chemistry, like other disciplines, has faced the challenge of investigating new routes fulfilling the requirements of green analytical chemistry. Nowadays, we can say that the researchers in this discipline are on the way to achieving this objective. Current and future investigations on electrochemical sensors and biosensors, the search for alternative green solvents and electrode materials, and the development of electroanalytical on-line methodologies with miniaturized systems making use of the recent advances in microfabrication technologies, offer novel and exciting working strategies closely adhering to the principles of green chemistry.

References

1. P. T. Anastas and J. C. Warner, *Green Chemistry: Theory and Practice*, Oxford University Press, New York, 1998.
2. M. Koel and M. Kaljurand, *Pure Appl. Chem.*, 2006, **78**, 1993.
3. J. Wang, *Electroanalysis*, 2005, **17**, 1341.
4. E. O. Jorge, M. M. M. Neto and M. M. Rocha, *Talanta*, 2007, **72**, 1392.
5. E. A. Hutton, S. B. Hocevar and M. Ogorevc, *Anal. Chim. Acta*, 2005, **537**, 285.
6. O. Zaouak, L. Authier, C. Cugnet, A. Castetbon and M. Potin-Gautier, *Electroanalysis*, 2009, **21**, 689.
7. M. Buckova, P. Grundler and G. U. Flechsig, *Electroanalysis*, 2005, **17**, 440.
8. D. Du, X. Ye, J. Zhang and D. Liu, *Electrochim. Acta*, 2008, **53**, 4478.
9. W. Li, G. Jin, H. Chen and J. Kong, *Talanta*, 2009, **78**, 717.
10. Y. Y. Li, Z. Y. Jiang, J. Q. Chen, H. L. Li and H. Q. Zhang, *Biomed. Environ. Sci.*, 2008, **21**, 479.
11. C. M. A. Brett, *Pure Appl. Chem.*, 2007, **79**, 1969.
12. F. Maya, J. M. Estela and V. Cerdà, *Talanta*, 2010, **81**, 1.
13. E. H. Hansen and M. Miró, *TrAC, Trends Anal. Chem.*, 2007, **26**, 18.
14. C. Zhao, J. F. Song and J. C. Zhang, *Anal. Bioanal. Chem.*, 2002, **374**, 874.
15. J. Li, L. Li, L. Zheng, Y. Xi an, S. Ai and L. Jin, *Anal. Chim. Acta*, 2005, **548**, 199.
16. J. Li, L. Zheng, L. Li, G. Shi, Y. Xian and L. Jin, *Talanta*, 2007, **72**, 1752.
17. H. B. Yu, C. J. Ma, X. Quan, S. Chen and H. M. Zhao, *Environ. Sci. Technol.*, 2009, **43**, 1935.
18. R. Pérez-Olmos, J. C. Soto, N. Zárate, A. N. Araújo and M. C. B. S. M. Montenegro, *Anal. Chim. Acta*, 2005, **554**, 1.

19. Y. Wang, Z. Liu, G. Yao, P. Zhu, X. Hu, Q. Xu and C. Yang, *Talanta*, 2010, **80**, 1959.
20. W. Siriangkhawut, S. Pencharee, K. Grudpan and J. Jakmunee, *Talanta*, 2009, **79**, 1118.
21. G. Kefala and A. Economou, *Anal. Chim. Acta*, 2006, **576**, 283.
22. J. Wang and Z. Taha, *Anal. Chem.*, 1991, **63**, 795.
23. M. S. M. Quintino, K. Araki and L. Angnes, *Talanta*, 2006, **68**, 1281.
24. F. R. Simoes, C. M. P. Vaz and C. M. A. Brett, *Anal. Lett.*, 2007, **40**, 1800.
25. I. Adraoui, M. E. Rhazi and A. Amine, *Anal. Lett.*, 2007, **40**, 349.
26. S. Alegret,*Integrated Analytical Systems*, Elsevier, Amsterdam, 2003.
27. Z. Zou, A. Jang, E. MacKnight, P. M. Wu, J. Do, P. L. Bishop and C. H. Ahn, *Sens. Actuators B*, 2008, **134**, 18.
28. A. Ríos, A. Escarpa, M. C. González and A. G. Crevillén, *TrAC, Trends Anal. Chem.*, 2006, **25**, 467.
29. J. Wang, M. P. Chatrathi and B. Tian, *Anal. Chim. Acta*, 2000, **416**, 9.
30. D. C. Shin, D. A. Tryk, A. Fujishima, A. Muck, G. Chen and J. Wang, *Electrophoresis*, 2004, **25**, 3017.
31. G. Chen, *Talanta*, 2007, **74**, 326.
32. M. Vlckova and M. A. Schwarz, *J. Chromatogr. A*, 2007, **1142**, 214.
33. X. Yao, H. X. Wu, J. Wang, S. Qu and G. Chen, *Chem. Eur. J.*, 2007, **13**, 846.
34. I. J. Chen and E. Lindner, *Anal. Chem.*, 2009, **81**, 9955.
35. S. Fan, F. Xiao, L. Liu, F. Zhao and B. Zeng, *Sens. Actuators B*, 2008, **132**, 34.
36. W. Sun, Y. Duan, Y. Li, H. Gao and K. Jiao, *Talanta*, 2009, **78**, 695.
37. J. B. Zheng, Y. Zhang and P. P. Yang, *Talanta*, 2007, **73**, 920.
38. W. Sun, Y. Z. Li, M. X. Yang, S. F. Liu and K. Jiao, *Electrochem. Commun.*, 2008, **10**, 298.
39. F. Xiao, F. Zhao, J. Li, L. Liu and B. Zeng, *Electrochim. Acta*, 2008, **53**, 7781.
40. Y. Shen, Y. Zhang, X. Qin, H. Guo and A. Ivaska, *Green Chem.*, 2007, **9**, 746.
41. C. Coll, R. H. Labrador, R. M. Manez, J. Soto, F. Sancenon, M. J. Segui and E. Sanchez, *Chem. Commun.*, 2005, **24**, 3033.
42. R. Maminska, A. Dybko and W. Wroblewski, *Sens. Actuators B*, 2006, **115**, 552.
43. J. Lu, F. Yan and J. Texter, *Prog. Polym. Sci.*, 2009, **34**, 431.
44. J. M. Pringle, J. Efthimiadis, P. C. Howlett, J. Efthimiadis, D. R. MacFarlane, A. B. Chaplin, S. B. Hall, D. L. Officer, G. G. Wallace and M. Forsyth, *Polymer*, 2004, **45**, 1447.
45. D. L. Goldfarb and H. R. Corti, *Electrochem. Commun.*, 2000, **2**, 663.
46. P. Munshi and S. Bhaduri, *Curr. Sci.*, 2009, **97**, 63.
47. E. F. Sullenberger, S. F. Dressman and A. C. Michael, *J. Phys. Chem.*, 1994, **98**, 5347.
48. S. F. Dressman and A. C. Michael, *Anal. Chem.*, 1995, **67**, 1339.

49. K. Byrappa, S. Ohara and T. Adschiri, *Adv. Drug Delivery Rev.*, 2008, **60**, 299.
50. J. Zhang, S. Ohara, M. Umetsu, T. Naka, Y. Hatakeyama and T. Adschiri, *Adv. Mater.*, 2007, **19**, 203.
51. G. M. An, N. Na, X. R. Zhang, Z. J. Miao, S. D. Miao, K. L. Ding and Z. M. Liu, *Nanotechnol.*, 2007, **18**, 435707.
52. Z. Li, J. Andzane, D. Erts, J. M. Tobin, K. Wang, M. A. Morris, G. Attard and J. D. Holmes, *Adv. Mater.*, 2007, **19**, 3043.
53. M. Jikei, S. Saitoh, H. Yasuda, H. Itoh, M. Sone, M. Kakimoto and H. Yoshida, *Polymer*, 2006, **47**, 1547.
54. W. Zhao, J. J. Xu, C. G. Shi and H. Y. Chen, *Electrochem. Commun.*, 2006, **8**, 773.
55. L. Ding, C. Hao, Y. Xue and H. Ju, *Biomacromolecules*, 2007, **8**, 1341.
56. K. Benaissi, L. Johnson, D. A. Walsh and W. Thielemans, *Green Chem.*, 2010, **12**, 220.
57. H. J. Jiang, Y. Zhao, H. Yang and D. L. Akins, *Mat. Chem. Phys.*, 2009, **114**, 879.
58. S. C. M. Fernandes, L. Oliveira, C. S. R. Freire, A. J. D. Silvestre, C. P. Neto and J. Desbrieres, *Green Chem.*, 2009, **11**, 2023.
59. I. Bilecka and M. Niederberger, *Electrochim. Acta*, 2010, **26**, 7717.
60. M. Niederberger, M. H. Bartl and G. D. Stucky, *Chem. Mater.*, 2002, **14**, 4364.
61. T. Brezesinski, J. Wang, J. Polleux, B. Dunn and S. H. Tolbert, *J. Am. Chem. Soc.*, 2009, **131**, 1802.
62. K. Yao, Z. Peng and X. Fan, *J. Environ. Sci. (China)*, 2009, **21**, 727.
63. Y. P. Wen, J. K. Xu, H. H. He, B. Y. Lu, Y. Z. Li and B. Dong, *J. Electroanal. Chem.*, 2009, **634**, 49.
64. I. Jiranek, K. Peckova, Z. Kralova, J. C. Moreira and J. Barek, *Electrochim. Acta*, 2009, **54**, 1939.
65. K. R. Rogers, *Anal. Chim. Acta*, 2006, **568**, 222.
66. A. Amine, H. Mohammadi, I. Bourais and G. Palleschi, *Biosens. Bioelectron.*, 2006, **21**, 1405.
67. W. A. Collier, M. A. Clear and A. L. Hart, *Biosens. Bioelectron.*, 2002, **17**, 815.
68. M. Del Carlo, V. Del Carlo, M. Mascini, A. Pepe, G. Diletti and G. D. Compagnone, *Food Chem.*, 2004, **84**, 651.
69. G. Bagni, D. Osella, E. Sturchio and M. Mascini, *Anal. Chim. Acta*, 2006, **81**, 573.
70. E. V. Rokhina, P. Lens and J. Vikutyte, *Trends Biotechnol.*, 2009, **27**, 298.
71. H. Zejli, J. L. Hidalgo-Hidalgo de Cisneros, I. Naranjo-Rodriguez, B. Liu, K. R. Temsamani and J. L. Marty, *Anal. Chim. Acta*, 2008, **612**, 198.
72. H. Zejli, J. L. Hidalgo-Hidalgo de Cisneros, I. Naranjo-Rodriguez, B. Liu, K. R. Temsamani and J. L. Marty, *Talanta*, 2008, **77**, 217.
73. R. Desai, M. M. Villalba, N. S. Lawrence and J. Davis, *Electroanalysis*, 2009, **21**, 789.
74. S. Da Silva, D. Shan and S. Cosnier, *Sens. Actuators B*, 2004, **103**, 397.

75. F. Patolski, E. Katz, V. Heleg-Shabtai and I. Willner, *Chem. Eur. J.*, 1998, **4**, 1068.
76. C. Ding, M. Zhang, F. Zhao and S. Zhang, *Anal. Biochem.*, 2008, **378**, 32.
77. J. Zhang, J. Lei, Y. Liu, J. Zhao and H. Ju, *Biosens. Bioelectron.*, 2009, **24**, 1858.
78. A. González-Cortés, P. Yáñez-Sedeño and J. M. Pingarrón, *Electrochim. Acta*, 2008, **53**, 5848.
79. A. S. Dios and M. E. Díaz-García, *Anal. Chim. Acta*, 2010, **666**, 1.
80. Y. Jiang, C. Guo, H. Xia, I. Mahmood, C. Liu and H. Liu, *J. Mol. Catal B: Enzym.*, 2009, **58**, 103.
81. L. Lu, T. Kang, S. Cheng and X. Guo, *Appl. Surf. Sci.*, 2009, **256**, 52.
82. R. Gao and J. Zheng, *Electrochem. Commun.*, 2009, **11**, 1527.
83. D. Brondani, J. Dupont, A. Spinelli and I. Cruz Vieira, *Sens. Actuators B*, 2009, **138**, 236.
84. D. Odaci, S. Timur and A. Telefoncu, *Sens. Actuators B*, 2008, **134**, 89.
85. J. Wang, M. Chatrathi and B. Tian, *Anal. Chem.*, 2001, **73**, 1296.
86. R. F. Carvalhal, D. S. Machado, R. K. Mendes, A. L. J. Almeida, N. H. Moreira, M. H. O. Piazetta, A. L. Gobbi and L. T. Kubota, *Biosens. Bioelectron.*, 2010, **25**, 2200.
87. Y. Jang, S. Y. Oh and J. K. Park, *Enzyme Microb. Technol.*, 2006, **39**, 1122.
88. H. Chen, C. Jiang, C. Yu, S. Zhang, B. Liu and J. Kong, *Biosens. Bioelectron.*, 2009, **24**, 3399.
89. J. Wu, Z. J. Zhang, Z. F. Fu and H. X. Ju, *Biosens. Bioelectron.*, 2007, **23**, 114.

CHAPTER 9
Green Analytical Chemistry in the Determination of Organic Pollutants in the Environment

SANDRA PÉREZ,[1] MARINELLA FARRÉ,[1]
CARLOS GONÇALVES,[2,3] JAUME ACEÑA,[1]
M. F. ALPENDURADA[2,3] AND DAMIÀ BARCELÓ[1,4]

[1] Department of Environmental Chemistry, IDAEA-CSIC, C/. Jordi Girona 18–26, 08034 Barcelona, Spain; [2] IAREN, Instituto da Água da Região Norte, Rua Dr. Eduardo Torres 229, 4450-113, Matosinhos, Portugal; [3] Laboratory of Hydrology, Faculty of Pharmacy, University of Porto, Rua Aníbal Cunha 164, 4050-047 Porto, Portugal; [4] Catalan Institute of Water Research, Catalan Institute for Water Research-ICRA, C/. Emili Grahit 101, Edifici H2O, Parc Científic i Tecnològic de la Universitat de Girona, 17003 Girona, Spain

9.1 Green Analytical Methodologies for the Analysis of Organic Pollutants

Green chemistry in environmental analysis has been discussed in detail in previous publications focused on environmental-friendly sample preparation techniques.[1] Thus, the present chapter deals with other aspects of green environmental analytical chemistry, emphasizing the increasing acceptance these approaches find worldwide, as exemplified by the number of applications that have been produced. Recent innovations, as well as techniques not covered in previous publications, will be discussed; these include green solvents for sample preparation, recent developments in chromatographic separations, improvements in greening the interfaces for mass spectrometric analysis, and biological techniques for sample preparation and detection.

Analytical schemes include different steps, most of which can be separated into two categories, the sample pretreatment and the detection step. Some detectors, such as mass spectrometers, are frequently coupled to a separation instrument including an additional step. Although an ideal green analysis tends to eliminate the preconcentration step, its elimination implies an increase in problems affecting analyte detection. In many cases, therefore, the preconcentration step should not be avoided. A variety of greener methods have been developed to preconcentrate and clean up the samples prior to the detection of the analytes.[2] Tobiszewski *et al.*[1] mention the elimination of the use of organic solvents in sample preparation as an important green approach, which can be achieved using solvent-free extraction on solid adsorbents, extraction with other types of solvents, assisted solvent extraction and miniaturized analytical systems.[1] This effort is needed to counteract the predicted growth of chemical production worldwide of 330% between 2000 and 2050 while population growth is estimated to increase by 47% in the same time period.[1]

Different steps of the global analytical process can make different contributions to greener chemistry. However, we should not forget that many auxiliary tasks, such as rinsing of glassware, dissolution of standards, cleaning of equipment parts, and chromatography mobile phases use organic solvents, and total reduction of solvent usage will not be effective if these tasks are not taken into account.

Regarding the separation and detection techniques, Tobiszewski *et al.*[1] suggest the use of gas chromatography (GC) as preferable to of liquid chromatography, whenever possible. However, currently the speed of ultra-performance liquid chromatography (UPLC) is captivating many analysts, partly because of the types of analytes that tend to be more amenable to liquid separations. A similarly rapid option is available for GC (fast GC) but as yet it has not gained wide acceptance,. It is true, however, that today's techniques use much less solvents and the throughput is maximized, but the number of determinations to perform has also increased enormously. For organic pollutant detection, mass spectrometry (MS) is the technique of choice. Some other detection techniques, Such as spectroscopy, electrochemistry and bioanalytical chemistry, are quite green already. However, new advances in sample introduction for MS are making this technique even greener.

Decontamination of glassware by oxidizing (metallic) agents and elimination of residues of dangerous substances are also important in this context. Fortunately, efforts have been made in recent years, including in our laboratories, to isolate these residues and send them for specialized treatment. As the main scientifically sound analytical processes have improved in environmental friendliness, these aspects gain greater relevance.

9.2 Sample Preparation

Sample treatment can be considered as the most polluting step of the whole analysis procedure, because it often requires the use of organic solvents. In truly green methods, this step should be avoided by choosing direct acquisition of the analytes. However, in most cases sample preparation remains essential. This section explains how to make the preconcentration and clean-up steps greener.

9.2.1 Solvent-Reduced Techniques

Green sample preparation techniques can be divided into those applicable to the extraction of environmental solids: *supercritical fluid extraction* (SFE), *subcritical water extraction* (SWE) and *pressurized liquid extraction* (PLE), and those applicable to liquid (aqueous) samples: *solid-phase microextraction* (SPME), *stir bar sorptive extraction* (SBSE), *liquid-phase microextraction* (LPME), *single-drop microextraction* (SDME), *dispersive liquid–liquid microextraction* (DLLME); distinguishing those used for gaseous samples: mainly purge and trap, and some of the above techniques carried out in the headspace. Atmospheric samples are not covered here.

The results of a search in the Scopus database aimed to grasp the acceptance and applications of these techniques in environmental (used as keyword) analytical chemistry in the period of 2004–09 are summarized in Figure 9.1. The first aspect to comment on is that the ranking is headed by three techniques that are each applicable to a different purpose: organics in aqueous samples

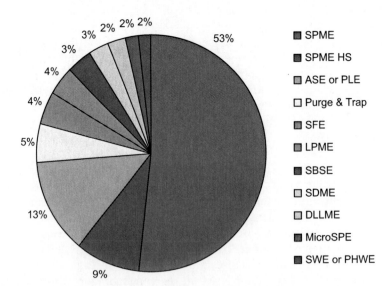

Figure 9.1 Relative distribution of publications indexed in Scopus in the period 2004–2009 reporting the use of various green techniques in environmental analytical chemistry.

(SPME), organics in solid samples (PLE) and volatile organics (purge and trap). The SPME technique is overwhelmingly preferred (62%) in environmental applications (not distinguishing the fundamental studies). The reason is many advantages of this technique, which in combination sum up the essence of a green technique: it is truly solvent-free, miniaturized and automated.[3] It must also be pointed out that SPME can be used in a wide range of matrices, from semivolatile to volatile analytes, hence the headspace (HS) SPME category (9%). Since the introduction of SPME, many substances traditionally analysed by purge and trap are being transferred to SPME methodologies, which are more straightforward and require less capital investment.

The three above mentioned techniques represent 79% of applications; the remaining 21% is distributed among seven techniques accounting for 2–4% each. Here we find some of the most innovative techniques that have in common the features of miniaturization and environmental friendliness, accomplished through reduction of sample size and reduction of solvent consumption. Except for SFE and SWE, a notable drawback that has not been overcome is the relatively low level of automation of these techniques. This requires operator intervention and manipulation, giving rise to additional variability of the results. The range of analytes that can be determined with these techniques is often limited to those that are not strongly polar in character, which constitutes an additional limitation especially considering the properties of the so-called emerging pollutants. These aspects clearly require further development. Present and future trends in each of these techniques will be detailed later.

9.2.1.1 Solid-Phase Microextraction

Because of the outstanding role SPME plays in green analytical chemistry we will broaden the scope of the discussion to demonstrate the potential it has found in other areas beyond the environmental field. The number of publications found in the literature involving SPME and various matrices (used as keywords) are presented in Figure 9.2. More than half of the applications deal with water, food and air; not surprisingly, as these are the matrices that our common well-being depends on. Furthermore, these matrices are very much regulated worldwide, so SPME is finding its way into compliance quality control. Because of its selectivity it is very suitable for enriching analytes from complex matrices such as food, biological fluids, forensic samples and soils, where it also works as a clean-up step.[3] The reduced volume of sample that is required also represents an advantage when working with biological fluids and forensic analysis.

Despite the many beneficial features of SPME and its wide implementation in routine analysis there are some opportunities for improvement when working with emerging pollutants, which deserve a closer look. All these improvements contribute to extending SPME's range of applications, which might ultimately replace old methodologies by greener ones.[4] Our approach is based on the

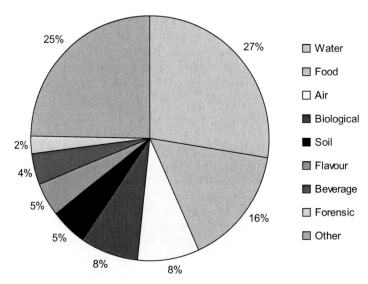

Figure 9.2 Range of matrices in which SPME has been used for analytical profiling of endogenous compounds or xenobiotic substances.

rationale that recent developments are presented along with the feature of analytical chemistry that is advanced.

SPME is known to be an equilibrium extraction technique that is as sensitive as the common exhaustive extraction techniques because the entire amount of analyte extracted is injected and analysed. Haddadi et al.[5] deployed an polydimethylsiloxane (PDMS) coated fibre internally cooled with carbon dioxide that allows nearly exhaustive extraction, and thus obtained improved sensitivity.

An internally cooled coated fibre has also been used for fast and exhaustive extraction of PAHs (polycyclic aromatic hydrocarbons) from soil samples. The efficiency of this arrangement is optimal since the sample can be heated to release the analytes and the fibre is cooled to increase the partition coefficients of the analytes on the fibre by creating a temperature gap between the coating and the headspace.[6,7] Chia et al. designed a system using chilled alcohol for cooling the fibre and applied it to the analysis of polychlorodibenzo-p-dioxins and polychlorodibenzofurans in contaminated soils.[8]

In an attempt to impart polar character to the coating and analyse polar organophosphorus pesticides in water, Bagheri et al. produced an amino-functionalized SPME fibre by using 3-(trimethoxysilylpropyl) amine as precursor.[9]

Sol–gel coatings have been developed to overcome the limitations of commercial phases in terms of solvent instability and swelling, low operating temperature, stripping of coating and inadequate polarity.[4] With the same aim of efficiently extracting polar analytes, Segro et al. produced an SPME coating composed of sol–gel titania poly(tetrahydrofuran) which proved very suitable for a wide range of analytes and resistant to extreme pH conditions.[10] A novel sol–gel coating based on the use of (3-aminopropyl)triethoxysilane and diethoxydiphenylsilane, with excellent thermal (400 °C) and chemical stability, has been applied to the determination of PAHs in water and in milk.[11] Polypyrrole is another new coating, inherently multifunctional and highly porous, that has turned out to be particularly suited for the extraction of polar, aromatic and anionic compounds, *e.g.* anatoxin, ochratoxin A, phenols and even nitrite, nitrate and chloride in water samples.[3] Azenha *et al.*[12] have produced SPME fibres with comparable performance to the commercial ones in the extraction of benzene, toluene, ethylbenzene and xylenes (BTEX), depositing a sol–gel polymer of PDMS-SiO_2 on to a titanium wire, which significantly benefited the fibre robustness. The amphiphilic and hydrophilic oligomers prepared by Basheer *et al.* provided extraction efficiencies towards organochlorine pesticides (OCPs) and triazines that were equal to or better than the PA and PDMS coatings.[13] Dong *et al.*[14] have demonstrated that the new calixarene coatings have better extraction capacity than OCPs despite being thinner. Cai *et al.*[15] have chosen to prepare crown ethers such as benzo-15-crown-5 as SPME coatings for the analysis of organophosphorus pesticides, on the basis of their polarity and highly porous structure.

Nanomaterials have also been investigated as adsorbents for micropollutants. Ji *et al.* have studied the adsorption mechanisms of oxidized multiwalled carbon nanotubes as a fibre coating for the extraction of pesticides from water.[16] The precision obtained was very good, but the sensitivity is insufficient to allow the determination of pesticides in water according to the legal requirements, probably because of the low sensitivity of the diode array detector, although this coating performed better than other home-made and commercial fibres. Rastkari *et al.* have evaluated single-walled carbon nanotubes for the analysis of gasoline additives (MTBE, ETBE and TAME) in human urine.[17] Tan *et al.* followed an alternative strategy preparing a molecularly imprinted polymer (MIP) as a coating for SPME fibres in order to analyse bisphenol A, an endocrine disruptor.[6] MIPs are synthetic mimics of biological antibodies that improve the selectivity of the extraction step by specific recognition of the analyte. This greatly simplifies the analysis of complex environmental and biological matrices, saving many fractionation and clean-up steps as well as equipment maintenance procedures, thus reducing labour and solvent consumption.[7] Li *et al.*[18] proposed the use of an organically modified silicate (ormosil) SPME stationary phase targeted for BDE-209 by molecular imprinting but also extracting small congeners from environmental samples. In 2001, Mullet prepared a MIP-coated silica fibre for the analysis of propanolol in urine and Turiel *et al.* built a monolithic MIP SPME fibre for the analysis of triazines in environmental and food samples.[3] In-tube SPME capillaries for the extraction of benzodiazepines and fluoxetine in human serum have been developed based on imunoaffinity extraction.[3] All these many approaches that make SPME so versatile and popular are, naturally, solvent-free.

Besides those applications mentioned above, SPME has been used for a range of other disparate applications: analysis of gasoline hydrocarbons in near-surface marine sediments,[19] analysis of methyl benzoate as a mould growth biomarker in classrooms of schools and universities[12] and study of the sorption of steroids on to environmental organic matter. *In vivo* sampling can be carried out, for instance introducing an SPME fibre into the bloodstream for studies of pharmacokinetics and other biochemical processes.[4] The use of SPME in degradation studies is also very interesting, since it reduces the amount of sample necessary and makes it possible to work with trace concentrations of pollutants. Gonçalves *et al.* studied the photodegradation kinetics of quinalphos in several water matrices,[20] and Llompart *et al.*[21] suggested photodegradation studies irradiating the pollutant adsorbed on to the fibre according to a procedure called photo-SPME. Mester and co-workers reviewed in depth the theoretical and practical aspects of the determination of inorganic analytes by SPME, which has gained wide acceptance.[21] Organometallic compounds such as organolead, organotin and organosulfur are among the preferred applications.[13]

A recent innovation is 96-fibre SPME devices that can be used for high-throughput bioanalytical and environmental applications, although most of these devices are home-made. PAS technology (Magdala) recently developed a multifibre SPME autosampler capable of controlling all steps of SPME. It was interfaced to a LC-MS/MS for the automated and high-throughput determination of ochratoxin A in human urine.[4] This approach is suitable for large-scale toxicological studies with a throughput of more than 1500 samples per day based on a solvent-free, miniaturized and automated technique. The amount of information gathered and the rapidity of the process relative to the resources involved are clearly ahead of today's state of the art, which proves that green chemistry does not force any trade-offs.

9.2.1.2 In-Tube Extraction

Nerin *et al.*[3] classify the in-tube SPME device in the same group as the needle trap device (NTD), in Pawliszyn's nomenclature. The former consists of an internally coated capillary or needle, while in the latter the lumen is packed with an extraction phase that is kept in place by some type of frit or plug.[22] Despite the similarity in shape, their applications are not identical: in-tube SPME is most useful for coupling and automation of high-pressure liquid chromatography (HPLC) methods dedicated to polar and thermolabile compounds while NTD is ideal for concentrating volatiles in gaseous samples or the headspace of liquid or solid samples followed by thermal desorption to a GC.

The NTD technique is more robust than SPME and the sorbent capacity is greater, making it possible to perform exhaustive extractions. Like SPME, NTD is totally solvent-free, but it also makes it possible to analyse pollutants bonded to particulate matter. Further details of this technique can be found elsewhere.[4]

In 2004, Kubinek and Berezkin[23] developed a NTD device for analysis of BTEX in aqueous samples, consisting of a needle filled with Porapak Q and an aluminium oxide bed.[3] An in-needle extraction device was also developed by Saito *et al.* in 2006 for analysis of VOCs using a specially designed needle packed with a copolymer of methacrylic acid and ethylene glycol dimethacrylate, which has thermal stability and high extraction efficiency.[24] Nowadays, the ITEX and INDEX devices are commercially available and are supplied with different sorbents.[3]

9.2.1.3 Thin-Film Extraction

Another recent tool is the thin-film microextraction (TFME) device.[25] It consists of a PDMS film that is cut in a funnel shape and held by a stainless steel wire in its centre allowing it to rotate around this axis. Due to its bigger surface area it possesses higher extraction capacity than the SPME fibre and the equilibration times are also faster. Although it is totally solvent-free, it remains quite a manual procedure and the desorption step is not so well established. TFME has been used both for grab sampling and also for on-site passive sampling of PAHs in Hamilton Harbor.[4]

9.2.1.4 Liquid-Phase Microextraction

Among the liquid–liquid microextraction techniques, single-drop microextraction (SDME) is a conceptually very interesting extraction and enrichment technique introduced in 1996. It is nearly solvent-free, since it requires just 1–3 µl of (typically) an organic solvent immiscible with water that is suspended from the tip of a microsyringe and held in contact with the sample or its headspace.[3] Its most attractive features are that it is inexpensive, does not require dedicate equipment, is easy in operation and provides an extract that can be directly analysed. To our knowledge it remains a quite manual technique, which generates increased variability in the results, although automation seems to be feasible. In 70% of applications the final analysis was carried out by GC because of the convenience of solvent injection, and also because the analytes that are more efficiently extracted by this technique are apolar, volatile or semivolatile. On these grounds, current research trends focus on extraction with new solvents for the enrichment of polar analytes and active extraction by chemical modification of the extracting drop.[3] The solvents used for extraction were originally organic, but today new greener solvents are being tested. Ionic liquids are employed for extraction of polar analytes, and water is used in headspace extraction. An interesting approach that opens the way to many possible applications consists of adding chemical reactivity to the extractive drop by doping it with a derivatization or complexation reagent, for instance. To circumvent some known operational problems of SDME, such as solvent drop instability and limited extracting surface area, Shen and Lee have developed the organic solvent film (OSF) approach, which might be called 'inside-needle solvent film extraction'.[26] This relies on a continuous renewal of the solvent film generated by plunger movements. Some modifications in the shape of the microsyringe tip have been introduced to increase the adhesion of the drop as well as to improve the extraction efficiency. Illustrative examples are the work of Ahmadi *et al.* and Ye *et al.*[3] The continuous-flow microextraction (CFME) technique first described by Liu and Lee[27] involves the introduction of a solvent drop into a chamber where the sample is pumped at a continuous flow rate.

Hollow-fibre liquid-phase microextraction (HFLPME) was originally developed to counteract the drop instability found in SDME. In the two-phase model a porous polypropylene capillary tube containing an organic solvent is placed in contact with the water sample. The analytes tend to partition into the solvent, which is then withdrawn from the hollow membrane and analysed. In the three-phase model an organic solvent is simply embedded in the pores of the semipermeable membrane and serves as a channel through which the analytes reach the third phase, where they concentrate. The liquid membrane is usually made of ethers with a long alkyl chain. The mass transport of analytes by simple passive diffusion can be low, but it can be enhanced if electrokinetic migration is used for ionizable molecules at an appropriate pH of the donor and acceptor phases.[28] The selectivity of the enrichment process is improved, thus the technique becomes particularly suitable for the analysis of biological samples.

Although the solvent is protected from evaporation and is physically constrained, handling the small hollow fibre and microlitre volumes of solvent requires skill and experience, with the risk of biased results. Automation of this technique is possible with gas and liquid chromatography as well as spectroscopy and electrophoresis, which eliminates these drawbacks. The implementation of HFLPME, an extremely versatile technique, makes a remarkable contribution to green chemistry since it is environmentally friendly, consumes little solvent, provides clean-up features due to its selectivity, has high enrichment factors and recoveries, and relies on low-cost materials.[29] Pedersen-Bjergaard and Rasmussen have dedicated much research effort to this technique and their paper[29] is a good sources for the fundamentals, history and applications of HFLPME.

Recent trends in this field involve the use of hydrophilic membranes like that used by Araujo *et al.*[30] to determine methanol in biodiesel or the solvent-resistant nylon-6 hydrophilic porous hollow-fibre membranes used by Kosaraju and Sirkar[31] to analyse acetic acid and phenol. Some authors have focused on the modification of existing membranes, but complete synthesis has been attempted by others. This is the case for the sol–gel zirconia hollow-fibre membrane produced by Xu and Lee[32] for the analysis of pinacolyl methylphosphonic acid, a nerve agent degradation product. Schellin and Popp propose the use of bag-shaped membranes to extract chlorinated volatiles from water. The extraction solvent is placed inside the bags, which are held in the headspace of a heated and stirred water sample.[33] Globally, the detection limits obtained by HFLPME analysing a wide diversity of analytes are in the sub-µg L^{-1} range, although two publications reported LODs (limits of detection) of pg L^{-1} in the determination of basic drugs, and polychlorinated biphenyls and polybrominated diphenyl ethers, respectively.[3]

Guo et al.[34] proposed a novel approach for direct determination of the n-octanol/water partition coefficients of organic substances using HFLPME enabling a more accurate, inexpensive, simple, and quick estimation than the conventional methods.[34]

For the interested reader, Nerin et al.[3] give a much more detailed discussion of these techniques.

9.2.1.5 Dispersive Liquid–Liquid Microextraction

Dispersive liquid–liquid microextraction (DLLME) is an interesting development that uses very small amounts of both solvents and sample. It was first introduced by Berijani and co-workers in 2006 and consists of three phases: the sample; the extraction solvent, with high density and immiscible with the sample; and a disperser solvent, miscible with both. The volume of disperser solvent normally used is around 1 ml whereas that of the extraction solvent is a few μl. A mixture of these last two in appropriate proportions is injected into the sample and instantaneously a cloudy solution made up of thousands of very tiny drops is formed. The intimate contact between the organic extraction solvent and the sample favours the transfer of the analytes by simple solvent partition.[35] The equilibrium state is achieved very quickly, resulting in a very short extraction time.[36] The extraction solvent is recovered by centrifugation and phase separation and then analysed. The parameters affecting the extraction efficiencies of DLLME are discussed by Zang et al.[35] The main advantages of the DLLME technique are simplicity of operation, rapidity, low cost, high recovery, high enrichment factor, and environmental friendliness.[36,37] However, DLLME is more suitable for the analytes with high or moderate lipophilicity ($K > 500$) and unsuitable for neutral analytes with high hydrophilicity.[35]

Among the liquid-phase microextractions, this technique finds great acceptance in trace analysis, as demonstrated by the applications presented below and those reviewed by Nerin et al.,[3] but most often the matrix is water of various environmental origins. It has been applied successfully, in environmental samples coupled to GC, LC and atomic absorption spectrometry (AAS), for the analysis of antimicrobials;[38] organophosphorous, organotin and pyrethroid pesticides; triazine and amide herbicides; fungicides; PAHs; organophosphorus flame retardants; polychlorinated biphenyls (PCBs); polybrominated diphenyl ethers (PBDEs); trichloromethane; chlorobenzenes; phthalate esters; pharmaceuticals; and heavy metals.[35] DLLME has been the technique of choice to analyse not only the conventional persistent organic pollutants but also emerging pollutants such as fragrances and antidepressant drugs.[39,40]

Often, the extract obtained is a liquid amenable for direct injection in the techniques above; otherwise, it can be evaporated to dryness and the residue dissolved in an appropriate solvent. The LODs obtained in the analysis of pesticides are in the low ng L^{-1}, suitable for regulatory compliance with the 0.1 μg L^{-1} limit enforced for drinking-water and groundwater. The LODs in the analysis of metals are also promising, being in the sub-μg L^{-1} or even sub-ng L^{-1} range.[35] Since the addition of disperser solvent increases the analyte's solubility in the water phase and diminishes the partition into the extraction phase, overall decreasing the extraction efficiency, Regueiro et al. propose a technique called *ultrasound-assisted-emulsification-microextraction* (USAEME).[39] Zhou et al.[40] developed a method for the analysis of aromatic amines by means of an ionic liquid, also based on the use of ultrasonic energy for dispersion of the extraction solvent (instead of rapid syringe injection). DLLME can also be used for the analysis of polar and non-volatile analytes using one-step derivatization and extraction, greatly simplifying the procedure and shortening the analysis time, *e.g.* in the analysis of chlorophenols.[39,41] Liu et al. propose a method of coupled SPE-DLLME-GC-ECD for the determination of PBDEs in water and plants.[42] Although the time required to process a sample must be more similar to that of a SPE procedure than a DLLME procedure, the authors state that ultra-trace concentrations of PBDEs can be reliably determined in water with LODs in the range 0.03–0.15 ng L^{-1}.

While maintaining all its green features, the future range of applications of DLLME is expected to broaden in three directions: (1) application to more complex matrices than environmental water; (2) wider selection of extraction solvents beyond the halogenated hydrocarbons; (3) coupling with sophisticated and highly sensitive detection instruments.[43]

Pena-Pereira published a very good review where DLLME as well as SDME and LPME are discussed with respect to their application in inorganic analysis including metals, metalloids, organometals, non-metals and speciation.[44]

When DLLME is coupled with graphite furnace atomic absorption spectrometry (GFAAS), direct injection analysis can be performed just after the removal of the extraction solvent.[43] Several chelating agents have been employed for the analysis of selenium, cadmium, gold, lead, cobalt, nickel and copper, providing an enrichment factor of from ten to several hundred-fold.

Interesting strategies are used in SDME such as continuous-flow microextraction, complexation, aqueous drops containing noble metal for enrichment of hydride-forming elements, and acidic and alkaline aqueous drops for extraction of ionizable inorganics like ammonia and cyanide.[44] Anthemidis and Ioannou reviewed the application of liquid–liquid microextraction techniques for inorganic analysis, focusing on *homogeneous liquid–liquid extraction* (HLLE) and DLLME.[45] The following phase separation mechanisms in HLLE and published applications are discussed: phase separation based on temperature; based on salting-out phenomenon with salt; based on salting-out effect with auxiliary solvent; pH-dependent phase separation with perfluorooctanoic acid (PFOA) pH-dependent ternary solvent systems and ion-pair formation.[45]

The automation of DLLME is still very limited and only one author has attempted sequential injection-DLLME for the analysis of copper and lead.[45] Although the overwhelming majority of applications relate to environmental water, Hu et al. propose the application of DLLME in the analysis of PCBs from soil samples.[46] The methodology relies on the same operational principles but the analytes are carried in the disperser solvent (acetone), previously used for the extraction of the soil sample, and are then transferred to the extraction solvent (chorobenzene), whereas the water phase only takes part in the ternary system required to produce a cloudy state. The hyphenation of the two techniques (liquid–solid extraction (LSE) and DLLME) increases the concentration factor and, although this is not mentioned by the authors, might provide a clean-up effect for the soil extract since the most polar matrix substances should remain in the aqueous phase.

9.2.1.6 Ionic Liquids as Green Extraction Solvents

Room-temperature ionic liquids (ILs) are a new sort of green solvents applied in DLLME and other techniques. They are generally composed of quaternary nitrogen cations, such as alkylammonium, dialkylimidazolium, and alkylpyridinium, and a suitable anion. They have attracted increasing interest in the chemistry community as green alternatives to the classical, environmentally damaging, medium of volatile organic solvents because of their remarkable properties, such as negligibly small vapour pressure, non-flammability, high thermal stability, tunable viscosity and miscibility with water and organic solvents.[35,44] ILs have been produced according to the needs of users. They have been employed in liquid–liquid extraction, stationary phases for GC, supported liquid membrane extraction, *etc*.[40] Liu published an excellent review on the application of ILs in analytical chemistry.[47] He updated this in 2009, together with other collaborators, with a list of imidazolium and non-imidazolium-based ILs, their characteristics and the main applications in sample preparation including the problems and challenges in this area.[48] Zhou and co-workers proposed an interesting green application of ILs called *temperature-controlled IL dispersive liquid–liquid microextraction*.[43] The temperature of the sample/extraction solvent IL mixture is used as the driving force for solubilization, extraction and cloud point formation. 1-hexyl-3-methylimidazolium hexafluorophosphate (C6MIMPF6) was used as the extraction solvent for the enrichment of pyrethroid pesticides from water. At 70 °C this IL is totally soluble in the water sample and extracts the pesticides; the solution then is cooled and becomes turbid, and the IL is recovered by centrifugation.[43] The recoveries were in the range of 77–136%. The authors extended the applicability of this method to analyse organophosphorus pesticides and atrazine in environmental samples. Liu *et al*. have directly extracted four heterocyclic pesticides from water using the same IL without the need for heating and cooling, which resulted in a faster procedure and is suitable for thermolabile compounds.[36] Liu *et al*. proved that the infinite subdivision of the extraction solvent in the cloudy state plays a pivotal role in the extraction efficiency. The recoveries were good, varying between 79% and 110%, and the precision was in the range of 3.5–10.7% RSD. However, the LODs were not as good as required for present standards, possibly because of the low sensitivity of the DAD detector.[36]

There are great expectations for ILs, since their unique properties will make it possible to enlarge the range of applicability of DLLME, hopefully to more polar compounds, and allow the development of new innovative approaches to DLLME suitable for environmental and other matrices.[43]

The applicability of ILs to the enrichment of heavy metals and subsequent analysis by AAS is conditioned by the previous addition of a chelating agent either directly to the water sample or to the extraction solvent/disperser mixture. At some point, the metal complex is extracted back into the fine droplets of the IL.

9.2.1.7 Stir Bar Sorptive Extraction

SBSE is a microextraction technique introduced in 1999 by Baltussen *et al.*[49] The extraction process takes place in a magnetic stir bar 1–2 cm in length coated with 25–125 µl of adsorbent polymer (a layer 0.3–1.0 mm thick). PDMS is the only polymer available as an extraction phase on commercial stir bars, so the large majority of applications use this coating.[50] The coated stir bar can be added to the sample for stirring and extraction (direct-SBSE) or exposed to the headspace of the sample (HS-SBSE). Once the extraction step is finished, the stir bar is removed from the sample and desorbed for the subsequent analysis. The desorption of the analytes can be accomplished by soaking in an appropriate solvent but the most straightforward option is thermal desorption in a special unit and direct coupling to GC. This set-up is more consistent with SBSE affinity for semivolatile and apolar compounds and subsequent analysis with highly sensitive GC-MS. Although the basic principles of SPME and SBSE are identical and the extraction phase is generally the same, the amount of PDMS is 50–250 times larger in SBSE, which increases the preconcentration efficiency.[50]

One decade after its introduction SBSE has gained wide acceptance in the field of environmental analysis, limited only by the type of extraction coating, which is unsuitable for direct preconcentration of polar pollutants.[51] The most widely used applications are for the analysis of persistent organic pollutants such as PAHs,[52–54] pesticides,[54,55] PCBs[54,56] and PBDEs.[56,57] Nevertheless, some methodologies have also been developed for determination of emerging environmental contaminants, such as alkylphenols,[56,58,59] sunscreen agents,[60] endocrine disruptors,[61,62] off-flavours in water[63] and various pharmaceuticals.[64] The reviews of Lancas *et al.* and Sanchez-Rojas are good sources for a great variety of applications in the environmental, clinical, forensic and biological fields.[50,51]

The present brief discussion focuses on the most innovative developments, although the original configuration of SBSE is preferred by most authors. Outstanding research is still scarce and focuses mainly on the development of new extraction phases and SBSE instrumentation.[50] Neng *et al.* proposed polyurethane (PU) foams as new polymeric phases for SBSE.[65] The authors demonstrated that these polymers present remarkable stability and excellent mechanical resistance for the enrichment of organic compounds from aqueous samples. PU can be regarded as a convenient alternative to the conventional

PDMS coating for the analysis of polar analytes, *e.g.* triazine herbicides.[66] PU foams doped with activated carbon, a mesoporous material and a calixarene did not show any advantages compared to the polymer alone.[65] One the other hand, Nogueira *et al.* developed and applied PU doped with activated carbon for the analysis of antidepressants, anticonvulsants, ivermectine and benzoimidazols at therapeutic levels with improved extraction efficiency.[50] This finding is explained by the double extraction mechanism: adsorption (activated carbon) and absorption (PU), as well as the special affinity of activated carbon for polar substances.[50] Silva *et al.* compared the PDMS and PU polymers and concluded that PU is a promising SBSE coating for the analysis of residues of several acidic non-steroidal anti-inflammatory drugs and lipid regulators in waters.[64] Melo *et al.* developed a new polymeric SBSE coating consisting of a dual-phase, polydimethylsiloxane (PDMS) and polypyrrole (PPY), for the analysis of antidepressants (mirtazapine, citalopram, paroxetine, duloxetine, fluoxetine and sertraline) from plasma samples.[67] The PDMS/PPY coated stir bar showed high extraction efficiency, sensitivity and selectivity toward the target analytes justified by the existence of both extraction mechanisms: adsorption (PPY) and sorption (PDMS).[67]

Huang *et al.* suggest the use of a monolithic material as SBSE coating.[68] The monolithic material was prepared by *in situ* copolymerization of octyl methacrylate and ethylene dimethacrylate in the presence of a porogen solvent containing 1-propanol, 1,4-butanediol and water, with azo-bisisobutyronitrile as the initiator. The adsorbent obtained is suitable for extraction of both polar and non-polar compounds and is characterized by good permeability, allowing fast adsorption and desorption of analytes. A novel poly(phthalazine ether sulfone ketone) (PPESK) SBSE coating has been described by Guan *et al.*[69] The stir bar showed a structure with a dense homogeneous porous surface and a sponge-like sublayer. The retention of analytes, such as organochlorine or organophosphorus pesticides used as model compounds, is due to an adsorption mechanism. PPESK exhibited better selectivity for polar compounds than PDMS; this was attributed to the interactions between compounds and the polar functional groups of PPESK.[69]

Zhu *et al.* prepared a 180 μm coating for SBSE with a MIP formed from a formic acid solution of nylon-6 polymer and targeted for monocrotophos.[70] Clean extracts and yields of 95% were obtained, demonstrating the suitability of this MIP SBSE coating for the analysis of environmental and biological samples. Additionally, the coating was useful for preconcentrating four structural analogues of monocrotophos. Huang *et al.* developed a new phase (poly(vinylpyridine-ethylene dimethacrylate)) to extract polar phenols, concluding that this phase can also be applied to analyse other groups of polar analytes.[71] Liu *et al.* described the use of a compact and thermally stable porous hydroxyterminated phase for the extraction of PAHs, n-alkanes, and phosphorus pesticides from water samples, and Lambert *et al.* prepared a biocompatible SBSE device using an alkyl-diol-silica (ADS) restricted access material (RAM).[50]

Bicchi *et al.* proposed an interesting arrangement for SBSE called 'dualphase twisters', composed of a hollow PDMS tube closed at both ends by magnetic stoppers.[72] The inner cavity of the tube can be filled with different adsorbents such as activated carbons. This special extraction device combines two extraction mechanisms, adsorption and absorption, and can be made suitable to enrich not only the traditional non-polar compounds (by the PDMS phase) but also polar ones.

Roy et al. have created an on-site extraction system for the analysis of 24 PAHs, including 15 EPA (U.S. Environmental Protection Agency) priority pollutants, composed of a SBSE stirrer and the EM 640 S field apparatus (Bruker).[52] According to the authors this device gave encouraging results, namely LODs around the sub-ppt level for most of the compounds, with the advantages of an on-site system such as unattended operation, minimized risk of cross-contamination and loss of analytes.

The range of application of SBSE has been further extended by using both pre-extraction derivatization reactions and *in situ* derivatization, typically with ethyl chloroformate and acetic acid anhydride as esterification and acetylation reagents, respectively.[50] Itoh et al. and Nakamura et al. used simultaneous derivatization (acetylation) and extraction to improve the preconcentration efficiency of hydroxy PAHs, particularly mono- and dihydroxy; and alkylphenols and bisphenol A, respectively.[53,59] Detection limits of a few ng L^{-1} were achieved, which is worth mentioning as these analytes are extremely toxic and occur at very low levels. Popp et al. have determined sub-ppt levels of PAHs from water samples and propose the rapid screening of highly contaminated waste material using silicone rods 1 mm in diameter and 10 mm long.[73] Lanças and co-workers have adapted a previously developed SPME-HPLC interface featuring temperature-controlled desorption, increasing the inner chamber to 100 μl in order to accommodate an SBSE stir bar.[50] This interface allows on-line coupling of SBSE-HPLC with a faster and improved desorption efficiency obtained in the analysis of antidepressants. A collaborator of the same group described a novel technique called *refrigerated sorptive extraction* (RSE), which allows heating the sample matrix and simultaneously cooling the bar coating.[50] This strategy increased the partition of the organochlorine pesticides (OCPs) used as model compounds on the PDMS adsorbent through a more favourable temperature gradient. It should be remembered that adsorption is an exothermic phenomenon, while diffusion within the sample matrix is faster at higher temperatures. Liquid desorption was chosen.

Despite the similarities with SPME, the use of SBSE in the inorganic field is scarce. In fact, Pena-Pereira et al. found only two articles where SBSE was used for the determination of organotin, methylmercury and butyltin species.[44] Prieto et al. determined organometallic compounds of mercury and tin in water, sediments and biological samples after derivatization with $NaBEt_4$ and collection of the volatile derivatives in the headspace.[74]

9.2.1.8 Microextraction in Packed Syringe

Microextraction in packed syringe (MEPS) is a new technique developed by the group of Abdel-Rehim for miniaturized solid-phase extraction, relying on the same concepts as SPE in cartridge format, which can be connected on-line to GC or LC without any modifications.[75,76] In MEPS, approximately 1 mg of the solid packing material is immobilized as a plug inside a syringe barrel (100–250 μl). For extraction, a 10–250 μl volume of an aqueous sample is withdrawn through the syringe by an autosampler.[77]

El-Beqqali et al. attribute the following main advantages to this technique: it is easy to use, fully automated, low cost and rapid in comparison with alternative techniques. Its suitability for environmental applications was demonstrated by the analysis of PAHs in water using a silica-C8 MEPS adsorbent. The adsorbent could be used for more than 400 repeated extractions before the syringe was discarded.[77] MEPS reduced the handling time by 30 and 100 times compared to SPME and SBSE, respectively, saving time and resources. Unlike conventional SPE, the sample is withdrawn into the syringe and this can be reversed several times, withdrawing several sample aliquots to increase sensitivity. The analytes are eluted with a very small amount of solvent or LC mobile phase (20–50 μl) directly into the instrument's injector, which gives the technique markedly green features. A sample volume as little as 1 ml was sufficient to obtain LODs of 5 ng L^{-1} or less, and the extraction time was about 2 min. At present, MEPS devices are commercially available from SGE Europe (Leicestershire, UK).

Abdel-Rehim et al. have analysed local anaesthetics in plasma (with extraction recoveries between 60 and 90%),[75] olomoucine[78] and β-blockers (acebutolol and metopropolol).[79] C2-silica was the adsorbent chosen to enrich the anaesthetics, whereas polystyrene was used for olomoucine and β-blockers. All sorts of adsorbents can be used in MEPS: reversed phase, normal phase, ion exchange, MIPs and RAM.[76,78]

Alternative syringe-based sample preparation techniques have been developed in the form of a needle-shaped extraction device. In such extraction devices, a section of the needle has a polymeric coating inside, or is packed with particulate materials or filled with a bundle of coated microfibres.[80] Several studies have shown that this sample preparation arrangement has the advantages of a SPME device without its disadvantages, such as fragility and low analyte capacity. These needle-shaped extraction devices are especially appropriate for the enrichment of volatile compounds since a continuous flow of the gaseous sample needs to be supplied through the extraction needle during the sampling process.[80] The ease of handling during the extraction and desorption processes, availability of a wide range of coatings and suitability for automation are additional attractive features when coupled to typical GC instruments on-line.[80]

Qi et al. have investigated the feasibility of a fibre-packed extraction device for the determination of aromatic pollutants in water samples (nitrobenzene, 2-naphthol, benzene, n-butyl-p-hydroxybenzoate, naphthalene, p-dichlorobenzene). Three nanofibres of poly(styrene-co-methacrylic acid), poly(styrene-co-styrenesulfonate) and polystyrene were employed as the extraction medium and packed into a section of pipette tip.[81] Zylon and Technora nanofibres have been chosen by some authors because of their excellent thermal stability and resistance to typical organic solvents.[82] HR-1 (100%-methyl-polysiloxane) and HR-17 (50%-phenyl-50%-methyl-polysiloxane were employed as the polymer coatings.[80] This type of extraction device, which works optimally for the enrichment of volatile analytes, can also be used for water samples. As in MEPS, the extraction takes place by pumping the aqueous sample on to the needle extraction device, and the subsequent desorption is achieved by flowing a desorption solvent through the needle, which can be aided by heating the syringe tip if a GC injector is used.[80] Ogawa *et al.* reported excellent thermal stability of these devices upon repeated use (up to 100 times) without any deterioration in extraction performance.[82] Saito *et al.* published a very good review on the principles, construction and special applications of fibre-packed needle-type extraction devices.[80] This group has long experience in the production of sample preconcentration and separation devices based on fibre-packed capillaries. A relevant green feature displayed by MEPS and fibre-packed needle devices, not highlighted until now but common to many other techniques previously discussed, is their 'reusability', meaning that the analytical costs are greatly reduced and wastes are minimized.

Another innovation is the packing of a common adsorbent or a monolithic methacrylate polymer, as reported by Blomberg, in a conventional micropipette tip or 96-well plate systems.[76] The latter allows a great improvement in analytical throughput. Such arrangements work in the same way as other micro solid-phase extractions described above; only the configuration and minor features vary. It is not surprising that different arrangements may be found advantageous for specific applications, but the general working principle and most of the advantages in terms of miniaturization, automation and environmental friendliness remain the same.

9.2.1.9 Passive Sampling

Passive sampling is another approach for environmental monitoring that follows green chemistry principles since it is carried out on site, avoiding sampling and sample pretreatment. Furthermore, passive sampling can give a measure of an organism's exposure to environmental xenobiotics and bioaccumulation potential, determining their time-weighted average concentration.[83–86] The ability to derive average concentration values is particularly attractive in environments subject to periodic peak pollution events such as industrial discharges, stormwater runoff or floods.[85]

Nowadays the methodologies employed for monitoring of environmental pollutants should not only give an indication of the pollutants' concentration and fate in the environment but also allow the evaluation of their effects and assessment of the potential hazard to human health.[83,87] Environmental contaminants cannot be seen strictly as individual threats; they interact with each other and the ecosystem conditions, to exert a biological effect that is difficult to model taking into account the individual inputs (concentration and toxicity). Furthermore, to assess the quality of our environment using conventional analytical chemistry a large number of samples must be analysed and the daily, monthly or annually time-weighted average levels determined.[83,87] Furthermore, the number of pollutants to be monitored is increasing every day. Thus, a large-scale environmental and human risk assessment can turn out to be prohibitively expensive. It is not only the financial constraints that can limit the wide implementation of this traditional approach, but also the resources involved. In addition, the toxic wastes produced by such activities, if not a threat in themselves, result in inefficient management of needs *versus* funds.

Passive sampling has great potential as a low-tech and cost-effective monitoring tool, avoiding almost every disadvantage of active sampling and/or sample preparation techniques.[83,87] Many advantages are attributed to passive sampling devices: simplicity, low cost, no need for expensive and complicated equipment, no power requirements, unattended operation, and the ability to produce accurate results.[83,86,87] Protection of the analytes against possible degradation and loss during collection, transport, storage and enrichment is an additional advantage. Although many desirable features of passive sampling are recognized, this approach must be seen as highly complementary to spot sampling and total replacement is not envisaged.[88]

Passive sampling has developed very fast over the past 20 years, and the present generation of passive samplers enables detection and analysis of bioavailable pollutants at low and very low concentrations. Organic and inorganic pollutants can be measured in air, water and soil, not only on the local scale but also on continental and global scales.[83,86,89] The use of passive samplers does not require isolating and controlling the environmental factors to produce reliable results; instead, the passive sampler is designed to sense and be influenced by them all: chemical composition, pressure, temperature, flow, biological activity, *etc*. Although passive sampling gives a more holistic view of environmental pollution, its design, calibration and quality assurance is a challenge that has been addressed in the last few years.

Environmental sciences is the largest application area of passive sampling, but new trends towards combining it with biological tests (to measure standard toxicity and genotoxicity assays) have emerged.[83,87] The main areas where passive sampling have proved useful are workplace exposure and monitoring; indoor and outdoor air-quality determination; aquatic sampling for ground and surface water pollution; and sediment and soil pollution monitoring. Water applications are predominant, accounting for around 50% of reports in the last 10 years.[83]

Linear uptake passive samplers and/or non-equilibrium samplers provide time-weighted average (TWA) concentrations of target analytes in the sample matrix averaged over a known period of time, whereas equilibrium passive samplers give an indication of bioconcentration of hydrophobic pollutants in living organisms.[86,87] The following equilibrium devices are examples of inherently green and resource-saving passive samplers: SPME, Empore disks, passive diffusion bag samplers (PDS), water-filled polyethylene bags (PE), diffusive multilayer sampler (DMLS), semipermeable membrane device (SPMD) and polar organic chemical integrative sampler (POCIS). Examples of non-equilibrium devices are: SPME, SPMD, passive *in situ* concentration/ extraction sampler (PISCES), POCIS, membrane-enclosed sorptive coating sampler (MESCO), passive samplers filled with sorbent or resin-based passive air sampler (Orsa 5, National Dräger, Radiello FS Maugeri, OVM 3500 3M), ceramic dosimeter, Chemcatcher, polyurethane foam disk (PUF).[83,86] Nowadays, passive samplers are used to monitor an extremely wide variety of pollutants, including pesticides, PAHs, PCBs, dioxins, furans and alkylphenols.[83,87]

Allan *et al.* have evaluated the performance of seven passive sampling devices for the monitoring of dissolved concentrations of PAHs, PCBs, hexachlorobenzene and p,p'-DDE through simultaneous field deployment for 7–28 days in the river Meuse (The Netherlands).[84] The passive samplers studied were Chemcatcher, low-density polyethylene membranes, two versions of the MESCO sampler, silicone rods, silicone strips and SPMDs. Passive samplers generally provided data for the target hydrophobic analytes that is less variable than that from 'whole water' analysis, since the latter may be strongly influenced by levels of suspended particulate matter. The TWA concentrations calculated varied by a factor of 2. The LODs provided by passive samplers with large surface area are likely to be well below typical concentrations encountered across Europe, making them suitable for monitoring tasks such as comparison with environmental quality standards (EQS) enforced by the water framework directive (WFD).[84]

Shaw and Mueller have simulated in the laboratory the effect of a flood, increasing the concentration of herbicides in waterways, and investigated the performance of the Chemcatcher to integrate pulsed concentrations.[85] This research aimed to offer an insight into the challenge of monitoring fluctuating pollutant concentrations that would be very intensive if grab samples were to be used. Under some circumstances the samplers under-predicted the average concentrations relative to a strictly integrative behaviour after a pulsed event. This was explained by the slow uptake rate of the analytes and a lag time to achieve steady-state conditions due to resistance to mass transfer across the membranes and the water boundary layer.

Bueno *et al.* have tested the usefulness of the POCIS sampler in the pharmaceutical configuration for 1-year monitoring of emerging contaminants, namely pharmaceuticals, biocides and some pesticides, in semienclosed coastal areas, where fish farms are installed.[90] Several pharmaceuticals and herbicides have been determined with adequate detection limits.

Soderstrom *et al.*[91] provided a comprehensive review of integrative passive sampling strategies for monitoring of emerging polar environmental pollutants, such as pharmaceuticals. There is now increased public awareness of the impact of these compounds on ecosystems, and monitoring tools are needed to face the expected growth in analytical determinations resulting from this awareness and the implementation of chemicals legislation such as the REACH procedure.

The interested reader looking for specific analytical solutions encompassing passive sampling is directed to the review of Zabiegala *et al.* which presents a comprehensive collection of applications targeting water and air compartments and is sufficiently up to date not to require any detailed amendment.[83]

The reviews of Greenwood et al.[86] and Seethapathy et al.[87] are very good sources for the fundamentals of passive sampling in environmental analysis, its virtues as an innovative green analytical concept and the configurations of numerous experimental and commercial devices.

9.2.1.10 Supercritical Fluid Extraction

When it was first invented in 1978, SFE was thought to have great potential, given the very convenient properties of supercritical fluids, particularly carbon dioxide: high diffusivity to extract organic compounds from solid matrices, non-flammable, moderate and easily accessible critical point conditions, low cost, decompressible directly to the atmosphere and harmless to the environment. SFE differs radically from conventional extraction techniques, since the analytes are in a concentrated form after extraction. Nevertheless, in practice SFE has never received the acceptance that was anticipated, since it has shown limitations in the extraction of polar compounds. It is possible to add an organic modifier or change the supercritical fluid, but some ideal properties of carbon dioxide are then lost. The high price of the equipment at that time and the almost simulataneous appearance of the PLE technique helped to restrain enthusiasm for SFE. However, its short development period leaves some hope that new improvements might appear in this field. This conviction is reinforced by the fact that all extraction techniques for solids make concessions to efficiency and environmental hazards. As can be seen below, SFE is often used for carrying out fundamental and kinetic studies as well as monitoring surveys. The kinetic studies are based on the assumption that only bioavailable pollutants can affect biodiversity; thus, although SFE sometimes does not provide exhaustive extraction of all analytes, it is able to give a measure of the amount of labile analytes in association with the matrix. On these grounds, Stroud et al. evaluated the bioacessibility of phenanthrene in two different soils using SFE.[92] Recently, SFE has been used for the monitoring of priority and emerging pollutants such as inorganic lead and organolead compounds in sand, urban dust and sediments;[93] pesticides in soil;[94] and polycyclic and nitro musks in raw sludge, primary sludge, waste activated sludge and aerobically/anaerobically digested biosolids.[95]

Lian et al. studied the occurrence of alkyl PAHs (APAHs) in the atmospheric environment of Shanghai, a highly polluted city,[96] using SFE to estimate the seasonal and spatial distribution of APAHs in dustfall. They found 41 APAHs, belonging to 6 groups: alkyl naphthalenes, alkyl phenanthrenes, alkyl anthracenes, alkyl fluoranthenes, alkyl pyrenes and alkyl chrysenes. Alkyl naphthalenes and alkyl phenanthrenes were the dominant APAHs (more than 50%). The source analysis indicated that dustfall APAHs derived mainly from vehicle emissions and used crankcase oil.[96] Yang et al. attempted to estimate the desorption rates of PAHs from river floodplain soils, which contain coal and coal-derived particles, using sequential SFE.[97] They demonstrated that, despite high soil PAH concentrations, the general environmental risk is reduced by the very slow desorption rates (ranging from decades for 2–4-ring PAHs to hundreds of years for 5–6-ring PAHs).[97] Schramm proposes an interesting and wise strategy for exposure risk assessment to air contaminants through hair biomonitoring of organic pollutants;[98] SFE was one of the techniques that proved successful for determination of contaminants in hair. Garcia-Rodriguez *et al.* reviewed and emphasized the great analytical potential of SFE in the analysis of polyhalogenated pollutants—polychlorinated biphenyls (PCBs), polybrominated biphenyls (PBBs), polybrominated diphenylethers (PBDEs), polychlorinated dibenzo-*p*-dioxin (PCDD)/Fs and OCPs—in aquaculture and marine samples, pointing out the integration of the extraction and clean-up steps for rapid sample processing, as well as the extraction efficiency.[99]

Kreitinger *et al.* determined total amounts of PAHs in 34 sediments between 4 and 5700 mg kg^{-1}, and realized that this data is totally unrelated to the survival and growth of *Hyalella azteca* (an amphipod crustacean).[100] Based on the results obtained, the authors concluded that the PAHs present in many sediments collected from sites of manufactured-gas plants have low bioavailability, and that both the measurement of the rapidly released PAH concentrations with mild SFE conditions and the dissolved porewater concentrations of PAHs are much more useful tools for estimating chronic toxicity to *Hyalella azteca*.

Jonker *et al.* also employed a mild SFE procedure, aimed to mimic desorption into water, to quantify the desorption kinetics of 13 native PAHs from pure charcoal, coal, and 4 types of soot, where the analytes are very strongly bonded.[101] The estimated times required for desorption into water of combustion-derived PAHs strongly sorbed to their particulate carrier are several millennia.

Zhang *et al.* proposed an analytical development of SFE, *on-line parallel-trap supercritical fluid extraction*, hyphenated to HPLC, which constitutes a higher degree of automation with consequent reduction of manipulation and thus less exposure of the analyst to toxic solvent vapours (which, although not often discussed, is also an objective of green chemistry). The interface consisted of automated 10-port valve and parallel traps that allowed continuous and alternate trapping and transferring of the sample extracted with supercritical carbon dioxide to the HPLC system.[102]

Rigou *et al.* built a field-compatible SFE device and proposed a suitable method for on-site PAH determination in contaminated soil in combination with field-based immunoassay.[103] Babel and del Mundo Dacera discussed the applicability of SFE for heavy metal removal from contaminated sewage sludge in order to minimize the prospective health risks of sludge during land application. The SFE process gave good results for removal of cadmium, copper, manganese and zinc.[104] Nevertheless, this method is limited by the complexity of the process and the cost of suitable ligands for effective metal extraction.[104]

From the interesting examples presented above it can be concluded that one of the strengths of SFE is its versatility in accomplishing very disparate objectives in environmental (green) chemistry.

9.2.1.11 Subcritical Water Extraction

SWE, also called *pressurized hot water extraction* (PHWE) ($T_c = $ 374 °C, $P_c = $ 218 atm), is environmentally friendly, since water is a non-flammable, non-toxic and easily available solvent.[1,3] The solubility of organic substances in the subcritical water is adjustable within a wide range by changing the conditions of pressure and temperature. The variable dielectric constant that can be obtained (<10 at 300 °C) provides the opportunity to extract polar compounds; nevertheless, this solvent is unsuitable for extracting very hydrophobic, thermolabile or hydrolysable compounds.[105] The great complexity of solid matrices (soils, sediments, sludge) tends to detract from the potential of this technique and requires great efforts to improve selectivity.

The most important area of application of SWE has been in the extraction and characterization of biological materials such as plants and foods; nevertheless, we focus here on its potential in the environmental field. It has been widely used in the extraction of soil samples for the purposes of studying soil composition in terms of environmental micropollutants, as well as the removal of chemical contaminants in remediation strategies. In addition to the diverse applications reported in the summary review of Nerin et al.,[3] we summarize those published up to now. Although, like SFE, it is most often used for kinetic studies, SWE has also found a legitimate application to bioavailability and bioaccessibility studies in support of risk assessment of contaminated land. One must remember that the extraction medium is water, which can be made to translate the equilibria of sorbed pollutants to the aquatic environmental compartment or transfer these pollutants to the aqueous body fluid. The need for accurate methods to incorporate biological exposure into risk-based policies for contaminated land has never been greater.[106] Latawiec and Reid evaluated two emerging non-exhaustive extraction techniques (SWE and Brij 700 extraction) to reflect bioaccessibility of PAHs to microorganisms, comparing them to formerly demonstrated methodologies. Although the use of cyclodextrins was the best predictor of the bioaccessible fraction for most compounds, other methods appeared more cost- and time-effective.[106] Fernandez-Gonzalez et al. combined two environmentally friendly techniques—PHWE followed by SPME—for the determination of PAHs in marine sediment.[107] The authors took advantage of the suitability of the PWHE extract for further processing by SPME to obtain a method with improved detection limits (0.4–15 μg kg^{-1}) and an intrinsic clean-up of the extract. Kalderis et al. used subcritical water at laboratory scale to give an insight into the degradation mechanism of TNT in contaminated soil.[108] Subcritical water was successful in remediating TNT on highly contaminated soils (12%), with destruction percentages of 98–100%. Haglund reviewed several remediation strategies for soil contaminated with polychlorinated dibenzo-p-dioxins/dibenzofurans (PCDD/Fs) including SWE.[109] Hashimoto et al. found that after 4 h of extraction, 99.4%, 94.5% and 60% of PCDDs were removed from the contaminated soil samples at 350, 300 and 150 °C, respectively, dechlorination being the main degradation pathway.[110] Dadkhah and Akgerman carried out a series of small-scale experiments (300 ml reactor) on soils polluted with PAHs, using subcritical water as extraction agent, with and without the addition of hydrogen peroxide for in situ wet oxidation.[111] In the combined extraction and oxidation flow experiments, PAHs were almost undetectable in the remaining soil. This can therefore be a feasible and cost-effective alternative approach for remediating PAH-contaminated soils and sediments. For sediments, Heltai et al. employed a sequential extraction scheme (supercritical carbon dioxide, subcritical water, and a subcritical mixture of 90% water and 10% carbon dioxide) to assess ecotoxicity due to heavy metals.[112] This procedure provided relevant information about environmentally mobile heavy metal fractions (water-soluble, bicarbonate-forming) of sediments. Morales-Riffo and Richter proposed an interesting method for determination of inorganic elements in airborne particulate matter (PM_{10}) by using acidified subcritical water and inductively coupled plasma-optical emission spectrometry (ICP-OES).[113] Elements such as aluminium, arsenic, boron, barium, cadmium, copper, iron, manganese, lead, selenium and zinc were rapidly and efficiently extracted with a solution of 0.1 mol L^{-1} HNO_3 under subcritical conditions. The green features of this procedure are exemplified by its use of just 5% of the amount of acid used in the standard microwave extraction procedure. Kronholm et al. analysed several phenolic compounds (phenol, 3-methylphenol, 4-chloro-3-methylphenol and 3,4-dichlorophenol) in sea sand and soil samples, exploiting the special affinity of the subcritical water for polar compounds (unsuitable for pure carbon dioxide SFE).[114] At 50 °C the phenols were selectively extracted, while at 300 °C the PAHs were extracted along with the phenols.

From the discussion above it is clear that finding environmentally friendly and cost-effective methods to remediate soils contaminated with persistent organic pollutants, such as PAHs and PCDD/Fs, is currently a major concern among researchers. SWE has not yet been used in the analysis of polar compounds, but the growing concern about emerging environmental pollutants is expected to increase interest in this technique.

9.2.1.12 Immunoaffinity Chromatography

This technique uses the binding properties between antigen and antibody for the selective extraction of an analyte from complex food or environmental matrices. The technique is especially useful for polar organic analytes. Immunoaffinity sorbents are used for preconcentration of closely related compounds, which are later eluted, separated and analysed; an example is the detection of isoproturon in water samples, allowing determination of 0.1 μg L^{-1},[115] or of bisphenol A.[116] A rabbit antibody immunoaffinity (IA) column procedure was evaluated as a clean-up method for the determination of atrazine in soil, sediments, and foodstuffs.[117] Polyclonal rabbit anti-atrazine antibody was immobilized in HiTrap Sepharose columns. The coupling efficiency (4.25 mg of antibody in 1 ml of resin bed) for the four IA columns ranged from 93% to 97% with an average of 96 ± 2% (2.1%). Recoveries of the 500, 50, and 5 ng ml^{-1} atrazine standard solutions from the four IA columns were 107 ± 7% (6.5%), 122 ± 14% (12%), and 114 ± 9% (8.0%) respectively, based on enzyme-linked immunosorbent assay (ELISA) data. The maximum loading was approximately 700 ng of atrazine for each IA column. Another example consisted of the extraction of PAHs[118] from water samples on to an antifluorene immunosorbent followed by LC with diode array detection. The reliability of the clean-up achieved by the IA was well demonstrated in the analysis of sediment and sludge complex samples containing the priority PAHs established by the U.S. EPA at concentrations ranging from $56\,\mu g\,kg^{-1}$ to $26\,mg\,kg^{-1}$. Results were compared to those obtained with conventional clean-up procedures, showing a better selectivity for PAHs. The main limitations of these procedures are the effects of the matrix on immunoreagents, the small volume admitted by the immunosorbents, and the sometimes difficult desorption step.

9.3 Greening Separation and Detection Techniques

The separation step prior to detection of the analytes is generally a crucial step in the analysis of complex samples. Recent advances in separation techniques decrease solvent use, analysis time and costs, making the procedures greener. In terms of the detection techniques they are quite green because limited quantities of hazardous chemicals are used. In this part of the chapter we therefore discuss some recent improvements in separation and detection (non-biological and biological approaches) that make these techniques greener.

9.3.1 Immunochemical Techniques

The binding properties of an antibody (Ab) to an antigen (Ag) have been used for the development of a wide variety of analytical techniques applicable in rapid environmental control for the measurement of both single and multiple analytes. Pesticides, PCBs, and surfactants are typical examples of organic pollutants that can be rapidly and efficiently determined by immunochemical methods. In addition, these technologies have also been exploited in sample preparation prior to analysis.

Different types of Abs are involved in immunoassays: polyclonal, monoclonal, and recombinant. Polyclonal Abs are obtained from the serum of animals immunized with a particular Ag. The Ab pool obtained from serum is the result of many B-cell clones; each secretes one specific Ab. Antiserum refers to a pool of serum containing all of the Ab fraction plus other serum proteins. Monoclonal Abs (MAbs) are produced following the fusion of myeloma cells with Ab-secreting B-cells. The resultant continuous cell line (hybridoma) produces large quantities of the homogeneous, well-defined, single-epitope Ab. The availability of large quantities of continuously produced Ab allows greater standardization and quality control of the Ab reagent. MAbs are therefore more precisely characterized, legally protected and more acceptable to regulatory agencies.

Preparation of recombinant Ab fragments with novel binding properties has been a primary goal of gene technologies. Their major asset lies in the possibility of focusing mutagenesis on that part of the gene that determines the structure and the affinity of the Ab binding site.

In green environmental analysis immunochemical methods offer a number of advantages over conventional methods, because they can provide fast, simple and cost-effective detection, with sensitivity in most of cases comparable to conventional techniques, requiring none or minimal quantities of solvent, with minimal sample pretreatment requisites (reducing wastes) and a reduced energy consumption during sample analysis. In addition, more advanced formats, such as immunoassays, can be designed to operate on-line in the field.

However, some limitations should be also pointed out, such as the lack of stability of immunoreagents, cross-reactivity between structurally related compounds and matrix effects. The lack of stability (especially thermal, and at extreme pH ranges) limits the use of immunochemical analysis in some cases.

Cross-reactivity is the lack of specificity of an immunoassay for an analyte when the immunoassay can react with other structurally related compounds; this is a characteristic that depends directly on the antibody. However, when the object of the analysis is the screening of a group of related substances, this property can be an advantage. Despite these limitations and the need to confirm positive results by means of conventional techniques, the use of immunoassays is justified, since samples can be selected, saving large amounts of solvents, energy and wastes in general.

Immunoassays may be carried out in various ways; Figure 9.3 shows the most common formats. Competitive assays are common in environmental

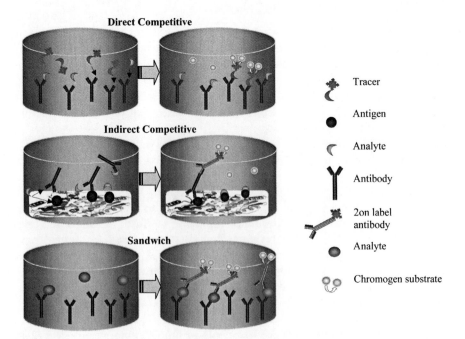

Figure 9.3 Summary of main immunoassay formats used in environmental analysis.

applications and can be performed in different ways, such as the analyte and the tracer competing for a limited number of binding sites (direct), or the analyte and the immobilized ligand (antigen) competing for a limited number of binding sites (indirect). Non-competitive immunoassays are also performed in sandwich-type techniques, in which the analyte is recognized by two different antibodies.

9.3.1.1 Enzyme-Based Immunoassays

Enzyme-based immunoassays (EIAs) typically use a change in colour, emission of light, or other signal to measure the concentration of an analyte. EIAs offer numerous advantages over other immunotechniques because their signal is amplified by forming a great number of product molecules, and they are widely used for rapid environmental analyses, especially those based on heterogeneous conditions, such as ELISAs. The main enzymes used are horseradish peroxidase (HRP), alkaline phosphatase (AP) and β-galactosidase. A large number of ELISAs to detect phenols,[119,120] surfactants,[121,122] and pesticides[123,124] have been reported during the last 15 years, and currently various well-implemented commercial kits are available on the market. Also, various immunoassays using dipstick formats have been developed for screening purposes.[125–128] Cho et al.[128] developed a direct competitive ELISA for fenthion in a microtitre plate and dipstick format. The microtitre plate ELISA showed an IC_{50} value of 1.2 μg L^{-1} with a detection limit of 0.1 μg L^{-1}. The antibodies showed negligible cross-reactivity with other organophosphorus pesticides. The use of the dipstick format using a support membrane allowed the quick visual detection of fenthion at concentrations greater than 10 μg L^{-1}. The IC_{50} value of the dipstick format using reflectance detection was 15 μg L^{-1} with a detection limit of 0.5 μg L^{-1}. The recoveries of fenthion from spiked vegetable samples using the two formats without any prior enrichment or clean-up steps were 87–116%. Recently, Lisa et al.[126] developed a gold nanoparticle (GNP)-based dipstick competitive immunoassay to detect organochlorine pesticides such as DDT at the nanogram level. GNPs of a specific size were synthesized and conjugated to anti-DDT antibodies. The intensity of colour development was inversely proportional to the DDT concentration. The lowest detection limit of DDT was determined to be 27 ng m L^{-1} under optimized conditions. Screening assays based on a dipstick format are a good example of a rapid and cost-effective screening methods, with reduced analysis time, wastes and solvents. Some examples of EIAs for environmental applications are summarized in Table 9.1.

9.3.1.2 Fluorescence Polarization Immunoassays

The fluorescence polarization immunoassays (FPIA) is a type of homogeneous competitive fluorescence immunoassay. With competitive binding, antigen from the specimen and antigen-fluorescein (AgF) labelled reagent competes for binding sites on the antibody. As a homogeneous immunoassay, the reaction is carried out in a single reaction solution, and the bound Ab–AgF complex does not require a washing step to separate it from 'free' labelled AgF. FPIA is used

Table 9.1 Examples of immunoassays.

Analytes	Type of antibody	Matrix	Limit of detection	References
Enzyme immunoassays				
Azinphos methyl	Monoclonal	Water and honeybee	0.6 ng L^{-1}	253
Bromopropylate	Polyclonal	Water	0.14 µg L^{-1}	254
Carbamazepine	Monoclonal	Wastewater	24 ng L^{-1}	255
Chlorpyrifos	Monoclonal	Water, fruit juice, honeybee	1–1.75 ng mL^{-1}	256
Oestrogens	Polyclonal and monoclonal	Wastewater		257
Glyphosate	Polyclonal	Surface water	1 µg L^{-1}	258
Imazalil	Monoclonal	Fruit juice	0.06 ng mL^{-1}	259
Linear alkylbenzene sulfonates	Polyclonal	Wastewater	1.8–0.6 µg L^{-1}	122
Parathion	Monoclonal	Water	0.7 ng ml^{-1}	260
PBDEs	Polyclonal	Fish and crabs		261
PBDES	Polyclonal	Water	100 ng L^{-1}	262
Triazines	Monoclonal	Water	0.11 µg L^{-1}	263
Triazines	Polyclonal	Water and food	0.1 µg L^{-1}	264
Triazophos	Monoclonal	Water		265
Triclosan	Polyclonal	Water and soil	10 ng L^{-1}	266
Triclosan and methyltriclosan	Polyclonal	Wastewater	12–25 ng L^{-1}	267
Fluorescent polarization immunoassays				
6-chloronocotinic acid	Polyclonal	Water, soil, urine and fruit	4 µg mL^{-1}	268
Linear alkylbenzene sulfonates	Polyclonal	Wastewater	0.03 mg L^{-1}	269
PAHs	Monoclonal and polyclonal	Water	0.9–3.4 ng mL^{-1}	270
Zearalenone	Monoclonal	Cereals	137 µg kg^{-1}	271

to provide accurate and sensitive measurement of small toxicology analytes such as therapeutic drugs, and some hormones. FPIA makes use of three key concepts to measure specific analytes in a homogeneous format:

- *Fluorescence*. Fluorescein is a fluorescent label. It absorbs light energy at 490 nm and releases this energy at a higher wavelength (520 nm) as fluorescent light.
- *Rotation of molecules in solution*. Larger molecules rotate more slowly in solution than do smaller molecules. This principle can be used to distinguish between the smaller antigen–fluorescein molecule, AgF, which rotates rapidly, and the larger Ab–AgF complexes, which rotate slowly in solution.
- *Polarized light*. Fluorescence polarization technology distinguishes AgF label from Ab–AgF by their different fluorescence polarization properties when exposed to polarized light. When polarized light is absorbed by the smaller AgF molecule, the AgF has the ability to rotate rapidly in solution

before the light is emitted as fluorescence. The emitted light will be released in a different plane from that in which it was absorbed, and is therefore unpolarized. With the larger Ab–AgF complex, the same absorbed polarized light is released as polarized fluorescence because the much larger Ab–AgF complex does not rotate as rapidly in solution. The light is released in the same plane as the absorbed light energy, and the detector can measure it.

A FPIA based on a monoclonal antibody for the detection of parathion-methyl was developed and optimized by Kolosova et al.,[129] with a linear range from 25 to 10 000 µg L^{-1}. The detection limit was 15 µg L^{-1}. Recovery in vegetable, fruit and soil samples averaged between 85 and 110%. The method developed showed a high specificity and reproducibility (CV (coefficient of variation) ranged from 1.5% to 9.1% for inter-assay and from 1.8 to 14.1% for intra-assay). Examples of applications to environmental monitoring using FPIAs are included in Table 9.1.

9.3.1.3 Chemiluminescent Magnetic Immunoassays

Chemiluminescent compounds can also be used to label analytes. A chemiluminescent label produces light when combined with a trigger reagent.

A chemiluminescent magnetic immunoassay (CMIA) for atrazine analysis was developed and evaluated by Fischer-Durand,[130] and this study demonstrates the feasibility of pesticide determination using CMIA, although the sensitivity of the current CMIA format does not reach the required levels.

9.3.1.4 Flow-Injection Immunoassays

Flow-injection immunoassays (FIIA) is based on the introduction of the sample into a carrier stream, which enters the reaction chamber where the immunoreaction takes place. In general in FIIA the antibodies are immobilized to form an affinity column and analyte is pumped over the column. The loading of the antibodies with analyte is followed by pumping over the column enzyme tracers that compete with the analyte for the limited binding sites of the antibodies. Generally, the indirect format produces a result inversely proportional to the analyte concentration. FIIA can be used with electrochemical, spectrophotometric, fluorimetric and chemiluminescence detection methods. Conventional FIIA has been successfully used for the detection of pesticides (*e.g.* triazines[131]). At present, FIIA is integrated into different immunosensors.

9.3.1.5 Immunosensors

Immunosensors are a class of biosensors which use immunoreagents as bioreceptors. They are therefore discussed in section 9.3.2.2.

9.3.2 Biosensors

A biosensor is defined by IUPAC as a self-contained integrated device that is capable of providing specific quantitative or semiquantitative analytical information using a biological recognition element (biochemical receptor), which is retained in direct spatial contact with a transduction element.

One of the major thrust of the green chemistry research activity is the development of new analytical methodologies.[132] New analytical tools are needed for real-time monitoring of industrial wastes.[133] Similarly, a real-time field measurement capability is desired for continuous environmental monitoring that would replace the common approach of sample collection and transport to a central laboratory.[134] Such monitoring offers a rapid control, while minimizing errors and costs associated with laboratory-based analyses. The development of greener analytical procedures, with negligible waste generation or non-toxic materials, is of considerable interest in connection to both centralized and field analyses. These green chemistry principles should thus be applied to all aspects of analytical science.

After many years of development, biosensors have begun to move out of the laboratory and into commercial applications. In spite of the slow and limited commercialization, which may be attributed to the costs of mass production and some key technical barriers, with the combination of advances in biotechnology, nanotechnology and information processing, biosensors promise to open the door to many exciting new environmental monitoring solutions. Based on these approaches, biosensors offer capabilities for rapid, miniaturized, on-line and at-site analysis, with minimal waste production and energy costs of analysis. The coupling of modern detection principles with recent advances in microelectronics has led to powerful and compact analytical devices for real-time, in process-monitoring. A vast array of devices for on-site and *in situ* environmental and industrial monitoring has been developed in recent years.[134,135]

Several reviews on biosensors and biological techniques for environmental analysis have been published during the last few years.[136–138]

9.3.2.1 Classes and Fundamentals

Biosensors have two main parts: a transduction element and a biological receptor. They can be classified according to the bioreceptor elements involved in the recognition process and according to the physicochemical transduction elements.

The main classes of bioreceptors applied in environmental analysis of organic pollution are enzymes, antibodies, DNA, and whole cells. Figure 9.4 summarizes the main classes of biosensors used in environmental analysis.

There are also four basic groups of transduction elements: electrochemical, optical, mass sensitive, and thermal sensors, with electrochemical transducers being the biosensors most commonly described in the literature.

9.3.2.1.1 Electrochemical transduction. Most of these devices fall into two major categories (according to the nature of the electrical signal): amperometric and potentiometric. In *amperometric biosensors* these changes describe

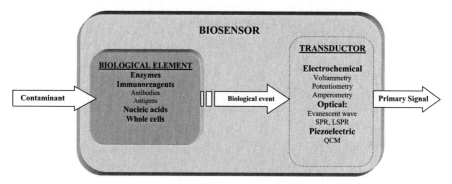

Figure 9.4 Biosensors for environmental analysis: general scheme.

the movement of electrons produced in a redox reaction (electrical current), whereas *potentiometric biosensors* respond to changes in the distribution of charges, causing an electrical potential (voltage) to be produced. Because of its simplicity, electrochemical transduction has given rise to a great variety of low-cost devices based on different formats.[139–141]

Enzyme electrodes have the longest tradition in the field of biosensors.[142–145] Such devices are usually prepared by attaching an enzyme layer to the electrode surface, which monitors changes occurring as a result of the biocatalytic reaction amperometrically or potentiometrically. Amperometric enzyme electrodes rely on the biocatalytic generation or consumption of electroactive species. A large number of hydrogen peroxide generating oxidases and NAD^+-dependent dehydrogenases have been particularly useful for the measurement of a wide range of substrates The liberated peroxide or NADH species can be readily detected at relatively modest potentials (0.5–0.8 V *vs* Ag/AgCl), depending upon the working electrode material. Lowering of these detection potentials is desirable in order to minimize interference from coexisting electroactive species. Potentiometric enzyme electrodes rely on the use of ion- or gas-selective electrode transducers, and thus allow the determination of substrates whose biocatalytic reaction results in local pH changes or the formation or consumption of ions or gas (*e.g.* NH_4^+ or CO_2). The resulting potential signal thus depends on the logarithm of the substrate concentration. The proper functioning of enzyme electrodes is greatly dependent on the immobilization procedure. The immobilization methods generally employed are physical adsorption at a solid surface, cross-linking between molecules, covalent binding to a surface, and entrapment within a membrane, surfactant matrix, polymer or microcapsule. In addition, sol–gel entrapment, Langmuir–Blodgett (LB) deposition, electropolymerization, self-assembled biomembranes and bulk modification have been widely used in recent years. However, new approaches have been described aimed at detecting biomolecule immobilization directly while retaining biological activity, and facilitating electron transfer between the immobilized proteins and electrode surfaces. This has led to an intensive use of nanomaterials for the construction of electrochemical biosensors with enhanced analytical performance.[139,146] The particular characteristics of carbon nanotubes make them an attractive material for use in electrochemical biosensors.[140]

9.3.2.1.2 Optical transducers. Optical transducers are based on the application of various optical phenomena, including adsorption, fluorescence, phosphorescence, polarization, rotation, interference and non-linear phenomena such as second harmonic generation. The choice of a particular optical method depends on the nature of the application and the desired sensitivities. In practice, fibre optics can be coupled with all optical techniques, thus increasing their versatility. Optical biosensor formats may involve direct detection of the analyte of interest or indirect detection through optically labelled probes. The main optical biosensor configurations used for environmental applications in recent years are *evanescent wave* (EW) and *surface plasmon resonance* (SPR).

EW is produced in the external media (with a different refractive index) of a waveguide by the electromagnetic field associated with the light guided by total internal refection (TIR). The electromagnetic field does not abruptly switch to zero at the interface between the two media, but decays exponentially with distance from the interface. The penetration depth of the evanescent field is defined as the distance where its strength is reduced to 1/e of its value at the interface and generally has a value around 100 nm. The penetration depth is dependent on the incident angle at the interface and is proportional to the wavelength of the excitation light. When molecules with an absorption spectrum including the excitation wavelength are located in the evanescent field, they absorb energy leading to attenuation in the reflected light of that wavelength. With the aim of improving detectability, various biosensors have been developed combining this principle with labelled molecules that can re-emit the absorbed evanescent photons at a longer wavelength as fluorescence. Part of this emission is coupled back to the waveguide and in this way it is transmitted to the receptor. This phenomenon is known as *total internal reflection fluorescence* (TIRF). TIRF has been used with planar and fibre-optic waveguides as signal transducers in a number of biosensors. In these devices, light is propagated down a waveguide, which generates an electromagnetic wave (the EW) at the surface of the optically denser medium of the waveguide and the adjacent less optically dense medium. The amplitude of the standing wave decreases exponentially with distance into the material of lower refractive index. The fluorescence of a fluorophore excited within the evanescent field can be collected either outside the waveguide, or by coupling the emission frequencies back into the waveguide. The biological sensing element is immobilized on the side of the waveguide rather than at the end. Based on the EW transducing principle, atrazine was detected at concentrations around 0.1 µg L^{-1},[147] and cyclodiene insecticides in the µg L^{-1} range.[148] TIRF is a fast, non-destructive, sensitive and versatile technique that is well suited for monitoring biomolecular interactions. TIRF allows the monitoring of conformational changes, orientation changes, and lateral mobility of biomolecules, and has been used in a high number of biosensors, for example River ANAlyser (RIANA) and Automated Water Analyser Computer Supported System (AWACSS) devices.[149–153] Hormones, pesticides, antibiotics, and endocrine-disrupting chemicals are among the organic pollutants that have been detected with this biosensor system.

Recent EW biosensors have been based in configurations such as grating couplers or Mach–Zehnder interferometers to enable direct measurements of small analytes without the use of fluorescent labels. In *Mach–Zehnder interferometry* (MZI), an optical waveguide is split into two arms which are recombined after a certain distance. The sensor arm is exposed to a variation in the refractive index due to a biorecognition reaction such as an immunoreaction in the sensor channel. Light travelling in the sensor arm will therefore experience a phase shift in comparison with light in the reference arm.[154]

Optical waveguide lightmode spectroscopy (OWLS) is a new sensing technique using an evanescent field for the *in situ* and label-free study of surface processes at molecular levels. It is based on the precise measurement of the resonance angle of linearly polarized laser light, diffracted by a grating and incoupled into a thin waveguide layer. The incoupling is a resonance phenomenon that occurs at a defined angle of incidence that depends on the refractive index of the medium covering the surface of the waveguide. In the waveguide layer, light is guided by TIR to the edges where it is detected by photodiodes. By varying the angle of incidence of the light, a spectrum is obtained from which the effective refractive indexes are calculated for both electrical and magnetic field waves.[155]

A *surface plasmon wave* can be described as a light-induced collective oscillation in electron density at the interface between a metal and a dielectric. At SPR, most incident photons are either absorbed or scattered at the metal/dielectric interface and, consequently, reflected light is greatly attenuated. The resonance wavelength and angle of incidence depend upon the permittivity of the metal and dielectric. SPR biosensors are based on special electromagnetic waves—surface plasmon polaritons—to probe interactions between an analyte in solution and a biomolecular recognition element immobilized on the SPR sensor surface. This principle has been extensively investigated for the monitoring of biological interaction in different types of configurations, and different SPR biosensors have been developed for environmental applications with the support of recent EU framework projects.

Since distinct SPR prototypes have appeared in the market (Biacore, IASys, Sensia, *etc.*), a significant number of applications of this principle have been reported in recent years. Recent papers reported the analysis of 17β-estradiol (E2) in sewage in the coastal marine environment.[156] Using the Biacore system, an LOD of 0.445 µg L^{-1} was achieved. Several papers have reported the analysis of pesticides in environmental samples with minimal or no sample preparation.[157–161] In another example an assay for bioeffect-related screening of chemicals with thyroid-disrupting activity was developed. In this assay two thyroid transport proteins, thyroxine-binding globulin and recombinant transtheryn, were applied in an inhibition assay format using a Biacore 3000 with CM5 biosensor chips coated with L-thyroxine.[162] An immunosensor chip utilizing SPR was made to detect carcinoembryonic antigen based on a protein A-conjugated surface.[163] In another example, a SPR immunosensor was developed for the detection of 2,4-dinitrophenol at ultra-low concentrations (immunoassay range between 1 ppt and 1 ppb). The sensor strategy was based on a competitive immunoreaction between 2,4-dinitrophenol and a 2,4-dinitrophenol-protein–bovine serum albumin conjugate (DNP-BSA). Anti-2,4-dinitrophenol monoclonal antibody was immobilized on a gold thin-film coated SPR sensor chip by means of a chemical coupling process.[164]

The development of large-scale biosensor arrays composed of highly miniaturized signal transducer elements that enable real-time, parallel monitoring of multiple species is an important driving force in biosensor research. This is particularly significant in high-throughput screening applications where many thousands of ligand–receptor or protein–protein interactions must be rapidly examined.

Recently, several research groups have begun to explore alternative strategies for the development of optical biosensors based on the extraordinary optical properties of noble metal nanoparticles, which exhibit a strong UV-vis absorption band that is not present in the spectrum of the bulk metal. This absorption band results when the incident photon frequency resonates with the collective oscillation of the conduction electrons and is known as the *localized surface plasmon resonance* (LSPR) (Figure 9.5). LSPR excitation results in wavelength selective absorption with extremely large molar extinction

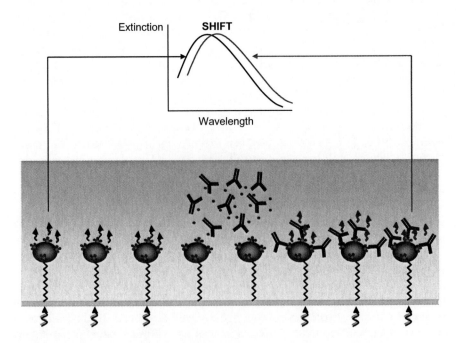

Figure 9.5 General scheme of a surface plasmon resonance immunosensor.

coefficients ($\sim 3.10^{11}\,\mathrm{M}^{-1}\,\mathrm{cm}^{-1}$), resonant Rayleigh scattering with an efficiency equivalent to that of 10^6 fluorophors, and enhanced local electromagnetic fields near the surface of the nanoparticle which are responsible for the intense signals observed in all surface-enhanced spectroscopies. It is well established that the peak extinction wavelength, λ_{max}, of the LSPR spectrum is dependent upon the size, shape, and interparticle spacing of the nanoparticle as well as its dielectric properties and those of the local environment. Consequently, there are at least four different nanoparticle-based sensing mechanisms that enable the transduction of macromolecular or chemical binding events into optical signals based on changes in the LSPR extinction or scattering intensity, shifts in LSPR λ_{max}, or both. These mechanisms are: (1) resonant Rayleigh scattering from nanoparticle labels in a manner analogous to fluorescent dye labels, (2) nanoparticle aggregation, (3) charge-transfer interactions at nanoparticle surfaces, and (4) local refractive index changes.

It has been demonstrated that nanoscale biosensors can be realized through shifts in the LSPR λ_{max} of triangular silver nanoparticles.[165] These wavelength shifts are caused by adsorbate-induced local refractive index changes in competition with charge-transfer interactions at the nanoparticle surface. Triangular silver nanoparticles have been shown to be unexpectedly sensitive to nanoparticle size, shape, and local dielectric environment.[166,167]

9.3.2.1.3 Mass-sensitive sensors. Measurements of small changes in mass is a form of transduction that has been used for biosensor development. Piezoelectric devices and surface acoustic wave devices can be classified in this category. Although this principle has been less reported for environmental applications, it is one of the most promising approaches because of the possibility of miniaturization and the high sensitivity and specificity achieved when coupled to the proper bioreceptor.

The vibration of piezoelectric crystals produces an oscillating electric field in which the resonant frequency of the crystal depends on its chemical nature, size, shape and mass. These crystals can be made to vibrate at a specific frequency of oscillation which is dependent on the electrical frequency. The frequency of oscillation is therefore dependent on the electrical frequency applied to the crystal as well as the crystal's mass. When the mass increases as a result of the binding of analytes, the oscillation frequency of the crystal changes and this change can be measured. The general equation of crystal microbalances can be summarized as follows when the change in mass (m) is very small compared to the total mass of the crystal:

$$\Delta f = Cf^2 \Delta m / A$$

where f is the vibration frequency of the crystal in the circuit, A is the area of the electrode and C is a constant determined in part by the crystal material and thickness. Piezoelectric crystals, sometimes referred to as *quartz crystal microbalances* (QCM), are typically made of quartz and operate at frequencies between 1 and 10 MHz. These devices can operate in liquids with a frequency determination limit of 0.1 Hz; the detection limit of mass bound to the electrode surface is about 10^{-10} to 10^{-11} g.

Acoustic wave devices made of piezoelectric materials are the most common sensors; they bend when a voltage is applied to the crystal. Acoustic wave sensors are operated by applying an oscillating voltage at the resonant frequency of the crystal, and measuring the change in resonant frequency when the target analyte interacts with the sensing surface.

9.3.2.2 Biosensors for Environmental Monitoring

This section presents and discusses the recent development of biosensors for the detection of organic pollution, classified according to the biological recognition elements.

9.3.2.2.1 Enzyme biosensors. Since the first biosensor described by Clarck and Lyons in 1962, traditional applications of enzymes have been linked to electrochemical transduction. Many examples of applications have been reported for the analysis of environmental pollutants, especially using oxidoreductases (*e.g.* tyrosinase, peroxidase and lactase),[168] and hydrolases (*e.g.* choline estearases).[145] Some examples of recent environmental applications of electrochemical enzyme biosensors are listed in Table 9.2.

In recent years the hugely increased interest in nanomaterials has been driven by their many desirable properties. In particular, the ability to tailor the size and structure and hence the properties of nanomaterials offers excellent prospects for designing novel sensing systems and enhancing the performance of bioanalytical assays.

An important design challenge in amperometric enzyme electrodes is the establishment of satisfactory electrical communication between the active site of the enzyme and the electrode surface. The redox centre of most oxidoreductases is electrically insulated by a protein shell and the enzyme thus cannot be oxidized or reduced at an electrode at any potential. In general, there is a need for co-substrates or mediators to achieve efficient transduction of the biorecognition event. Aligned carbon nanotubes (CNTs), prepared by self-assembly, can act as molecular wires to allow electrical communication between the underlying electrode and redox proteins (covalently attached to the ends of the single-walled CNTs).[169]

The deposition of platinum nanoparticles on CNTs has led to further improvements in the detection of enzymatically liberated peroxide species.[170] The same group of researchers reported the oxidative determination of arsenite (As(III)) using boron-doped diamond (BDD) macro- and microelectrodes modified by electrodeposition of platinum nanoparticles.[171] The formation of platinum nanoparticles was proved using cyclic voltammetry measurement, and atomic force microscopy (AFM) and scanning electron microscopy (SEM)

Table 9.2 Examples of electrochemical biosensors reported for environmental analysis.

Analyte/effects	Biological receptor/transducer	Matrix	Limit of detection	References
Bisphenol A	Tyrosinase/amperometric detection	Drinking and surface water	0.02 μM	272
Bisphenol A	Antibody/potentiometric	Water	0.6 μg L^{-1}	273
Bisphenol A, 17-β-estradiol	Tyrosinase/amperometric detection	Water	10^{-6} M for bisphenol A	274
Carbamates	AChE, BChE/potentiometric detection	Aqueous synthetic samples	1.5.10^{-5}–2.5.10^{-3} M	275
Catechol, p-cresol, phenol, p-chlorophenol, and p-methylcatechol	Tyrosinase/amperometric detection	Spring and surface water	9×10^{-8} M	276
Fenitrothion and ethyl p-nitrophenol thiobenzene phosphonate	Whole-cell biosensor/amperometric detection	Water	1.4 μg L^{-1} of fenitrothion and 1.6 μg L^{-1} of EPN	213
Genotoxicity	DNA/amperometric detection	Wastewater		193
Genotoxicity	DNA electrochemical biosensors	Water		277
Heavy metals	DNA/amperometric detection	Water	4.0×10^{-11}, 1.0×10^{-10}, 1.0×10^{-9}, and 5.0×10^{-9} M for Cu(II), Pb(II), Cd(II), and Fe(III), respectively	196
Organophosphates	Whole-cell biosensor/amperometric detection	Water	Phosphate levels in the marine system were 0.04 mg L^{-1}	278
PAT gene	DNA/impedance measurement	Water	1.0×10^{-11} M	279
Phenol, o-cresol, p-cresol, m-cresol, catechol, dopamine, adrenaline (ephinephrine)	Tyrosinase/amperometric detection	Water solution	303 μA mM^{-1}	280
Surfactants	Bacteria/amperometric detection	Water	0.25 mg L^{-1}	281
Toxicity	Bacteria/amperometric detection	Water and wastewater		217

revealed the size and size distribution of the deposited nanoparticles. With linearity up to 100 ppb and a detection limit of 0.5 ppb, the electrochemical system could be used for processing tap and river water samples. Over 150 repetitive runs could be performed, and electrochemical etching of platinum allowed the reuse of the BDD microelectrode. The presence of copper and chloride ions, the two most severe interferents at levels commonly found in groundwater, did not interfere with the assay.

In addition to CNT films, it is possible to use CNT-based inks.[172] The excellent electrocatalytic properties of metal nanoparticles (compared to bulk metal electrodes) can also benefit amperometric enzyme electrodes. For example, You *et al.*[173] dispersed iridium nanoparticles in graphite-like carbon for improved amperometric biosensing of glutamate.

Another promising and controllable route for preparing conducting-polymer nanowire enzyme sensors involves electrodeposition within the channel between electrodes.[174] The catalytic properties of metal nanoparticles have also facilitated electrical contact between redox centres of proteins and electrode surfaces. For example, gold nanoparticles were shown to be extremely useful as electron relays for the alignment of glucose oxidase on conducting supports and wiring its redox centre.[175]

The development of electrochemical DNA and protein sensors on screen-printed electrodes based on the catalytic activity of hydrazine has been described by Shiddiky *et al.*[176]

A wide range of enzyme electrodes based on dehydrogenase or oxidase enzymes rely on amperometric monitoring of the liberated NADH or hydrogen peroxide products. The anodic detection of these species is often hampered by the large overvoltage encountered for their oxidation. It has been proven that enhancement of the redox activity of hydrogen peroxide[177] and NADH[178] using CNT-modified electrodes helps to overcome voltage limitations.

The ability of CNTs to promote electron transfer reactions is due to the presence of edge plane defects at their end caps. Yang *et al.*[179] reported a lactate electrochemical biosensor with a titanate nanotube as direct electron transfer promoter. The nanotubes offer the pathway for direct electron transfer between the electrode surface and the active redox centres of lactate oxidase, which enables the biosensor to operate at a low working potential and to avoid the influence of the presence of O_2 on the amperometric current response.

Gong *et al.*[180] reported a nanosized calcium carbonate–chitosan (nano-$CaCO_3$–chi) composite film with a three-dimensional porous, network-like structure, providing a favourable and biocompatible microenvironment to immobilize an enzyme. By using such a composite film as the enzyme immobilization matrix, a highly sensitive and stable acetylcholine esterase (AChE) sensor was achieved for determination of methyl parathion as a model of organophosphate pesticides. The inhibition of methyl parathion was proportional to its concentration ranging from 0.005–0.2 to 0.75–3.75 µg mL^{-1}. The detection limit was found to be as low as 1 ng mL^{-1}.

Recently an electrochemical biosensor for the determination of organophosphorus insecticides in vegetable crops was described. The self-assembled monolayers (SAMs) of single-walled CNTs wrapped by thiol-terminated single-strand oligonucleotide (ssDNA) on gold was used to prepare a nanosized polyaniline matrix for AChE enzyme immobilization.[181] The detection limit of the biosensor for methyl parathion and chlorpyrifos pesticides was found to be 1×10^{-12} M in both cases.

9.3.2.2.2 Immunosensors.
Antibody–antigen interactions have been exploited in many biosensors for environmental analysis. Table 9.3 summarizes some successful applications for environmental analysis. Electrochemical immunosensors have been used for environmental analysis in amperometric, potentiometric, and conductimetric configurations.

Amperometric immunosensors measure the current generated by oxidation or reduction of redox substances at the electrode surface, which is held at an appropriate electrical potential. One-use screen-printed electrodes have been developed: examples are the detection of PAHs[182] and the quantitative

Table 9.3 Immunosensors for environmental applications.

Analytes	Transducer	Matrix	Limit of detection	References
2,4-Dichloro-phenoxyacetic acid	SPR	Water	0.1 ng mL^{-1}	282
2,4-Dichloro-phenoxyacetic acid	Fluorescence/chemiluminescence	Water	0.02 ng mL^{-1}	263
Atrazine	SPR	River water	20 ng L^{-1}	161
Atrazine	TIRF	Water	0.155 ng mL^{-1}	283
Atrazine	SPR	Water	1 µg mL^{-1}	284
Atrazine	Impedimetric	Rain, surface, marine and ground water	20 ng L^{-1}	285
Chlorpyrifos	SPR	Surface, ground and drinking water	55 ng L^{-1}	286
Dioxins		Flyash	0.1–0.01 ng mL^{-1}	287
Hormones	TIRF	Water	1.4 ng L^{-1}	151
Isoproturon	Fluorescence	Water	9.7 ng L^{-1}	288
Monocrotophos	Electrochemical	Water	0.3 ng mL^{-1}	289
Organophosphorus and carbamate	QCM	Water	1 ng mL^{-1}	290
Pesticides	TIRF	Water	—	151
Pesticides	Electromagneto-immunosensor	Water	—	264
Stanozolol	LSPR	Water	—	167
Sulphamethoxazole	Piezoelectric	Soil, water	0.15 ng mL^{-1}	291

LSPR, localized surface plasmon resonance; QCM, quartz crystal microbalance; SPR, surface plasmon resonance; TIRF, total internal reflection.

detection of 2,4,6-trichloroanisole.[183] Other approaches explored the use of recombinant single-chain antibody (scAb) fragments; this approach was used for atrazine determination.[184] Recently, an amperometric immunosensor for ricin in water using CNT and graphite paste electrodes has been described by Suresh et al.[185]

A nanostructured progesterone immunosensor using a tyrosinase–colloidal gold–graphite–Teflon biosensor as amperometric transducer has been developed by Pingarron's group.[186] A QCM has been used for dioxins detection.[138]

TIRF has been used with planar and fibre-optic waveguides as signal transducers in a number of biosensors that have been reported. Based on the EW transducing principle, the simultaneous analysis of relevant environmental organic pollutants, such as atrazine, isoproturon and oestrone, in water samples was reported using optical biosensors (RIANA[187,188] and AWACSS[149,189]).

A large number of SPR immunoassays have been developed for environmental analysis, as been presented in the previous section. Currently, novel SPR based configurations have been gaining attention, e.g. SPR imaging (SPRi) which is at the forefront of optical label-free and real-time detection. It offers the possibility of monitoring hundreds of biological interactions simultaneously and, from the binding profiles, allows the estimation of the kinetic parameters of the interactions between the immobilized probes and the ligands in solution. The current state of development of SPRi technology and its application, including commercially available SPRi instruments, has recently been reviewed by Scarano et al.[190]

9.3.2.2.3 Nucleic acid and biosensors.
Due to their wide range of physical, chemical and biological properties, nucleic acids have been incorporated into a wide range of biosensors and bioanalytical assays, many of which possess interesting features for environmental applications.

Several DNA biosensors and bioassays have been reported for the detection of chemically induced DNA damage. The structure of DNA is very sensitive to the influence of environmental pollutants, such as PAHs.[191] Different groups of organic pollutants are characterized by a great affinity to DNA, causing mutagenesis and carcinogenesis. It is therefore very convenient to use DNA-containing systems, e.g. DNA-based biosensors,[192–194] to perform genotoxic assays, or for rapid testing of pollutants for mutagenic and carcinogenic activity. Several researchers have developed electrochemical DNA systems for environmental analysis. Mascini's group[195] determined genotoxic compounds by their effect on the oxidation signal of the guanine peak of calf thymus DNA immobilized on the surface of disposable electrodes and investigated by chronopotentiometric analysis. This type of DNA biosensor was able to detect known intercalating compounds, such as aromatic amines, and its applicability to river and waste water samples was also demonstrated.[196] However, many different configurations have been applied in the rapid screening of genotoxic compounds using the molecular interaction between surface-linked DNA and the target pollutants or drugs, using electrochemical as well as optical and mass transducers.[142,197]

There is also an ongoing effort in the area of biosensor technology for measuring nucleic acid hybridization. In this case a sequence-specific probe (usually a short synthetic oligonucleotide) is integrated within a signal transducer. The probe, immobilized onto the transducer surface, acts as the biorecognition molecule and recognizes the target DNA or RNA. Optical, electrochemical and mass detectors are mainly used for transducing the biorecognition event into an analytical signal. However, such nucleic acid sensing applications require high sensitivity and substantial efforts have been devoted to amplifying the transduction of the oligonucleotide interaction.[198]

Nanoparticle-based amplification schemes have led to the sensitivity of bioelectronic assays being improved by several orders of magnitude. In 2001 Wang's group reported[199] the use of colloidal gold tags for the electronic detection of DNA hybridization. The coupling of the biorecognition element to surfaces of magnetic beads offers effective minimization of non-specific binding. The hybridization of probe-coated magnetic beads with the metal-tagged targets results in three-dimensional network structures of magnetic beads, cross-linked together through the DNA and gold nanoparticles. Several amplification processes can be used to dramatically enhancing the sensitivity of particle-based bioelectronic assays. The concept of using inorganic-nanocrystal tracers for multitarget electronic detection of DNA[200] can be scaled up and multi-plexed. One-dimensional nanowires can also be used to bridge two closely spaced electrodes for label-free DNA detection. For example, a p-type silicon nanowire-functionalized with peptide nucleic acid (PNA) probes has been shown to be extremely useful for real-time label-free conductometric monitoring of the hybridization event.[201] This relies on the binding of the negatively charged DNA target, which leads to an increase in conductance, reflecting the increased surface charge. CNTs can also lead to ultrasensitive bioelectronic detection of DNA hybridization. For example, CNTs can be used as carriers for several thousand enzyme tags and for accumulating the α-naphthol product of the enzymatic reaction. Such a CNT-derived double-step amplification pathway (of both the recognition and transduction events) allows the detection of DNA down to the 1.3 zmol level. Recently, Feng et al.[202] reported a gold nanoparticle/polyaniline nanotube membranes on a glassy carbon electrode (Au/nanoPAN/GCE) for the DNA sensing. The properties of the Au/nano-PAN/GCE and the characteristics of the immobilization and hybridization of DNA were studied by cyclic voltammetry, differential pulse voltammetry and electrochemical impedance spectroscopy. The synergistic effect of the two kinds of nanomaterials, nanogold and nanoPAN, could dramatically enhance the sensitivity for recognition of DNA hybridization. A simple and sensitive electrochemical DNA biosensor based on *in situ* DNA amplification with nanosilver as label and horseradish peroxide (HRP) as enhancer has been designed by Fu et al.[203]

The highly sensitive and sequence-specific detection of single-stranded oligonucleotides using nonoxidized silicon nanowires (SiNWs) has been demonstrated by Zhang et al.[204] To maximize device sensitivity, the surface of the SiNWs was functionalized with a densely packed organic monolayer via hydrosilylation, subsequently immobilized with PNA capable of recognizing the label-free complementary target DNA. Because of the selective functionalization of the SiNWs, binding competition between the nanowire and the underlying oxide is avoided.

Thus, as reported above, the promising performance of NA (nucleic acid) biosensors for hybridization studies has been demonstrated in several studies.

9.3.2.2.4 Nuclear receptors.
The oestrogen receptor (ER) has been the basis of several biosensors to assess the oestrogenic potency of complex matrices polluted with endocrine disruptor compounds (EDC).

Steroid hormones induce different effects in mammalian cells after binding to specific intercellular receptors, which are ligand-dependent transcription factors. Many endocrine disruptors can bind to ERs either as agonists or as antagonists. Thus, the binding ability of the chemicals towards the ER is used to test their potential environmental toxicity. The advantage of receptor assays is that they are quite simple to perform and allow the identification of all endocrine disrupters that act through the ER. The natural sensing element most commonly used is the human ER.

Using SPR, Usami et al.[205] developed a simple competitive assay for the evaluation of different chemicals using human recombinant ER. The system measured the binding between oestradiol immobilized on the sensor chip and an injected human recombinant ER.

In another example using an optical SPR sensor and the human ER-α, Hock et al.[206] performed binding studies with oestradiol and xenoestrogens. Butala et al.[207] described the use of a fractal analysis to model the binding and dissociation kinetics between analytes in solution and the ER immobilized on a SPR sensor chip.

An evanescent-based biosensor[208] was developed with laser-based fibre optics using fluorescent dye-labelled recombinant human ER-α (rhERα) and hERβ as probes. A three-tiered approach evaluating various steps in the formation of the ER complex and its subsequent activity was developed for instrument calibration to detect oestrogen mimics in biological samples, water and soil. Using this approach, binding affinities and activities of certain known oestrogen mimics were determined for their use as calibrator molecules. Results indicated rhERα and rhERβ may be employed as probes to distinguish oestrogen mimics with a broad range of affinities.

9.3.2.2.5 Whole-cell biosensors.
The main classes are bacterial, algal, fungal, and cell-based biosensors. Fabrication of a whole-cell biosensor requires immobilization of microorganisms on transducers. Since whole-cell biosensor response, operational stability and long-term use are, to some extent, a function of the immobilization strategy used, immobilization technology plays a very important role and the choice of immobilization technique is critical.

Microorganisms can be immobilized on transducer or support matrices by chemical or physical methods.[209]

Chemical methods of bacterial immobilization include covalent binding and cross-linking.[210] Covalent binding methods are based on a covalent bond between functional groups of the microorganism cell wall components, such as amine, carboxylic or sulphydryl, and the transducer, such as amine, carboxylic, epoxy or tosyl. To achieve this goal, whole cells are exposed to harmful chemicals and harsh reaction conditions, which may damage the cell membrane and decrease the biological activity. For this reason, this method has not been successful for immobilization of viable microbial cells.[209] Cross-linking involves bridging between functional groups on the outer membrane of the cells by multifunctional reagents such as glutaraldehyde or cyanuric chloride, to form a network. This method has found wide acceptance for the immobilization of microorganisms. The cells may be cross-linked directly on to the transducer surface or on to a removable support membrane, which can then be placed on the transducer. The ability to replace the membrane with the immobilized cells is an advantage of the latter approach. Although cross-linking has advantages over covalent binding, the cell viability and/or the cell membrane biomolecules can be affected by the cross-linking agents. Thus cross-linking is suitable in constructing microbial biosensors where cell viability is not important and only the intracellular enzymes are involved in the detection.[210]

Adsorption and entrapment are the two widely used physical methods for microbial immobilization, because they produce a relatively small perturbation of the microorganism's native structure and function.[211] Physical adsorption is the simplest. Typically, a microbial suspension is incubated with the electrode or an immobilization matrix, such as alumina and glass beads,[210,212] followed by rinsing with buffer to remove unabsorbed cells. The microbes are immobilized as a result of adsorptive interactions such as ionic, polar or hydrogen bonding and hydrophobic interaction. However, immobilization using adsorption alone generally leads to poor long-term stability because of desorption of microbes.

The immobilization of microorganisms by entrapment can be achieved either by the retention of the cells in close proximity to the transducer surface using a dialysis or filter membrane or in chemical/biological polymers/gels (*e.g.* alginate, carrageenan, agarose, chitosan, collagen, polyacrylamide, polyvinylachohol, poly(ethylene glycol), polyurethane).[213] A major disadvantage of entrapment immobilization is the additional diffusion resistance offered by the entrapment material, which will result in lower sensitivity and detection limits.

One of the most relevant applications of *bacterial biosensors* has been the determination of the biochemical oxygen demand (BOD), toxicity, genotoxicity and oestrogenicity. In recent years, several publications have reviewed technology based on microbial biosensors for ecotoxicity assessment. An example of a recent development is a BOD biosensor using salt-tolerant *Bacillus licheniformis* for use in seawater.[214] In another recent work, a novel reactor-type biosensor for rapid measurement of BOD was developed, based on using immobilized microbial cell (IMC) beads as recognition bioelement in a completely mixed reactor which was used as determining chamber, replacing the membrane traditionally used as a recognition bioelement. This novel kind of BOD biosensor significantly increased the sensitivity of the response and the detecting precision, and prolonged the lifetime of the recognition element.[215]

In addition to BOD determination, bacterial biosensors have been developed to detect different groups of organic pollutants. For example, amperometric biosensors based on genetically engineered *Moraxella* sp. and *Pseudomonas putida* with surface-expressed organophosphorus hydrolase (OPH) have been developed by Lei et al.[213] for the detection of organophosphorus pesticides. The biosensor consisted of recombinant PNP-degrading/oxidizing bacteria *P. putida* JS444, anchoring and displaying organophosphorus hydrolase (OPH) on its cell surface as biological sensing element, and a dissolved oxygen electrode as the transducer. Surface-expressed OPH catalysed the hydrolysis of fenitrothion and ethyl *p*-nitrophenol thiobenzenephosphonate (EPN) to release 3-methyl-4-nitrophenol and *p*-nitrophenol, respectively, which were oxidized by the enzymatic machinery of *P. putida* JS444 to carbon dioxide while consuming oxygen, which was measured and correlated to the concentration of organophosphates. Under the optimum operating conditions, the biosensor was able to measure as little as 277 ppb of fenitrothion.

Another relevant application of bacterial biosensors is the toxicity assessment of wastewater. A whole-cell bacterial biosensor using screen-printed electrodes with *E. coli* was reported by Farré et al.[216] In another study, the same group determined the acute toxicity of wastewaters using *P. putida* electrodes, in conjunction with chemical analysis.[217] In another example, a biodetector was designed based on a bioluminescence test using the bacterium *Vibrio fischeri*, performed in a liquid continuous flow-through system. In this system a new flow-cell holder and a new case including a top cover to connect the flow cell with the waste and the incubation capillary in a light-proof manner are described.[218]

Different examples of the application of biosensors based on bioluminescent bacteria for wastewater monitoring have been reported. Whole-cell, genetically modified, bioluminescent biosensors and their immobilization on thin films of poly(vinyl alcohol) cryogels have been constructed out by Philp et al.[219] The biosensor was designed for use in monitoring the toxicity of industrial wastewaters containing phenolic materials. It has been proved to operate predictably with pure toxicants, within the wastewater treatment plant. Several types of recombinant bioluminescent bacteria have been employed to set up a multi-channel, continuous, water toxicity monitoring system. The one developed by Kim's research group was based on channels that each hosted a different recombinant bacterial strain, and consisted of two mini-bioreactors to enable continuous operation, *i.e.* without system interruption due to highly toxic samples.[220] The luminescent strains were DPD2540 (*fabA::luxCDABE*), DPD2794 (*recA::luxCDABE*), and TV1061 (*grpE::luxCDABE*), induced by cell membrane-, DNA-, and protein-damaging agents; GC2 (lac::luxCDABE) was a constitutive strain. The field samples were waters discharged from a nuclear power plant and a thermoelectronic power plant. Each channel showed specific luminescent response profiles, and by comparing the luminescent bacteria signals of standard chemicals with those of discharged water samples, the equivalent toxicity of the field water could be estimated.[220]

Another important group of whole-cell biosensors is those based on *algal* immobilization. Fluorescence transduction has been widely applied in ecotoxicity biosensors. For example, a multiple-strain algal biosensor was constructed for the detection of herbicides inhibiting photosynthesis. In this work,[221] nine different microalgal strains were immobilized on an array biochip using permeable membranes. The biosensor allowed on-line measurements of aqueous solutions passing through a flow cell, using chlorophyll fluorescence as the biosensor response signal. The herbicides atrazine, simazine, diuron, isoproturon and paraquat were detectable within minutes. Another example, a miniaturized biosensor-based optical instrument using the algal photosystem II (PSII) system has been designed and fabricated for multiarray fluorescence measurements of several biomediators in series, with applications in environmental monitoring and agrofood analysis.[222] A recent paper[223] described the development of an optical biosensor for environmental monitoring based on computational and biotechnological tools for engineering the photosynthetic D1 protein of *Chlamydomonas reinhardtii*. The fluorescence properties were exploited and a reusable and portable multiarray optical biosensor for environmental monitoring was developed with LODs between 0.8×10^{-11} and 3.0×10^{-9} M for triazine herbicides.

Electrochemical transduction has also been used in some examples. Amperometic algal-based sensors have been designed by Shitanda *et al.*,[224] taking advantage of variations in the photosynthetically produced oxygen.

Fungal cells can provide all the advantages of bacterial cells, but in addition they can provide information that is more relevant to other eukaryote organisms. These cells are easy to cultivate and to manipulate for sensor configurations, and are amenable to a wide range of transducer methodologies. There are several reports on the use of wild-type fungal cells and yeast to detect toxicity. The use of eukaryote cells as the detection element is preferable because their response to toxins more accurately predicts the response in plant and animal cells than does the prokaryote response. The naturally bioluminescent fungi *Armillaria mellea* and *Mycena citricolor* were used as the sensing element in toxicity bioassays,[225] and as the basis for the development of toxicity biosensors. The bioluminescence inhibition during 60 min exposure to the toxin was monitored to provide EC_{50} values.

Immobilized *Sacharomices cerevisiae* was used to assess the toxicity of four free and three conjugated cholanic acids, and a toxicity scale was constructed from the results.[226] The authors concluded that this sensor could therefore be considered a valid instrument for the preliminary evaluation of the toxicity of organic compounds or drugs. Campanella *et al.* also used *S. cerevisiae* as the sensing element of four general toxicity biosensors. Toxicity was determined by detecting the retardation of metabolic activity.

Many of the current yeast sensors are for genotoxicity and are based on genetically modified *S. cerevisiae*. Other genetically modified yeast sensors have been constructed to detect specific molecules or groups of molecules. Some modified cell sensors function as model eukaryote cells to detect toxicity and in particular genotoxicity, either of single specific molecules or of mixtures of molecules such as in environmental samples. Although their use in the laboratory generally not problematic, the use of genetically modified sensors outside the laboratory, for on-site environmental monitoring for example, may be restricted by biosafety requirements.

Cell cultures, in particular those derived from fish, have been successfully employed as a biological alternative to the use of whole animals in ecotoxicity studies. The use of fish cell lines in conventional bioassays such as neutral red retention assays is, however, labour intensive, lengthy and costly. To date several transduction element have been explored, but most biosensors schemes are based on optical transduction. Although vertebrate cell lines in biosensor configurations can be monitored electrochemically,[227] stability was initially poor compared with microbial cells. Alternative approaches for immobilization or transduction have been developed. One alternative approach to the monitoring of cellular activity has been based on the use of luminescence reporter genes and several works describe biosensors using this strategy. Fibroblastic cells of blue gill sunfish (BF-2) have been transfected with a plasmid containing the luciferase gene *Luc*, which allows the luminescence of the transgenic cells following environmental challenge to be monitored via an optical transducer.[228–230] Other transduction elements have also been used; for example, an impedimetric biosensor for drinking-water toxicity assessment was presented by Curtis *et al.*[231]

In recent years mammalian cell-based biosensors have emerged as powerful functional tools for the rapid detection of hazards and threats associated with food, agriculture, environment and biosecurity. Assessing hazard-induced physiological responses, such as receptor–ligand interactions, signal transduction, gene expression, membrane damage, apoptosis and oncosis of living sensing organisms can provide insight into the basis of toxicity for a particular hazard. Banerjee *et al.*[232] reviewed mammalian cell-based biosensors for pathogen and toxin detection.

9.3.2.3 *Autonomous Biosensor Wireless Networks*

The concept of autonomous wireless networks is an attractive futuristic idea which envisages the widespread deployment of networked sensors for continuous environmental monitoring.

Nowadays, molecular and biological events can be monitored and converted to digital signals by means of biosensors and this fact opens up new opportunities for analytical devices to dramatically enhance the future of environmental monitoring over large geographical areas. However, the cost of reliable autonomous biosensing is still far too high for massively scaled-up deployments. Some of the drawbacks that must first be overcome are the complexity of biosensing processes and the need for regular recalibration; energy consumption and waste production should also be minimized. There is also a need for more stable biological reagents that can be stored locally (*i.e.* on/in the device) for extended periods of time.

Considerable advances have been made in recent years in hardware and software for building wireless biosensor networks. However, to ensure effective data gathering by biosensor networks for monitoring remote outdoor environments, the following problems remain:

- *Reactivity:* the ability of the network to react to its environment, and provide only relevant data to users;
- *Robustness:* the ability of network nodes to function correctly in harsh outdoor environments;
- *Network lifetime:* maximizing the length of time during which the network is able to deliver data before batteries or other consumables are exhausted.

The fundamental building block of all sensor networks is the so-called *sensor node*. A sensor node is the smallest component of a sensor network that has integrated sensing and communication capabilities. It contains basic networking capabilities through wireless communications with other nodes as well as some data storage capacity and a microcontroller that performs basic processing operations. Sensor nodes usually come with a sensor board that slots onto the controller board, which allows the user to interface other sensors, provided the signal is presented in an appropriate form for the controller. They also include a power supply, usually provided by an on-board battery.

In recent years, there has been a focus on integrating a local energy scavenging capability, as the small lithium button batteries commonly employed have limited lifetime and regular manual replacement is unrealistic. The sensor nodes within a wireless sensor network are also commonly referred to as '*motes*'. During the past few years, the most widely used motes have been those provided by Crossbow Technologies Inc., based in San Jose, CA, which is a spin-off from the University of California Berkeley.

The hardware requirements for wireless sensors include robust radio technology, a low-cost and energy-efficient processor, flexible signal inputs/outputs for linking a variety of sensors, a long-lifetime energy source, and a flexible, open-source development platform. Additional constraints for wireless sensor nodes include a small physical footprint capable of running on low-power processors, small memory requirement, and high modularity to aid software rescue. Thus, the basic components of a sensor node are a microcontroller, radio transceiver, set of transducers, and power source, and the software that runs on these nodes must be small and allow for efficient energy use.

In the simplest case, motes are programmed before deployment to perform measurements at a particular sampling rate and return the captured data in a prearranged format. In more sophisticated deployments, the motes are programmed to facilitate sampling rates that adapt to external events and function cooperatively in terms of finding the optimum route for returning data to remote base stations.

Because of the limited computing power of sensor motes, they often employ TinyOS, an operating system written in the nesC programming language, which is a dialect of C specifically designed for restricted operating environments such as exists on sensor motes where limited memory and processor power are available.[233] More recently products that are fully C compliant have become available, and these are generally preferred by experienced programmers.

However, sensor and biosensor networks are in their infancy. A wireless magnetoelastic biosensor for the detection of acid phosphatase was reported by Wu et al.[234]

Bendikov et al.[235] reported millimetre scale sensors integrated with a sampling capability. Under the EU framework programmes, the AWACSS project developed an optical sensor to detect river pollution. These projects address the problems of mobility, scale, and sensing in water as well as the development of new sensors.

9.3.3 Non-Biological Techniques

9.3.3.1 Separation Techniques: Greening Liquid Chromatography

HPLC and GC are often used prior to the detection systems for analysis of organic pollutants. GC, as the name suggests, does not use solvents but relies on the separation of the analytes in the gas phase. However, HPLC, a proven technique that has been used in laboratories worldwide for the past 30-plus years,[236] uses large amounts of solvents and has become one of the preferred techniques for analysing polar contaminants. However, advances in HPLC have led to increased interest in comparing the ultimate performance limits of methodologies, with the aim of increasing the resolving power per unit time. The use of nanoflows for HPLC is one example,[237,238] but it is not commonly used in environmental analysis. Recently, UPLC has been introduced to enhance sample throughput and reduce analysis time and ultimately mobile phase consumption. UPLC uses HPLC columns packed with sub-2 μm particles, implying operation at high back-pressures, to significantly reduce analytical run times. By decreasing the particle size of the packing material, the analyst can reduce the height of a theoretical plate, making shorter column lengths possible and widening the range of usable flow rates without sacrificing separation power. Consequently, the analysis time is shortened without losing separation quality.[239]

Also, increasing the temperature of the chromatographic mobile phase above ambient temperature reduces the time necessary for the chromatogram and consequently reduces solvent use.[240] This kind of application is called *high-temperature liquid chromatography* (HTLC) and can be of great value since the concomitant decrease of the mobile phase viscosity and increase in solute diffusivity with temperature allows the use of high flow rates, leading to low analysis time without loss of efficiency or significant increase in column back-pressure.[240] As the dielectric constant and surface tension of water decrease with increasing temperature, a large proportion of organic solvent in the mobile phase is replaced by water.[240] An increases in the temperature of the mobile phase (water) by 5 °C is comparable to an increase of 1% acetonitrile,[241] and an increase of 3.75 °C is equivalent to an increase of 1% of methanol.[242] Therefore water above its boiling point at atmospheric pressure, up to its supercritical temperature, exhibits reduced viscosity, increased capability to dissolve non-polar compounds, and decreased polarity.[243] Several authors have studied the use of heated water in HPLC separations,[244,245] but high-temperature chromatography using UPLC columns (HT-UPLC) is relatively new.[240,243] Nguyen et al.[240] evaluated the combination of high-temperature and UPLC for the first time for the separation of several doping agents and a cocktail of pharmaceuticals. Both separations were obtained in less than 1 min at 90 °C and pressures higher than 400 bar. In addition, a shorter column (30 mm length) was used at high temperature and allowed a separation of eight pharmaceuticals in only 40 s.[240] Two chromatographic methods were compared for the separation of eight drugs at 60 °C and 180 °C.[243] The authors evaluated UPLC Aquity columns 10 cm and 30 cm in length. The latter was constructed by linking three 10 cm columns together in series, using zero dead volume fittings in order to maximize the separation. The authors calculated the flow rate with an experimental model developed on the basis of data obtained using a range of model drugs, which demonstrated the relationship between temperature, flow and pressure. For the chromatographic separation of the target analytes, they determined that the maximum flow rate for a back-pressure of 15 000 psi was 0.4 ml min^{-1} at 60 °C and 1 ml min^{-1} at 180 °C. It was shown that there was an improvement in efficiency using the longer columns, and the retention time was dramatically reduced by the use of the higher temperatures. The estimated back-pressure of the 30 cm column at room temperature was 65 000 psi, which is impractical with any commercially available chromatographic system.[243]

Another green chromatographic technique for the separation of organic pollutants is planar chromatography, particularly *high performance thin-layer chromatography* (HPTLC). This technique is a simpler, more cost-effective method, using less solvent than HPLC for the chromatographic separation of a wide spectrum of substances.[246,247] However, because of the advance of HPLC, HPTLC has taken a back seat, in spite of its multiple advantages, such as the possibility of performing parallel analysis, high flexibility in chromatography and detection, and tolerance for samples highly loaded with matrix. HTPLC is a well-known off-line separation approach, but it is underestimated. For some applications it has been shown to be superior to HPLC systems. For instance, the rapid analysis of bioactive secondary metabolites in marine sponges by coupling HPTLC with bioluminescence and MS showed the utility of this technique.[248] HPTLC gives a very fast response and the organic mobile phase evaporates from the plate, so that the solvent cannot impede direct detection with the bioassay. If the separated compounds were bioactive, they inhibited bacterial luminescence and could be identified as dark zones on the luminescent background. The microbiological detection revealed new compounds. Then the mass spectra of the bioactive zones were recorded within seconds by *direct analysis in real time* (DART) or with an exact mass analyser (LTQ Orbitrap XL hybrid FT-MS), making it possible to identify avarone as a bioactive metabolite of the sponge *Dysidea avara*. This methodology proved to be very effective not only for the detection but also for the identification of unknown bioactive metabolites in sponges.[248] Another feature of the technique is the novel *in situ* pre-and post-chromatographic derivatization of the target analytes. For example, the derivatization of the metabolites from the sponge *Dysidea avara*, by dipping the plate into sulphuric acid dissolved in water (cooled) and then in methanol, took 1 min.[248] Another example is the rapid and sensitive determination of acrylamide in drinking-water by HPTLC with fluorescence detection after derivatization with dansulfinic acid.[249] After SPE, aliquots of 100 µl of water extracts were oversprayed on a HPTLC silica gel plate at the starting zone and covered *in situ* with the derivatization agent dansulfinic acid by heating the plate for different periods. Subsequently, the products containing fluorescent cromophores were analysed by fluorescent detection.[249] Coupling HPTLC with UV or MS detectors, the quantification of target analytes in the plate is also feasible. Aranda and Morlock[250] developed a method for the quantification of caffeine in pharmaceutical and energy drink samples using stable isotope dilution. After sample preparation, samples and caffeine standard were applied on silica gel 60 F_{254} HPTLC plates and overspotted with caffeine-d_3 used for correction of the plunger positioning. In this work, it was shown that isotope dilution was a good option for the correction of plunger positioning because the stable isotopic internal standard showed the same chromatographic behaviour (same hR_f value) as the analyte and thus the positioning error could be nullified. The calibration showed a linear regression with a determination coefficient of 0.9998.[250]

9.3.3.2 Detection Techniques

In spectroscopic and electrochemistry techniques the size of the sample is relatively small and there is no sample preparation step. These kinds of techniques are green in their basis. However, detection techniques like MS require sample preparation and frequently chromatographic separation of the sample prior to detection. Developments in MS systems have aimed almost exclusively at improving atmospheric pressure instrumentation interfaces with systems where the samples are introduced in a liquid stream, namely *atmospheric pressure chemical ionization* (APCI) and *electrospray ionization* (ESI). Some efforts have been made to reduce the introduction of solvent in the electrospray ion source. The developments in the submicrolitre per minute flow regime as referred to as *nanoelectrospray*. However, new achievements in ion source developments, many of which deal with surface analysis and solid sample introduction, eliminate the use of solvent because no sample preparation is needed and no solvent is used for sample introduction into the MS system.

All the traditional atmospheric pressure ionization sources like ESI, *matrix-assisted laser desorption ionization* (MALDI), APCI or *atmospheric pressure photoionization* (APPI) still require extensive sample preparation steps before the sample can be introduced into the MS instrument. With the introduction of *desorption electrospray* (DESI) and DART, it became possible, for the first time, to analyse samples directly in their native condition without sample preparation steps.[251] These techniques are relatively new, but they have been applied successfully for the detection of organic pollutants in the environment. Saturated non-functionalized hydrocarbons in petroleum distillates derived from crude oils were determined by DESI. These compounds are difficult to analyse by MS using atmospheric pressure ionization methods. In this method the hydrocarbons were oxidized *in situ* by initiating an electrical discharge during DESI to generate the corresponding alcohols and ketones.[252] This kind of reactive DESI experiment can be used as an *in situ* derivatization method for rapid and direct analysis of alkanes at atmospheric pressure without sample preparation. The limit of detection for alkanes from pure samples was 20 ng, and the technique was successfully implemented using exact mass measurements.

9.4 Future Trends in Organic Pollutants Analysis

Miniaturization, automation and environmental friendliness are the three vectors that drive present research in green analytical chemistry. Without any doubt, microextraction techniques represent a step in this direction. The additional advantages of portability and sustainability make microextraction techniques very attractive for any laboratory.[7] Miniaturization in analytical systems according to the μ-TAS concept is still underdeveloped. Other authors place great hopes on this approach since it can bring great advantages in terms of reduced chemical consumption, faster analysis, smaller sample volumes, increased separation efficiency and great automation potential, as well as the possibility to operate on-site and the corresponding reduced costs.[1]

Regarding the extraction and the purification of samples, the range of potential applications of automated SPME, for instance, will be extended to clinical studies and routine therapeutic drug monitoring in the near future. Currently, the main limitations of automated multifibre SPME are the lack of commercially available coatings suitable for use with the system and its unsuitability for the analysis of volatile compounds.[4] Far more microextraction techniques are applicable to liquid samples than for solid samples. Similarly, the range of applications reported is enormously greater for water samples than for other matrices encompassing analytes from all sorts of chemical families. Nevertheless, persistent organic pollutants such as pesticides continue to be a major concern as compared to the emerging pollutants such as pharmaceuticals.

In the field of the chromatographic separation, many efforts have been made to reduce time and organic solvents, as in UPLC or the reintroduced HPLTC.

These techniques minimize the use of solvent consumption by performing rapid separations.

Also, several approaches have been recently emerged for eliminating the use of organic solvents in ionization techniques for MS. These ambient desorption ionization techniques, in a single operational step, successfully analyse samples directly in their native form without the use of solvents for sample preparation and introduction to MS.

Other techniques, such as bioanalysis, have always been green, and there are a few opportunities to improve them in this respect.

Together with many other industrial sources, analytical chemistry itself was a cause of environmental pollution by using unfriendly solvents and reagents and wasting energy resources. However, it was also responsible for the identification of the problem generated by the whole of society, and is trying to develop analytical strategies that have minimal environmental impact.

References

1. M. Tobiszewski, A. Mechlinska, B. Zygmunt and J. Namiesnik, *TrAC. Trends Anal. Chem.*, 2009, **28**, 943.
2. L. H. Keith, L. U. Gron and J. L. Young, *Chem. Rev.*, 2007, **107**, 2695.
3. C. Nerín, J. Salafranca, M. Aznar and R. Batlle, *Anal. Bioanal. Chem.*, 2009, **393**, 809.
4. S. Risticevic, V. Niri, D. Vuckovic and J. Pawliszyn, *Anal. Bioanal. Chem.*, 2009, **393**, 781.
5. S. H. Haddadi, V. H. Niri and J. Pawliszyn, *Anal. Chim. Acta*, 2009, **652**, 224.
6. K. J. Chia, T. Y. Lee and S. D. Huang, *Anal. Chim. Acta*, 2004, **527**, 157.
7. H. Bagheri, Z. Ayazi and E. Babanezhad, *Microchem. J.*, 2010, **94**, 1.
8. S. S. Segro, Y. Cabezas and A. Malik, *J. Chromatogr. A*, 2009, **1216**, 4329.
9. F. Bianchi, F. Bisceglie, M. Careri, S. Di Berardino, A. Mangia and M. Musci, *J. Chromatogr. A*, 2008, **15**, 1196–1197.
10. M. A. Azenha, P. J. Nogueira and A. F. Silva, *Anal. Chem.*, 2006, **78**, 2071.
11. C. Dong, Z. Zeng and X. Li, *Talanta*, 2005, **66**, 721.
12. L. Cai, S. Gong, M. Chen and C. Wu, *Anal. Chim. Acta*, 2006, **559**, 89.
13. Y. Ji, X. Liu, X. Jiang, H. Huang, Z. Xu, H. Zhang and C. Wang, *Chromatographia*, 2009, **70**, 753.
14. N. Rastkari, R. Ahmadkhaniha and M. Yunesian, *J. Chromatogr. B*, 2009, **877**, 1568.
15. F. Tan, H. Zhao, X. Li, X. Quan, J. Chen, X. Xiang and X. Zhang, *J. Chromatogr. A*, 2009, **1216**, 5647.
16. J. Haginaka, *J. Sep. Sci.*, 2009, **32**, 1548.
17. M. K. Y. Li, N. Y. Lei, C. Gong, Y. Yu, K. H. Lam, M. H. W. Lam, H. Yu and P. K. S. Lam, *Anal. Chim. Acta*, 2009, **633**, 197.

18. M. A. Abrams, N. F. Dahdah and E. Francu, *Appl. Geochem.*, 2009, **24**, 1951.
19. C. Gonçalves, A. Dimou, V. Sakkas, M. F. Alpendurada and T. A. Albanis, *Chemosphere*, 2006, **64**, 1375.
20. M. Llompart, M. Lores, M. Lourido, L. Sánchez-Prado and R. Cela, *J. Chromatogr. A*, 2003, **985**, 175.
21. Z. Mester and R. Sturgeon, *Spectrochim. Acta Part B: Atom. Spectr.*, 2005, **60**, 1243.
22. V. Niri, H. E. In-Yong, K. Farhad Riazi and P. Janusz, *J. Sep. Sci.*, 2009, **32**, 1075.
23. R. Kubinec, V. G. Berezkin, R. Górová, G. Addová, H. Mracnová and L. Soják, *J. Chromatogr. B*, 2004, **800**, 295.
24. Y. Saito, I. Ueta, K. Kotera, M. Ogawa, H. Wada and K. Jinno, *J. Chromatogr. A*, 2006, **1106**, 190.
25. I. Bruheim, X. Liu and J. Pawliszyn, *Anal. Chem.*, 2003, **75**, 1002.
26. G. Shen and H. K. Lee, *Anal. Chem.*, 2003, **75**, 98.
27. W. Liu and H. K. Lee, *Anal. Chem.*, 2000, **72**, 4462.
28. S. Pedersen-Bjergaard and K. E. Rasmussen, *J. Chromatogr. A*, 2006, **1109**, 183.
29. S. Pedersen-Bjergaard and K. E. Rasmussen, *J. Chromatogr. A*, 2008, **1184**, 132.
30. A. R. T. S. Araujo, M. L. M. F. S. Saraiva, J. L. F. C. Lima and M. G. A. Korn, *Anal. Chim. Acta*, 2008, **613**, 177.
31. P. B. Kosaraju and K. K. Sirkar, *J. Membr. Sci.*, 2007, **288**, 41.
32. L. Xu, C. Basheer and H. K. Lee, *J. Chromatogr. A*, 2007, **1152**, 184.
33. M. Schellin and P. Popp, *J. Chromatogr. A*, 2006, **1103**, 211.
34. Y. Guo, J. Zhang, D. Liu and H. Fu, *Anal. Bioanal. Chem.*, 2006, **386**, 2193.
35. X. H. Zang, Q. H. Wu, M. Y. Zhang, G. H. Xi and Z. Wang, *Chin. J. Anal. Chem.*, 2009, **37**, 161.
36. Y. Liu, E. Zhao, W. Zhu, H. Gao and Z. Zhou, *J. Chromatogr. A*, 2009, **1216**, 885.
37. M. Rezaee, Y. Assadi, M. R. M. Hosseini, E. Aghaee, F. Ahmadi and S. J. Berijani, *J. Chromatogr. A*, 2006, **1116**, 1.
38. J. H. Guo, X. H. Li, X. L. Cao, Y. Li, X. Z. Wang and X. B. Xu, *J. Chromatogr. A*, 2009, **1216**, 3038.
39. J. Regueiro, M. Llompart, C. Garcia-Jares, J. C. Garcia-Monteagudo and R. Cela, *J. Chromatogr. A*, 2008, **1190**, 27.
40. Q. Zhou, X. Zhang and J. Xiao, *J. Chromatogr. A*, 2009, **1216**, 4361.
41. N. Fattahi, Y. Assadi, M. R. M. Hosseini and E. Z. Jahromi, *J. Chromatogr. A*, 2007, **1157**, 23.
42. X. Liu, J. Li, Z. Zhao, W. Zhang, K. Lin, C. Huang and X. Wang, *J. Chromatogr. A*, 2009, **1216**, 2220.
43. S. Pérez, P. Eichhorn, V. Ceballos and D. Barceló, *J. Mass Spectrom.*, 2009, **44**, 1308.
44. F. Pena-Pereira, I. Lavilla and C. Bendicho, *Spectrochim. Acta Part B: Atom. Spectr.*, 2009, **64**, 1.

45. A. N. Anthemidis and K. I. G. Ioannou, *Talanta*, 2009, **80**, 413.
46. J. Hu, L. Fu, X. Zhao, X. Liu, H. Wang, X. Wang and L. Dai, *Anal. Chim. Acta*, 2009, **640**, 100.
47. J. F. Liu, G. B. Jiang, J. F. Liu and J. A. Jönsson, *TrAC. Trends Anal. Chem.*, 2005, **24**, 20.
48. R. Liu, J. F. Liu, Y. G. Yin, X. L. Hu and G. B. Jiang, *Anal. Bioanal. Chem.*, 2009, **393**, 871.
49. E. Baltussen, P. Sandra, F. David and C. Cramers, *J. Microcolumn Sep.*, 1999, **11**, 737.
50. F. M. Lancas, M. E. C. Queiroz, P. Grossi and I. R. B. Olivares, *J. Sep. Sci.*, 2009, **32**, 813.
51. F. Sanchez-Rojas, C. Bosch-Ojeda and J. M. Cano-Pavon, *Chromatographia*, 2009, **69**, s79.
52. G. Roy, R. Vuillemin and J. Guyomarch, *Talanta*, 2005, **66**, 540.
53. N. Itoh, H. Tao and T. Ibusuki, *Anal. Chim. Acta*, 2005, **535**, 243.
54. E. Pérez-Carrera, V. M. L. León, A. G. Parra and E. González-Mazo, *J. Chromatogr. A*, 2007, **1170**, 82.
55. A. Giordano, M. Fernández-Franzón, M. Ruiz, G. Font and Y. Picó, *Anal. Bioanal. Chem.*, 2009, **393**, 1733.
56. J. Sánchez-Avila, J. Quintana, F. Ventura, R. Tauler, C. M. Duarte and S. Lacorte, *Mar. Pollut. Bull.*, **60**, 103.
57. P. Serôdio, M. S. Cabral and J. M. F. Nogueira, *J. Chromatogr. A*, 2007, **1141**, 259.
58. M. Kawaguchi, R. Ito, N. Sakui, N. Okanouchi, K. Saito, Y. Seto and H. Nakazawa, *Anal. Bioanal. Chem.*, 2007, **388**, 391.
59. S. Nakamura and S. Daishima, *J. Chromatogr. A*, 2004, **1038**, 291.
60. R. Rodil and M. Moeder, *J. Chromatogr. A*, 2008, **1179**, 81.
61. B. L. L. Tan, D. W. Hawker, J. F. Müller, L. A. Tremblay and H. F. Chapman, *Wat. Res.*, 2008, **42**, 404.
62. C. Almeida and J. M. F. Nogueira, *J. Pharm. Biomed. Anal.*, 2006, **41**, 1303.
63. T. Bauld, P. Teasdale, H. Stratton and H. Uwins, *Wat. Sci. Technol.*, 2007, **55**, 59.
64. A. R. M. Silva, F. C. M. Portugal and J. M. F. Nogueira, *J. Chromatogr. A*, 2008, **1209**, 10.
65. N. R. Neng, M. L. Pinto, J. Pires, P. M. Marcos and J. M. F. Nogueira, *J. Chromatogr. A*, 2007, **1171**, 8.
66. F. C. M. Portugal, M. L. Pinto and J. M. F. Nogueira, *Talanta*, 2008, **77**, 765.
67. L. P. Melo, A. M. Nogueira, F. M. Lanças and M. E. C. Queiroz, *Anal. Chim. Acta* 2009, **633**, 57.
68. X. Huang and D. Yuan, *J. Chromatogr. A*, 2007, **1154**, 152.
69. W. Guan, Y. Wang, F. Xu and Y. Guan, *J. Chromatogr. A*, 2008, **1177**, 28.
70. X. Zhu, J. Cai, J. Yang, Q. Su and Y. Gao, *J. Chromatogr. A*, 2006, **1131**, 37.
71. X. Huang, N. Qiu and D. Yuan, *J. Chromatogr. A*, 2008, **1194**, 134.

72. C. Bicchi, C. Cordero, E. Liberto, P. Rubiolo, B. Sgorbini, F. David and P. Sandra, *J. Chromatogr. A*, 2005, **1094**, 9.
73. P. Popp, C. Bauer, A. Paschke and L. Montero, *Anal. Chim. Acta*, 2004, **504**, 307.
74. A. Prieto, O. Zuloaga, A. Usobiaga, N. Etxebarria, L. A. Fernández, C. Marcic and A. de Diego, *J. Chromatogr. A*, 2008, **1185**, 130.
75. M. Abdel-Rehim, *J. Chromatogr. B*, 2004, **801**, 317.
76. L. G. Blomberg, *Anal. Bioanal. Chem.*, 2009, **393**, 797.
77. A. El-Beqqali, A. Kussak and M. Abdel-Rehim, *J. Chromatogr. A*, 2006, **1114**, 234.
78. M. Abdel-Rehim, P. Skansen, M. Vita, Z. Hassan, L. Blomberg and M. Hassan, *Anal. Chim. Acta*, 2005, **539**, 35.
79. A. El-Beqqali, A. Kussak, L. Blomberg and M. Abdel-Rehim, *J. Liq. Chromatogr. Related Technol.*, 2007, **30**, 575.
80. Y. Saito, I. Ueta, M. Ogawa, A. Abe, K. Yogo, S. Shirai and K. Jinno, *Anal. Bioanal. Chem.*, 2009, **393**, 861.
81. D. Qi, X. Kang, L. Chen, Y. Zhang, H. Wei and Z. Gu, *Anal. Bioanal. Chem.*, 2008, **390**, 929.
82. M. Ogawa, Y. Saito, I. Ueta and K. Jinno, *Anal. Bioanal. Chem.*, 2007, **388**, 619.
83. B. Zabiegala, A. Kot-Wasik, M. Urbanowicz and J. Namiesnik, *Anal. Bioanal. Chem.*, 2010, **396**, 273.
84. I. J. Allan, K. Booij, A. Paschke, B. Vrana, G. A. Mills and R. Greenwood, *Environ. Sci. Technol.*, 2009, **43**, 5383.
85. M. Shaw and J. F. Mueller, *Environ. Sci. Technol.*, 2009, **43**, 1443.
86. R. Greenwood, G. A. Mills and B. Vrana, *J. Chromatogr. A*, 2009, **1216**, 631.
87. S. Seethapathy, T. Górecki and X. Li, *J. Chromatogr. A*, 2008, **1184**, 234.
88. Z. Zhang, A. Hibberd and J. L. Zhou, *Anal. Chim. Acta*, 2008, **607**, 37.
89. I. J. Allan, J. Knutsson, N. Guigues, G. A. Mills, A. M. Fouillac and R. Greenwood, *J. Environ. Monit.*, 2008, **10**, 821.
90. M. J. Martínez Bueno, M. D. Hernando, A. Agüera and A. R. Fernández-Alba, *Talanta*, 2009, **77**, 1518.
91. H. Söderström, R. H. Lindberg and J. Fick, *J. Chromatogr. A*, 2009, **1216**, 623.
92. J. L. Stroud, A. H. Rhodes, K. T. Semple, Z. Simek and J. Hofman, *Environ. Pollut.*, 2008, **156**, 664.
93. M. C. Zúñiga, E. Jover, V. Arancibia and J. M. Bayona, *Talanta*, 2009, **80**, 504.
94. C. Gonçalves, J. J. Carvalho, M. A. Azenha and M. F. Alpendurada, *J. Chromatogr. A*, 2006, **1110**, 6.
95. S. A. Smyth, L. Lishman, M. Alaee, S. Kleywegt, L. Svoboda, J. J. Yang, H. B. Lee and P. Seto, *Chemosphere*, 2007, **67**, 267.
96. J. J. Lian, Y. Ren, J. M. Chen, T. Wang and T. T. Cheng, *J. Environ. Monit.*, 2009, **11**, 187.
97. Y. Yang, T. Cajthaml and T. Hofmann, *Environ. Pollut.*, 2008, **156**, 745.
98. K. W. Schramm, *Chemosphere*, 2008, **72**, 1103.

99. D. García-Rodríguez, A. M. Carro-Díaz and R. A. Lorenzo-Ferreira, *J. Sep. Sci.*, 2008, **31**, 1333.
100. J. P. Kreitinger, E. F. Neuhauser, F. G. Doherty and S. B. Hawthorne, *Environ. Toxicol. Chem.*, 2007, **26**, 1146.
101. M. T. O. Jonker, S. B. Hawthorne and A. A. Koelmans, *Environ. Sci. Technol.*, 2005, **39**, 7889.
102. J. Zhang, Z. Liang, L. H. Zhang, W. B. Zhang, Y. S. Huo, Y. K. Zhang, G. Xuexiao and H. Xuebao, *Chemical J. Chin. Universit.*, 2006, **27**, 2291.
103. P. Rigou, S. Saini and S. J. Setford, *Int. J. Environ. Anal. Chem.* 2004, **84**, 979.
104. S. Babel and D. del Mundo Dacera, *Waste Manage.*, 2006, **26**, 988.
105. V. Andreu and Y. Picó, *TrAC Trends Anal. Chem.*, 2004, **23**, 772.
106. A. E. Latawiec and B. J. Reid, *Environ. Int.*, 2009, **35**, 911.
107. V. Fernández-González, E. Concha-Graña, S. Muniategui-Lorenzo, P. López-Mahía and D. Prada-Rodríguez, *J. Chromatogr. A*, 2008, **65**, 1196–1197.
108. D. Kalderis, S. B. Hawthorne, A. A. Clifford and E. Gidarakos, *J. Hazard. Mater.*, 2008, **159**, 329.
109. P. Haglund, *Ambio*, 2007, **36**, 467.
110. S. Hashimoto, K. Watanabe, K. Nose and M. Morita, *Chemosphere*, 2004, **54**, 89.
111. A. A. Dadkhah and A. Akgerman, *J. Hazard. Mater.*, 2006, **137**, 518.
112. G. Heltai, K. Percsich, G. Halász, K. Jung and I. Fekete, *Microchem. J.*, 2005, **79**, 231.
113. J. J. Morales-Riffo and P. Richter, *Anal. Bioanal. Chem.*, 2004, **380**, 129.
114. J. Kronholm, P. Revilla-Ruiz, S. P. Porras, K. Hartonen, R. Carabias-Martínez and M. L. Riekkola, *J. Chromatogr. A*, 2004, **1022**, 9.
115. I. Ferrer, V. Pichon, M. C. Hennion and D. Barceló, *J. Chromatogr. A*, 1997, **777**, 91.
116. M. Zhao, Y. Liu, Y. Li, X. Zhang and W. Chang, *J. Chromatogr. B*, 2003, **783**, 401.
117. J. C. Chuang, J. M. Van Emon, R. Jones, J. Durnford and R. A. Lordo, *Anal. Chim. Acta*, 2007, **583**, 32.
118. S. Pérez, I. Ferrer, M. C. Hennion and D. Barceló, *Anal. Chem.*, 1998, **70**, 4996.
119. R. Galve, F. Camps, F. Sanchez-Baeza and M. P. Marco, *Anal. Chem.*, 2000, **72**, 2237.
120. A. Oubiña, B. Ballesteros, R. Galve, D. Barcelo and M. P. Marco, *Anal. Chim. Acta*, 1999, **387**, 255.
121. J. Ramón-Azcón, R. Galve, F. Sánchez-Baeza and M. P. Marco, *Anal. Chem.*, 2006, **78**, 71.
122. M. Farré, J. Ramón, R. Galve, M. F. Marco and D. Barceló, *Environ. Sci. Technol.*, 2006, **40**, 5064.
123. L. Bruun, C. Koch, M. H. Jakobsen and J. Aamand, *Anal. Chim. Acta*, 2000, **423**, 205.

124. F. A. Esteve-Turrillas, A. Abad-Fuentes and J. V. Mercader, *Food Chem.*, 2011, **124**, 1727.
125. J. Kaur, K. V. Singh, R. Boro, K. R. Thampi, M. Raje, G. C. Varshney and C. R. Suri, *Environ. Sci. Technol.*, 2007, **41**, 5028.
126. M. Lisa, R. S. Chouhan, A. C. Vinayaka, H. K. Manonmani and M. S. Thakur, *Biosens. Bioelectron.*, 2009, **25**, 224.
127. M. J. Kim, J. Y. Shim, Y. T. Lee and H. S. Lee, *Int. J. Food Sci. Technol.*, 2006, **41**, 927.
128. Y. A. Cho, Y. J. Kim, B. D. Hammock, Y. T. Lee and H. S. Lee, *J. Agric. Food Chem.*, 2003, **51**, 7854.
129. A. Y. Kolosova, J. H. Park, S. A. Eremin, S. J. Kang and D. H. Chung, *J. Agric. Food Chem.*, 2003, **51**, 1107.
130. N. Fischer-Durand, A. Vessières, J. M. Heldt, F. le Bideau and G. Jaouen, *J. Organomet. Chem.*, 2003, **668**, 59.
131. M. Wortberg, C. Midendorf, A. Katerkamp, T. Rump, J. Krause and K. Cammann, *Anal. Chim. Acta*, 1994, **289**, 177.
132. P. T. Anastas, *Crit. Rev. Anal. Chem.*, 1999, **29**, 167.
133. J. Workman Jr, K. E. Creasy, S. Doherty, L. Bond, M. Koch, A. Ullman and D. J. Veltkamp, *Anal. Chem.*, 2001, **73**, 2705.
134. J. Buffle and G. Horvai, *In situ Monitoring of Aquatic Systems: Chemical Analysis and Speciation*, 2000, John Wiley & Sons, Chichester, p. 279.
135. J. Wang, *TrAC, Trends Anal. Chem.*, 1997, **16**, 84.
136. M. Farré, R. Brix and D. Barceló, *TrAC, Trends Anal. Chem.*, 2005, **24**, 532.
137. M. Farré and D. Barceló, *TrAC, Trends Anal. Chem.*, 2003, **22**, 299.
138. S. Kurosawa, J. W. Park, H. Aizawa, S. I. Wakida, H. Tao and K. Ishihara, *Biosens. Bioelectron.*, 2006, **22**, 473.
139. M. Pumera, S. Sánchez, I. Ichinose and J. Tang, *Sens. Actuators, B: Chemical*, 2007, **123**, 1195.
140. G. A. Rivas, M. D. Rubianes, M. C. Rodríguez, N. F. Ferreyra, G. L. Luque, M. L. Pedano, S. A. Miscoria and C. Parrado, *Talanta*, 2007, **74**, 291.
141. M. Badihi-Mossberg, V. Buchner and J. Rishpon, *Electroanalysis*, 2007, **19**, 2015.
142. I. Palchetti, S. Laschi and M. Mascini, *Met. Mol. Biol. (Clifton, N.J.)*, 2009, **504**, 115.
143. M. A. Rahman, P. Kumar, D. S. Park and Y. B. Shim, *Sensors*, 2008, **8**, 118.
144. T. Ahuja, I. A. Mir, D. Kumar and Rajesh, *Biomaterials*, 2007, **28**, 791.
145. S. Andreescu and J. L. Marty, *Biomol. Eng.*, 2006, **23**, 1.
146. J. M. Pingarrón, P. Yañez-Sedeño and A. González-Cortés, *Electrochim. Acta*, 2008, **53**, 5848.
147. E. F. Schipper, S. Rauchalles, R. P. H. Kooyman, B. Hock and J. Greve, *Anal. Chem.*, 1998, **70**, 1192.
148. K. E. Brummel, J. Wright and M. E. Eldefrawi, *J. Agric. Food Chem.*, 1997, **45**, 3292.

149. J. Tschmelak, M. Kumpf, N. Kappel, G. Proll and G. Gauglitz, *Talanta*, 2006, **69**, 343.
150. J. Tschmelak, G. Proll and G. Gauglitz, *Biosens. Bioelectron.*, 2004, **20**, 743.
151. J. Tschmelak, G. Proll and G. Gauglitz, *Talanta*, 2005, **65**, 313.
152. E. Mallat, C. Barzen, R. Abuknesha, G. Gauglitz and D. Barceló, *Anal. Chim. Acta*, 2001, **426**, 209.
153. E. Mallat, C. Barzen, A. Klotz, A. Brecht, G. Gauglitz and D. Barceló, *Environ. Sci. Technol.*, 1999, **33**, 965.
154. F. Prieto, B. Sepúlveda, A. Calle, A. Llobera, C. Domínguez and L. M. Lechuga, *Sens. Actuators, B: Chemical*, 2003, **92**, 151.
155. P. B. Luppa, L. J. Sokoll and D. W. Chan, *Clin. Chim. Acta*, 2001, **314**, 1.
156. W. w. Zhang, Y. c. Chen, Z. f. Luo, J. y. Wang and D. y. Ma, *Chem. Res. Chin. Univ.*, 2007, **23**, 404.
157. E. Mauriz, A. Calle, L. M. Lechuga, J. Quintana, A. Montoya and J. J. Manclús, *Anal. Chim. Acta*, 2006, **561**, 40.
158. E. Mauriz, A. Calle, J. J. Manclús, A. Montoya, A. M. Escuela, J. R. Sendra and L. M. Lechuga, *Sens. Actuators, B: Chemical*, 2006, **118**, 399.
159. E. Mauriz, A. Calle, J. J. Manclús, A. Montoya, A. Hildebrandt, D. Barceló and L. M. Lechuga, *Biosens. Bioelectron.*, 2007, **22**, 1410.
160. E. Mauriz, A. Calle, J. J. Manclús, A. Montoya and L. M. Lechuga, *Anal. Bioanal. Chem.*, 2007, **387**, 2757.
161. M. Farré, E. Martínez, J. Ramón, A. Navarro, J. Radjenovic, E. Mauriz, L. Lechuga, M. P. Marco and D. Barceló, *Anal. Bioanal. Chem.*, 2007, **388**, 207.
162. G. R. Marchesini, E. Meulenberg, W. Haasnoot, M. Mizuguchi and H. Irth, *Anal. Chem.*, 2006, **78**, 1107.
163. D. P. Tang, R. Yuan and Y. Q. Chai, *Bioprocess Biosyst. Eng.*, 2006, **28**, 315.
164. H. Aizawa, M. Tozuka, S. Kurosawa, K. Kobayashi, S. M. Reddy and M. Higuchi, *Anal. Chim. Acta*, 2007, **591**, 191.
165. A. J. Haes and R. P. Van Duyne, *J. Am. Chem. Soc.*, 2002, **124**, 10596.
166. J. N. Anker, W. P. Hall, O. Lyandres, N. C. Shah, J. Zhao and R. P. Van Duyne, *Nat. Mater.*, 2008, **7**, 442.
167. M. P. Kreuzer, R. Quidant, J. P. Salvador, M. P. Marco and G. Badenes, *Anal. Bioanal. Chem.*, 2008, **391**, 1813.
168. J. Kulys, R. Vidziunaite, R. Janciene and A. Palaima, *Electroanalysis*, 2006, **18**, 1771.
169. X. Yu, D. Xu and Q. Cheng, *Proteomics*, 2006, **6**, 5493.
170. S. Hrapovic, Y. Liu, K. B. Male and J. H. T. Luong, *Anal. Chem.*, 2004, **76**, 1083.
171. S. Hrapovic, Y. Liu and J. H. T. Luong, *Anal. Chem.*, 2007, **79**, 500.
172. J. Wang, J. Dai and T. Yarlagadda, *Langmuir*, 2005, **21**, 9.
173. T. You, O. Niwa, R. Kurita, Y. Iwasaki, K. Hayashi, K. Suzuki and S. Hirono, *Electroanalysis*, 2004, **16**, 54.

174. K. Ramanathan, M. A. Bangar, M. Yun, W. Chen, A. Mulchandani and N. V. Myung, *Nano Lett.*, 2004, **4**, 1237.
175. Y. Xiao, F. Patolsky, E. Katz, J. F. Hainfeld and I. Willner, *Science*, 2003, **299**, 1877.
176. M. J. A. Shiddiky, M. A. Rahman, C. S. Cheol and Y. B. Shim, *Anal. Biochem.*, 2008, **379**, 170.
177. J. Wang, M. Musameh and Y. Lin, *J. Am. Chem. Soc.*, 2003, **125**, 2408.
178. M. Musameh, J. Wang, A. Merkoci and Y. Lin, *Electrochem. Commun.*, 2002, **4**, 743.
179. M. L. Yang, J. Wang, H. Q. Li, J. G. Zheng and N. Q. N. Wu, *Nanotechnology*, 2008, **19**.
180. J. Gong, T. Liu, D. Song, X. Zhang and L. Zhang, *Electrochem. Commun.*, 2009, **11**, 1873.
181. S. Viswanathan, H. Radecka and J. Radecki, *Biosens. Bioelectron.*, 2009, **24**, 2772.
182. S. Ayers, K. D. Gillis, M. Lindau and B. A. Minch, *IEEE Trans. Circuits Sys. I*, 2007, **54**, 736.
183. E. Moore, M. Pravda and G. G. Guilbault, *Anal. Chim. Acta*, 2003, **484**, 15.
184. K. Grennan, G. Strachan, A. J. Porter, A. J. Killard and M. R. Smyth, *Anal. Chim. Acta*, 2003, **500**, 287.
185. S. Suresh, A. K. Gupta, V. K. Rao, k. Om and R. Vijayaraghavan, *Talanta*, 2010, **81**, 703.
186. V. Carralero, A. González-Cortés, P. Yañez-Sedeño and J. M. Pingarrón, *Anal. Chim. Acta*, 2007, **596**, 86.
187. S. Rodriguez-Mozaz, M. J. López de Alda and D. Barceló, *Talanta*, 2006, **69**, 377.
188. E. Mallat, D. Barceló, C. Barzen, G. Gauglitz and R. Abuknesha, *TrAC, Trends Anal. Chem.*, 2001, **20**, 124.
189. J. Tschmelak, G. Proll, J. Riedt, J. Kaiser, P. Kraemmer, L. Bárzaga, J. S. Wilkinson, P. Hua, J. P. Hole, R. Nudd, M. Jackson, R. Abuknesha, D. Barceló, S. Rodriguez-Mozaz, M. J. López De Alda, F. Sacher, J. Stien, J. Slobodnik, P. Oswald, H. Kozmenko, E. Korenková, L. Tóthová, Z. Krascsenits and G. Gauglitz, *Biosens. Bioelectron.*, 2005, **20**, 1509.
190. S. Scarano, M. Mascini, A. P. F. Turner and M. Minunni, *Biosens. Bioelectron.*, **25**, 957.
191. R. A. Doong, H. M. Shih and S. H. Lee, *Sens. Actuators, B*, 2005, **323**, 111–112.
192. F. Lucarelli, A. Kicela, I. Palchetti, G. Marrazza and M. Mascini, *Bioelectrochemistry*, 2002, **58**, 113.
193. F. Lucarelli, I. Palchetti, G. Marrazza and M. Mascini, *Talanta*, 2002, **56**, 949.
194. E. Bakker, *Clin. Chem.*, 2006, **52**, 557.
195. F. Lucarelli, S. Tombelli, M. Minunni, G. Marrazza and M. Mascini, *Anal. Chim. Acta*, 2008, **609**, 139.

196. S. S. Babkina and N. A. Ulakhovich, *Anal. Chem.*, 2005, **77**, 5678.
197. I. Palchetti and M. Mascini, *Analyst*, 2008, **133**, 846.
198. J. J. Gooding, *Electroanalysis*, 2002, **14**, 1149.
199. J. Wang, D. Xu, A. N. Kawde and R. Polsky, *Anal. Chem.*, 2001, **73**, 5576.
200. A. N. Kawde and J. Wang, *Electroanalysis*, 2004, **16**, 101.
201. J. I. Hahm and C. M. Lieber, *Nano Lett.*, 2004, **4**, 51.
202. Y. Feng, T. Yang, W. Zhang, C. Jiang and K. Jiao, *Anal. Chim. Acta*, 2008, **616**, 144.
203. X. H. Fu, *Bioprocess Biosyst. Eng.*, 2008, **31**, 69.
204. Q. Zhang, Q. Sun, B. Hu, Q. Shen, G. Yang, X. Liang, X. Sun and F. Liu, *Food Chem.*, 2008, **106**, 1278.
205. M. Usami, K. Mitsunaga and Y. Ohno, *J. Steroid Biochem. Mol. Biol.*, 2002, **81**, 47.
206. B. Hock, M. Seifert and K. Kramer, *Biosens. Bioelectron.*, 2002, **17**, 239.
207. H. D. Butala and A. Sadana, *J. Colloid Interface Sci.*, 2003, **263**, 420.
208. J. L. Wittliff, S. A. Andres, T. L. Kruer, D. A. Kerr, I. A. Smolenkova and J. L. Erb, *Adv. Exp. Med. Biol.*, 2008, **614**, 315.
209. Y. Lei, W. Chen and A. Mulchandani, *Anal. Chim. Acta*, 2006, **568**, 200.
210. S. F. D'Souza, *Appl. Biochem. Biotechnol. Part A*, 2001, **96**, 225.
211. R. E. Ionescu, N. Jaffrezic-Renault, L. Bouffier, C. Gondran, S. Cosnier, D. G. Pinacho, M. P. Marco, F. J. Sánchez-Baeza, T. Healy and C. Martelet, *Biosens. Bioelectron.*, 2007, **23**, 549.
212. V. Nanduri, A. K. Bhunia, S. I. Tu, G. C. Paoli and J. D. Brewster, *Biosens. Bioelectron.*, 2007, **23**, 248.
213. Y. Lei, P. Mulchandani, W. Chen and A. Mulchandani, *Appl. Biochem. Biotechnol.*, 2007, **136**, 243.
214. J. Cui, X. Wang, G. Wang and L. Ma, *3rd International Conference on Bioinformatics and Biomedical Engineering, iCBBE* 2009, 2009.
215. J. Wang, Y. Zhang, Y. Wang, R. Xu, Z. Sun and Z. Jie, *Biosens. Bioelectron.*, 2010, **25**, 1705.
216. M. Farré and D. Barceló, *Anal. Bioanal. Chem.*, 2001, **371**, 467.
217. M. Farré, O. Pasini, M. Carmen Alonso, M. Castillo and D. Barceló, *Anal. Chim. Acta*, 2001, **426**, 155.
218. P. Stolper, S. Fabel, M. G. Weller, D. Knopp and R. Niessner, *Anal. Bioanal. Chem.*, 2008, **390**, 1181.
219. J. C. Philp, S. Balmand, E. Hajto, M. J. Bailey, S. Wiles, A. S. Whiteley, A. K. Lilley, J. Hajto and S. A. Dunbar, *Anal. Chim. Acta*, 2003, **487**, 61.
220. B. C. Kim and M. B. Gu, *Environ. Monit. Assess.*, 2005, **109**, 123.
221. B. Podola and M. Melkonian, *J. Appl. Phycol.*, 2005, **17**, 261.
222. A. Tibuzzi, G. Rea, G. Pezzotti, D. Esposito, U. Johanningmeier and M. T. Giardi, *J. Phy. Conden. Matt.*, 2007, 19.
223. M. T. Giardi, V. Scognamiglio, G. Rea, G. Rodio, A. Antonacci, M. Lambreva, G. Pezzotti and U. Johanningmeier, *Biosens. Bioelectron.*, 2009, **25**, 294.

224. I. Shitanda, K. Takada, Y. Sakai and T. Tatsuma, *Anal. Chim. Acta*, 2005, **530**, 191.
225. H. J. Weitz, C. D. Campbell and K. Killham, *Environ. Microbiol.*, 2002, **4**, 422.
226. L. Campanella, G. Favero, D. Mastrofini and M. Tomassetti, *J. Pharm. Biomed. Anal.*, 1996, **14**, 1007.
227. M. E. Polak, D. M. Rawson and B. G. D. Haggett, *Biosens. Bioelectron.*, 1996, **11**, 1253.
228. A. Bentley, A. Atkinson, J. Jezek and D. M. Rawson, *Toxicol. In Vitro*, 2001, **15**, 469.
229. M. P. Zhao, Y. Z. Li, Z. Q. Guo, X. X. Zhang and W. B. Chang, *Talanta*, 2002, **57**, 1205.
230. K. P. Dierksen, L. Mojovic, B. A. Caldwell, R. R. Preston, R. Upson, J. Lawrence, P. N. McFadden and J. E. Trempy, *J. Appl. Toxicol.*, 2004, **24**, 363.
231. T. M. Curtis, M. W. Widder, L. M. Brennan, S. J. Schwager, W. H. Van Der Schalie, J. Fey and N. Salazar, *Lab Chip*, 2009, **9**, 2176.
232. P. Banerjee, D. Lenz, J. P. Robinson, J. L. Rickus and A. K. Bhunia, *Lab. Invest.*, 2008, **88**, 196.
233. D. Diamond, *Talanta*, 2008, **75**, 605.
234. S. Wu, X. Gao, Q. Cai and C. A. Grimes, *Sens. Actuators B*, 2007, **123**, 856.
235. T. A. Bendikov, J. Kim and T. C. Harmon, *Sens. Actuators, B*, 2005, **106**, 512.
236. M. E. Swartz, *Sep. Sci. Redif.*, 2005, 8. www.chromatographyonline.com.
237. G. R. Marchesini, J. Buijs, W. Haasnoot, D. Hooijerink, O. Jansson and M. W. F. Nielen, *Anal. Chem.*, 2008, **80**, 1159.
238. G. R. Marchesini, H. Hooijerink, W. Haasnoot, M. W. F. Nielen, J. Buijs, K. Campbell, C. T. Elliott and M. W. F. Nielen, *TrAC, Trends Anal. Chem.*, 2009, **28**, 792.
239. A. L. Batt, M. S. Kostich and J. M. Lazorchak, *Anal. Chem.*, 2008, **80**, 5021.
240. D. T. T. Nguyen, D. Guillarme, S. Heinisch, M. P. Barrioulet, J. L. Rocca, S. Rudaz and J. L. Veuthey, *J. Chromatogr. A*, 2007, **1167**, 76.
241. M. H. Chen and C. Horváth, *J. Chromatogr. A*, 1997, **788**, 50.
242. J. Bowermaster, McNair and M. Harold, *J. Chromatogr. Sci.*, 1984, **22**, 165.
243. A. M. Edge, S. Shillingford, C. Smith, R. Payne and I. D. Wilson, *J. Chromatogr. A*, 2006, **1132**, 206.
244. I. D. Wilson, *Chromatographia*, 2000, **52**, S28.
245. R. M. Smith, R. J. Burgess, O. Chienthavorn and J. R. Stuttard, *LC-GC Europe*, 1999, **12**, 30.
246. A. Alpmann and G. Morlock, *Anal. Bioanal. Chem.*, 2006, **386**, 1543.
247. G. S. Morlock and W. Schwack, *LC-GC Europe*, 2008, **21**, 336.
248. A. Klöppel, W. Grasse, F. Brümmer and G. E. Morlock, *J. Planar. Chromatogr.*, **21**, 431.

249. A. Alpmann and G. Morlock, *J. Sep. Sci.*, 2008, **31**, 71.
250. M. Aranda and G. Morlock, *Rapid Commun. Mass Spectrom.*, 2007, **21**, 1297.
251. A. Venter, M. Nefliu and R. Graham Cooks, *TrAC, Trends Anal. Chem.*, 2008, **27**, 284.
252. C. Wu, K. Qian, M. Nefliu and R. G. Cooks, *J. Am. Soc. Mass Spectrom.*, 2010, **21**, 261.
253. A. Ivanov, G. Evtugyn, H. Budnikov, S. Girotti, S. Ghini, E. Ferri, A. Montoya and J. V. Mercader, *Anal. Lett.*, 2008, **41**, 392.
254. J. Ramón-Azcón, F. Sánchez-Baeza, N. Sanvicens and M. P. Marco, *J. Agric. Food Chem.*, 2009, **57**, 375.
255. A. Bahlmann, M. G. Weller, U. Panne and R. J. Schneider, *Anal. Bioanal. Chem.*, 2009, **395**, 1809.
256. C. Soler, S. Girotti, S. Ghini, F. Fini, A. Montoya, J. J. Manclús and J. Mañas, *Anal. Lett.*, 2008, **41**, 2539.
257. M. Farré, M. Kuster, R. Brix, F. Rubio, M. J. L. de Alda and D. Barceló, *J. Chrom. A*, 2007, **1160**, 166.
258. J. D. Byer, J. Struger, P. Klawunn, A. Todd and E. D. Sverko, *Environ. Sci. Technol.*, 2008, **42**, 6052.
259. M. J. Moreno, A. Abad, R. Pelegrí, M. I. Martínez, A. Sáez, M. Gamán and A. Montoya, *J. Agric. Food Chem.*, 2001, **49**, 1713.
260. C. Wang, Y. Liu, Y. Guo, C. Liang, X. Li and G. Zhu, *Food Chem.*, 2009, **115**, 365.
261. T. Xu, I. K. Cho, D. Wang, F. M. Rubio, W. L. Shelver, A. M. E. Gasc, J. Li and Q. X. Li, *Environ. Poll.*, 2009, **157**, 417.
262. W. L. Shelver, C. D. Parrotta, R. Slawecki, Q. X. Li, M. G. Ikonomou, D. Barcelo, S. Lacorte and F. M. Rubio, *Chemosphere*, 2008, **73**, 18.
263. S. Herranz, J. Ramón-Azcón, E. Benito-Peña, M. D. Marazuela, M. P. Marco and M. C. Moreno-Bondi, *Anal. Bioanal. Chem.*, 2008, **391**, 1801.
264. E. Zacco, M. I. Pividori, S. Alegret, R. Galve and M. P. Marco, *Anal. Chem.*, 2006, **78**, 1780.
265. W. Gui, C. Liang, Y. Guo and G. Zhu, *Anal. Lett.*, 2010, **43**, 487.
266. W. L. Shelver, L. M. Kamp, J. L. Church and F. M. Rubio, *J. Agric. Food Chem.*, 2007, **55**, 3758.
267. L. Kantiani, M. Farré, D. Asperger, F. Rubio, S. González, M. J. López de Alda, M. Petrovic, W. L. Shelver and D. Barceló, *J. Hydrology*, 2008, **361**, 1.
268. W. B. Shim, M. E. Yakovleva, K. Y. Kim, B. R. Nam, E. S. Vylegzhanina, A. A. Komarov, S. A. Eremin and D. H. Chung, *J. Agric. Food Chem.*, 2009, **57**, 791.
269. M. L. Sánchez-Martínez, M. P. Aguilar-Caballos, S. A. Eremin and A. Gómez-Hens, *Talanta*, 2007, **72**, 243.
270. I. Y. Goryacheva, S. A. Eremin, E. A. Shutaleva, M. Suchanek, R. Niessner and D. Knopp, *Anal. Lett.*, 2007, **40**, 1445.
271. H. S. Chun, E. H. Choi, H. J. Chang, S. W. Choi and S. A. Eremin, *Anal. Chim Acta*, 2009, **639**, 83.

272. D. G. Mita, A. Attanasio, F. Arduini, N. Diano, V. Grano, U. Bencivenga, S. Rossi, A. Amine and D. Moscone, *Biosens. Bioelectron.*, 2007, **23**, 60.
273. M. H. Piao, H. B. Noh, M. A. Rahman, M. S. Won and Y. B. Shim, *Electroanal.*, 2008, **20**, 30.
274. H. Notsu, T. Tatsuma and A. Fujishima, *J. Electroanal. Chem.*, 2002, **523**, 86.
275. A. N. Ivanov, G. A. Evtugyn, R. E. Gyurcsanyi, K. Toth and H. C. Budnikov, *Anal. Chim. Acta*, 2000, **404**, 55.
276. J. Kochana, A. Gala, A. Parczewski and J. Adamski, *Anal. Bioanal. Chem.*, 2008, **391**, 1275.
277. M. Del Carlo, M. Di Marcello, M. Perugini, V. Ponzielli, M. Sergi, M. Mascini and D. Compagnone, *Microchim. Acta*, 2008, **163**, 163.
278. N. Liu, X. Cai, Y. Lei, Q. Zhang, M. B. Chan-Park, C. Li, W. Chen and A. Mulchandani, *Electroanal.*, 2007, **19**, 616.
279. Y. Ma, K. Jiao, T. Yang and D. Sun, *Sens. Actuators B*, 2008, **131**, 565.
280. Y. C. Tsai and C. C. Chiu, *Sens. Actuators B*, 2007, **125**, 10.
281. L. A. Taranova, A. P. Fesay, G. V. Ivashchenko, A. N. Reshetilov, A. Winther-Nielsen and J. Emneus, *Appl. Biochem. Microbiol.*, 2004, **40**, 472.
282. J. K. Sook, K. V. Gobi, H. Tanaka, Y. Shoyama and N. Miura, *Chem. Lett.*, 2006, **35**, 1132.
283. S. Rodriguez-Mozaz, S. Reder, M. Lopez De Alda, G. Gauglitz and D. Barceló, *Biosens. Bioelectron.*, 2004, **19**, 633.
284. C. Nakamura, M. Hasegawa, N. Nakamura and J. Miyake, *Biosens. Bioelectron.*, 2003, **18**, 599.
285. S. Hleli, C. Martelet, A. Abdelghani, N. Burais and N. Jaffrezic-Renault, *Sens. Actuators B*, 2006, **113**, 711.
286. E. Mauriz, A. Calle, L. M. Lechuga, J. Quintana, A. Montoya and J. J. Manclús, *Anal. Chim. Acta*, 2006, **561**, 40.
287. S. Kurosawa, H. Aizawa and J. W. Park, *Analyst*, 2005, **130**, 1495.
288. P. Pulido-Tofiño, J. M. Barrero-Moreno and M. C. Pérez-Conde, *Anal. Chim. Acta*, 2006, **562**, 122.
289. D. Dan, C. Shizhen, S. Dandan, L. Haibing and X. Chen, *Biosens. Bioelectron.*, 2008, **23**, 1864.
290. N. G. Karousos, S. Aouabdi, A. S. Way and S. M. Reddy, *Anal. Chim. Acta*, 2002, **469**, 189.
291. E. V. Melikhova, E. N. Kalmykova, S. A. Eremin and T. N. Ermolaeva, *J. Anal. Chem.*, 2006, **61**, 687.

CHAPTER 10
On-line Decontamination of Analytical Wastes

SERGIO ARMENTA AND MIGUEL DE LA GUARDIA

Departamento de Química Analítica, Edificio de Investigación, Universidad de Valencia, C/. Dr. Moliner 50, 46100 Burjassot, Valencia, Spain

10.1 Introduction

Analytical chemistry can be considered as a tool for problem solving, made up of several steps, from sampling, sample pretreatment (including extraction, digestion, dissolution and preconcentration) to sample analysis and data treatment. Traditionally, the development of an analytical procedure focused on the improvement of analytical properties such as accuracy, precision, selectivity, sensitivity and robustness, also taking into consideration other aspects such as costs, time and the safety of the method. All these parameters were improved by optimizing the various steps of the analytical procedures and developing new technologies.

In the last decade of the 20th century the new concept of 'green chemistry' appeared, which had as its main aim to reduce or mitigate the side effects and the negative impact of chemistry in the environment. Taking this new paradigm into consideration, a new step had to be considered in analytical chemistry: waste management. Since then, many efforts have been dedicated to reducing the side effects of each step of the analytical chemistry process. For instance, new screening methodologies have developed to reduce the number of samples to be analysed, classical methods have been replaced by new procedures which provide direct and non-destructive analysis of untreated samples; new sample treatment methods have been applied for extraction or digestion of samples, including microwave assisted extraction or digestion, ultrasound extraction and pressurized solvent extraction; and new preconcentration techniques which reduce to the minimum the use of organic solvents, such as solid phase microextraction (SPME), liquid phase microextraction (LPME) and membrane assisted extraction, are now available, together with a reduction of the size of analytical systems to the micrometre scale and automation of the whole process which reduces the consume of solvents and reagents. In addition, the replacement of traditional, widely used solvents and chemicals by new ones that are innocuous or, at least, less toxic, also provides greener alternatives.

Unfortunately, it is not always possible to avoid the use of hazardous chemicals. In such cases, an additional effort must be made to add a last step in the analytical process to decontaminate the analytical wastes, or at least reduce their amount and toxicity, providing on-line recycling or recovery of solvents and reagents, mineralization of organic compounds and passivation of toxic species. In short, it is important to incorporate chemical solutions for the problems relating to waste generation, with on-line waste decontamination being preferred to the external treatment.

Decontamination of analytical wastes is important because reducing the amount of hazardous waste means lower costs and increased safety for workers and the environment. The reduction of costs incurred by green analytical chemistry justifies the selection of these methodologies from an economic perspective. The second advantage of green methods is that associated with operator and environmental safety; green alternatives can greatly reduce the risks of causing terrible human and economic damage.

In this sense, different methodologies have been developed to reduce the toxicity of analytical laboratory wastes, including recycling, degradation and passivation. Through the use of those methodologies, preferably in on-line mode, the amount of toxic wastes can be substantially reduced, avoiding safety problems relating to the storage of large quantities of residues and reducing the cost of their external management.

10.2 Recycling of Analytical Wastes: Solvents and Reagents

Reducing solvent wastes is currently one of the most important elements of pollution prevention programmes worldwide. The use of solvents is a major contributor to air and water pollution, and is a leading source of hazardous analytical wastes. Reducing solvent usage also reduces the cost of purchasing virgin solvents as well as reducing storage and management costs.

It is clear that when implementing a green analytical method, aimed at reducing organic solvent volume, the first alternative should be to reduce or eliminate the use of solvents at the source. Replacement of toxic solvents by aqueous solutions, or by innocuous, or at least, less toxic solvents is the direct way of eliminating the source of organic solvent wastes. It means modifying procedures to reduce the release of solvents and pollutants, and the costs associated with those releases. On the other hand, if solvents cannot be eliminated from the process, the second option is to recycle or reuse the generated wastes.

Recovery of organic solvents is especially significant in some areas of analytical chemistry such as liquid chromatography (LC), where the large amounts of solvents consumed lead to high purchase and disposal costs, and also in infrared spectroscopy where chlorinated solvents are commonly used because of their high transparency in the mid infrared spectral region.

The volume of LC mobile phase purchased and discarded each year is about 20 million litres.[1] The purchase cost of solvents is about $20 per litre. Disposal, assuming it is done properly, adds at least $30 per litre. Thus, the total annual cost to global society of high-pressure liquid chromatography (HPLC) mobile phase is $1 billion. Recycling the mobile phases reduces these costs by 50–90% and additionally reduces the environmental side effects. Dolan,[2] writing in 2007, states that for more than 10 years solvent recycling has been one of the main subjects of his 'LC troubleshooting' column,[3] and indicates different ways to reduce the volume of solvent waste from LC analyses, including (1) reusing the mobile phase, (2) automated recycling of mobile phase, (3) distillation, and (4) modifying the size of columns and particles used in the process.

As Dolan indicates, the most direct way to reduce solvent consumption in LC analysis is to reuse the mobile phase. Total recycling recirculates the mobile phase from the waste line to the solvent reservoir: it is only applicable to isocratic methodologies where the composition of the mobile phase is constant and does not vary with time. In this case, recycling the mobile phase implies its contamination with a small amount of the injected sample or standard. If the volume of the mobile phase is relatively large, this contamination can be considered as negligible. However, the analyte will slowly accumulate in the mobile phase. This method has become frequently used in LC, but its effects on sample quantification are not yet well documented.[4] Abreu and Lawrence[5] spiked mobile phase with three different concentrations of two analytes, tartaric acid and sodium nitrate, to simulate mobile phase recycling, and used these mobile phases to analyse these two analytes. In analysis with mobile phase recycling, three different cases can be observed: (1) samples with analyte concentrations greater than those in the recycled mobile phase, resulting in positive peaks; (2) concentration of the analyte in the sample lower than that of the recycled mobile phase, resulting in negative peaks; (3) concentration of the analyte in the sample equal to that of the recycled mobile phase, so no peaks are observed. The conclusion of this paper is that the mobile phase should be replaced when the analyte concentration in the recycled mobile phase approximates to the lowest concentration of analyte in samples. Using this methodology, Yokoyama *et al.* determined 17 amino acids in beverages.[6] Fifty consecutive separations could be achieved without significant changes of peak height and retention times, reducing mobile phase volume used by a factor of 60 compared to that used for conventional reversed-phase LC (RPLC).

However, to avoid problems related to mysterious peaks (often negative) that arise from the vacancy effect, total recycling is rarely used today. An alternative is the so-called *closed-loop eluent recycling system*. In this system, the mobile phase is recirculated to the solvent reservoir after removal of dissolved sample constituents by adsorption on to activated carbon,[7] restoring the purity of the mobile phase. Separation of anions has been also achieved by anion chromatography using recycled mobile phase[8] after removing analyte ions with mixed cation- and anion-exchange resins. The mobile phase consumption was reduced to one tenth that used in conventional anion chromatography. In a similar way, carboxylates have been determined by LC using recycled mobile phase,[9] attaching a graphitic carbon column to remove the analytes from the mobile phase. The LC system needs only 5 ml mobile phase for 50 consecutive analyses of citrate. However, the residue of carboxylates injected as samples could not be removed from the mobile phase.

Another way of significantly reducing mobile phase consumption is automated recycling of the mobile phase. Those recyclers split the column effluent between waste and the mobile phase reservoir. The first generation of recyclers, developed in 1970, used timers that relied on the run-to-run reproducibility of retention time windows. To prevent contamination of the recycled mobile phase by sample constituents due to variations in retention time, the window for elution of the peak needed to be about $\pm 15\%$ wider than the peak width at baseline. This was a wide window for long runs, which reduced the potential savings.

Second-generation mobile phase recyclers generally employed threshold peak sensing. These recyclers often successfully recovered most of the pure mobile phase. However, some problems can arise when the threshold value is too high, such as peak detection being delayed, sending the leading edge of the peak back to the reservoir, and small peaks not being detected. When the threshold is too low the savings are never realized because the recycling valve never resets. Additional problems can arise when the baseline of the chromatogram drifts (see Figure 10.1).

Third-generation solvent recyclers contain a microprocessor which uses a sensitive level-sensing circuit to direct the mobile phase to waste whenever the output from the system detector exceeds a level set by the user. When the output from the system detector drops below the programmed level, the mobile phase is redirected to the solvent reservoir to be used again, reducing both solvent disposal and purchasing costs (see Figure 10.2). Nowadays, there are various commercially available devices on the market that can be used for solvent recycling (see Table 10.1). Some instruments can compensate for non-ideal shapes using an algorithm, such as that incorporated in the SolventTrak mobile phase recycler (Axxiom Chromatography, Moorpark, CA), which permits the user to select a return-to-baseline delay time to make sure that the baseline is really achieved. One study demonstrated savings of 88–96% of mobile phase and a dollar savings of 81% ($\sim$ \$1000 per year).[10] But the potential savings are greatest when an instrument has a low number of peaks per hour or per day, due either to very simple chromatograms or to low sample

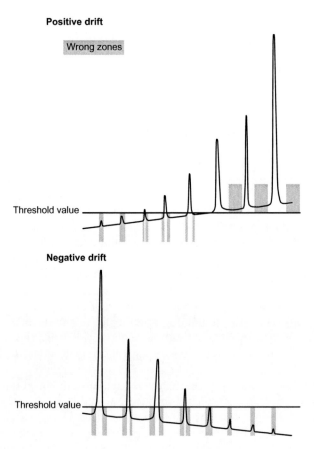

Figure 10.1 Problems found in LC solvent recycling when the baseline of the chromatogram drifts.

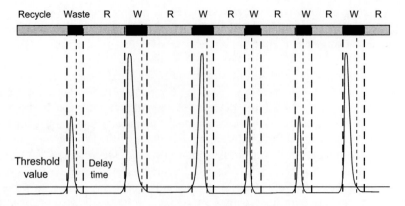

Figure 10.2 Solvent recycling scheme of the third generation of recyclers.

Table 10.1 Commercially available devices for solvent recycling.

Automatic solvent recycling systems
7206 Solvent Recycler Micro Solv
LA2890 Solvent Recycler Laserchrom
S-3 HPLC Solvent Recycler
SolventTrak Antech solutions
SolventTrak phenomenex
SolventTrak II SMI-LabHut
Solvent Recycler 2000 Alltech Associates
Solvent Recycler 3000 Alltech Associates
Thermo Scientific SRS Pro Solvent Recycling System

Distillation systems for HPLC solvents
9600 solvent recycling B/R Corporation

load. However, mobile phase recyclers are not the universal solution. Recycling makes sense only for isocratic separations, and there are some exceptions where mobile phase recycling is not the best solution, such as separations with small baseline periods, ion chromatography with suppression which modifies the mobile phase composition to improve detection, and the use of capillary columns where mobile phase recovery does not to justify cost reduction.

Recovery of mobile phase from gradient mode runs involves distillation of the organic solvent.[11,12] There are three types of apparatus for this: simple distillation, fractional distillation units and thin film evaporators. The most common method is simple distillation which heats solvent wastes: the vapour reverts back to liquid form in a condenser and is collected. The waste remaining in the bottom of the still is collected and disposed of. Fractional stills produce a higher purity of recycled product but they are generally more expensive to operate and are generally better suited to larger volumes. Thin film evaporators distil by running a thin film of dirty solvent down a heated cylindrical vessel where it is vaporized. The vapours are collected and condensed back into liquid form for reuse. Thin film evaporators are generally suited for use in high-volume, continuous processes and require the dirty solvent to have a low content of suspended solids in order to work well.

In 2002 Stepnowski *et al.* developed a methodology for methanol/water[13] and acetonitrile/water[14] LC solvent recycling based on a batch distillation system combined with microbial biodegradation. The biodegradation is used for the first cuts and heavy ends of the solvent waste, which are not suitable for further recovery. Additionally, B/R Instrument Corporation (Maryland, USA) developed and sells a fractional distillation system specifically designed for recovery of LC solvents in an automated or manual process. However, solvent distillation is not a panacea, since the distillate is of slightly lower purity than the original feedstock. Moreover, it should be considered that common solvents like ethanol and acetonitrile distillates are azeotropes.

As already mentioned, solvent recycling is not a perfect solution and analytical laboratories should consider its possible limitations before investing in any solvent recycling equipment: (1) the costs of the recycled solvent should justify the costs of purchasing and maintaining recycling equipment, together with the costs of training operators, and (2) the properties of the recycled solvent (purity, stability, etc.) should be appropriate for its use. For instance, recycling could alter the composition and usefulness of an organic solvent. An example of an organic solvent which cannot be recycled is 1,1,1-trichloroethane, which can break down during distillation and become acidic. Additionally, stabilizers and other additives may be required to make the recycled product stable.

Another analytical chemistry research area where the recovery of solvents is of practical relevance is infrared spectroscopy, especially in the mid infrared range, where chlorinated hydrocarbons are still the preferred solvents because of their high transparency. Replacement of chlorinated solvents has been extensively discussed in the literature since J.D.R. Thomas raised the matter in a paper entitled 'Away with chlorinated solvents: how will analytical chemistry cope?'[15] One approach employed to avoid the use of chlorinated hydrocarbons is their replacement by more innocuous solvents in infrared determinations. An example of this is the replacement of CFC-113 for determining oil and grease in wastewater by different solvents such as the dimer/trimer of chlorotrifluoroethylene (S-316)[16] or carbon disulfide.[17] However, it is not always so easy to find a solvent with appropriate properties to replace chlorinated ones. In those cases, the reduction of the volume of chlorinated solvent consumed in infrared determinations is based on the recycling of the solvents by on-line distillation. Using a closed *flow injection analysis* (FIA) manifold, which includes a distillation unit incorporated on-line after the infrared measurements to recover chlorinated solvents, a drastic reduction of the volume of wastes was achieved. This system has been successfully applied in the determination of ketoprofen[18] and propyphenazone and caffeine[19] in pharmaceuticals.

FIA is an approach widely used in analytical measurements, its main advantages being the reduction of reagents and sample consumption and their high degree of automation compared to batchwise methods. However, the carrier and reagent solutions are fed continuously into the flow system, being converted into toxic waste after the detection step. Cyclic FIA modifies the manifold by recirculation of the reagent solution, leading to an appreciable reduction of reagent consumption[20] (see Figure 10.3). This concept has been well known since the early 1970s and several examples of cyclic FIA methods have been described in the literature for enzyme recycling and metal determination.[21,22] Other examples employ chemiluminescence detection for the determination of copper,[23] thiamine[24] and H_2O_2,[25] using the reagent solution as carrier solution because of the short lifetime of the chemiluminescence effect. UV spectrometers have also been successfully used as detectors in cyclic FIA methods for the determination of chloride,[26] ascorbic acid[27] and lead, using an ion exchange column to regenerate the colorimetric reagent,[28] for calcium ion with regeneration of the reagent with chlorophosphonazo III and an ion exchanger,[29] and for methamphetamine with solvent extraction.[30]

Classical FIA manifold

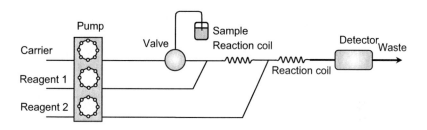

Cyclic FIA manifold

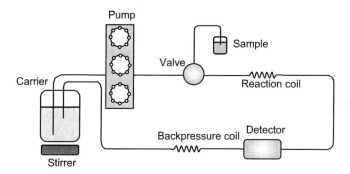

Figure 10.3 Typical configurations of FIA and cyclic FIA systems to reduce solvent consumption.

The usefulness of a continuous circulating multicommutated flow system, based on an open-loop configuration and continuous renewal of the carrier solution, for the determination of chloride in parenteral solutions through the formation of the iron(III) thiocyanate complex after displacement of thiocyanate by chloride[31] has been also demonstrated, reducing reagent consumption and consequently waste generation.

10.3 Degradation of Wastes

As mentioned earlier, solvent and reagent recycling is not a panacea and cannot always be applied to minimize the side effects of analytical laboratories. Thus, the development of degradation methods to minimize waste generation is a major concern of scientists around the world. It should be noted that most of the research activity in this topic focuses on the development of new or more efficient treatment technologies for industrial and municipal wastewater. However, these techniques may be adapted to the treatment of analytical laboratory residues, reducing the environmental impact and the economic costs of external treatment of residues.

10.3.1 Thermal Degradation

Thermal decomposition is defined as a chemical reaction in which a chemical substance breaks up into at least two chemical substances when heated. The reaction is endothermic as heat is required to break chemical bonds in the compound undergoing decomposition. Until the mid 1980s, waste combustion was widely considered as the basic method to eliminate toxic chemical wastes. Nowadays, it is mainly used to decompose solid organic matter in the food, petrochemical and polymer industries.[32]

10.3.2 Chemical Oxidation

Chemical oxidation is normally applied in the treatment of wastewater to eliminate toxic or hazardous compounds and can be combined with biological oxidation to reduce the total organic carbon content of wastewaters or improve their biodegradability. Potassium permanganate has been used as oxidant, particularly for the destruction of hazardous compounds.[33] In order to enhance oxidizing power, there is much interest nowadays in the use of *advanced oxidation technologies* (AOTs). In these processes, contaminants are oxidized by different reagents such as ozone, hydrogen peroxide, Fenton's reagent and combinations of these. The procedures may also be combined with UV irradiation and ultrasound. The photo-Fenton reaction uses UV light, Fe^{2+} and H_2O_2 and has been successfully used for the degradation of many different organic pollutants in aqueous solution.[34,35] A modification of the photo-Fenton reaction is based on the system hv–ferrioxolate–H_2O_2 which is a highly efficient and powerful oxidant for the destruction of organic pollutants in contaminated groundwater and wastewaters.[36] As the review by Gogate and Pandit indicates,[37] these AOTs have been commonly applied for industrial wastewater treatment and their use for the *in situ* destruction of aqueous laboratory wastes is still rare.[38]

The main problems of these of waste degradation techniques are that they do not totally mineralize organic contaminants, because of the need for combination with biological treatments,[39] and oxidants are continuously consumed during the reaction because it takes place in homogeneous solution.

10.3.3 Photocatalytic Degradation

Heterogeneous photocatalysis, as one of the most promising AOPs, has attracted much attention in the past two decades because this process can be operated at mild conditions of temperature and pressure and can completely mineralize toxic organic compounds to carbon dioxide, water and mineral acids.[40,41]

The ideal photocatalyst should possess the following properties (1) photoactivity, (2) biological and chemical inertness, (3) stability toward photocorrosion, (4) suitability towards visible or near UV light, (5) low cost, and (6) lack of toxicity. A wide range of semiconductors provide those properties and may be used for heterogeneous photocatalysis, such as the oxides TiO_2, ZnO, MgO, WO_3, ZrO_2, CeO_2, Fe_2O_3, or sulfides CdS, ZnS. Titanium dioxide (TiO_2) powder is one of the most popular photocatalysts due to its nontoxicity, chemical inertness, and low costs. The main advantage of heterogeneous photodegradation systems is the possibility of recovering and reusing the catalyst, resulting in considerable savings and simpler operation of the equipment involved.

Reactors using heterogeneous catalysis can be briefly classified into two types: those that use the catalyst as suspension form and those that immobilize it on a substrate. The main drawback of the first type is the difficulty of separating the catalyst from the solution because of the micrometric size of the particles. To facilitate the recovery of the photocatalyst, TiO_2 has been retained on silica gel,[42] polymeric membranes,[43] porous alumina-silica ceramic,[44] zeolite,[45] activated carbon[46] and the inner walls of a glass reactor.[47]

The photocatalytic degradation principle has been successfully applied in several fields such as drinking-water, wastewater and treatment of industrial and analytical laboratory wastes. Applications of heterogeneous photocatalysis include removing heavy metals such as Hg, Cr, Pb, Cd, As, Ni and Cu,[48,49] and destruction of organic compounds[50–52] and inorganic species such as nitrate,[53] sulfide and sulphite,[54] and cyanide.[55] Recently, microwave irradiation has been used to improve degradation rate of photocatalytic reactions.[56]

10.3.4 Biodegradation

Biological processes use the natural metabolism of living cells to degrade chemicals by means of a sequence of reactions catalysed by enzymes. Municipal and industrial wastewaters have traditionally been treated with biological processes, typically aerobic organisms in suspended systems. Nowadays, a process known as *membrane bioreactor* (MBR) is widely used for municipal and industrial wastewater treatment (see Figure 10.4). MBR is the combination of a membrane process like ultrafiltration with a suspended growth bioreactor.[57] The performance of MBR filtration inevitably decreases with time due to the deposition of soluble and particulate materials on to and into the membrane.[58]

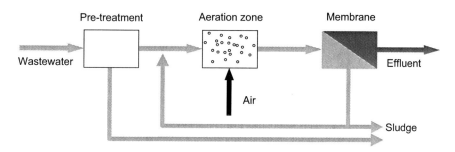

Figure 10.4 Scheme of a membrane bioreactor (MBR) process used for municipal and industrial wastewater treatment.

However, biodegradation of analytical laboratory wastes is more challenging because they contain less biodegradable and often toxic pollutants.[59] Moreover, pH, redox potential, moisture, and other characteristics of analytical residues influence the proliferation and activity of microorganisms in degrading toxic compounds.[60] This is probably the reason for the limited applicability of toxic waste biodegradation in analytical laboratories.

10.4 Passivation of Toxic Wastes

Sometimes degradation of the pollutants is not possible, or is an expensive or slow process. In such cases, passivation is an alternative way to reduce the toxicity and amount of analytical wastes. The main objective of passivation methods is the effective removal of pollutants from aqueous streams, reducing the amount of residues from several litres to a few milligrams. The use of this type of methodologies widely implemented in industrial waste treatment but its application to analytical laboratories is still limited.

Conventional methods for removal of metal ions are precipitation, co-precipitation, adsorption, chelation and surfactant-based treatments. The traditional methodology used for the removal of metal ions from wastewater is based on their precipitation, typically involving an increase in pH. However, this method can be time consuming and inefficient. For instance, cadmium and chromium may precipitate out of solution under alkaline conditions, while metals such as molybdenum will not precipitate as easily and will typically remain in solution at levels above environmental regulatory discharge limits. The destabilization of the surface charge of the particles can be accomplished either by dosing the coagulants (chemical coagulation) or by generating them electrochemically *in situ* (electrocoagulation).

The addition to the wastewater of a metal sulfate, such as alum, $Al_2(SO_4)_3 \cdot 18H_2O$, makes the precipitation of the metal ions more effective even under acidic conditions. These processes, however, tend to generate large volumes of sludge with high bound water content that can be slow to filter and difficult to dewater.[61] Electrocoagulation processes have attracted significant attention for inorganic ion removal, as an alternative to the addition of metal salts for breaking stable emulsions and suspensions.[62] The procedure introduces highly charged polymeric metal hydroxide species to remove metals, colloidal solids and particles, and soluble inorganic pollutants from wastewater streams. It has been found that anodized aluminium is more effective than the aluminium ion introduced in the form of aluminium sulfate solution.[63] However, the main limitation of the method is that the hydrogen gas produced at the aluminium cathode in an electrocoagulation cell prevents the flocs from settling properly.[64] To overcome this problem, an electrocoagulation–electroflotation combination has been used to float the precipitate and to separate it from the treated water.[65] The main advantages of this methodology over conventional treatments include versatility, energy efficiency, safety, selectivity, possibility of automation and cost-effectiveness.[66]

The co-precipitation process is one of the usual methods of removing heavy metals from wastewater. The method comprises co-precipitating the heavy metal ions with a carrier precipitate which is formed *in situ* within the aqueous solution; it was patented in the USA in 1991 by W.T. Douglas.[67] Removal of heavy metals by co-precipitation with ferrite has been well known for more than 40 years[68] and the co-precipitation of metals in laboratory wastes using aluminium hydroxide has been also successfully developed.[69,70]

Recently, the use of low-cost natural adsorbents including carbonaceous materials, agricultural products and waste by-products has been investigated for removal of heavy metals and organic compounds from wastewater as potential alternative to the conventional processes. Plant residues, which are mainly lignocellulosic materials, can inherently adsorb waste chemicals such as dyes and cations in water due to the coulombic interaction between the two substrates and physical absorption. Moreover, they are renewable agricultural wastes, available abundantly at no or low cost. Thus, *Eucalyptus camaldulensis* leaves,[71] roadside tree leaves,[72] sunflower stalks[73] and chemically modified plant wastes[74] have been successfully employed for wastewater treatment.

Another type of metal ion passivation is based on the effective chelation of the ions by an organic molecule. The chelating agent can combine with heavy metal ions through coordinate links to form a stable insoluble chelate complex. Dithiocarbamate, alone or anchored on to polymer composites,[75] and chitosan[76] have been widely used as organic heavy metal chelating agents for wastewater treatment. The metal chelate species can be removed by ion exchange, reverse osmosis or nanofiltration. These techniques generate concentrates of the metal form which will require further treatment. Ion exchange (chelating) resins, based on cross-linked polystyrene, are also used widely in industry for metal ion removal. However, the main disadvantages of chelating resins are their low selectivity between heavy metals, and their elastic nature which limits their use in on-line packed column systems. Thus, the modification of chelator resins through the bonding of dyes or other chelating agents on to polymer microspheres or hydrophobic polymer resins to be used in on-line packed systems has been intensively studied in the last two decades.[77,78] Another area of research focuses on developing new biopolymer ion exchange resins from inexpensive natural materials, such as crustacean shells,[79] seaweed[80] and corn.[81]

During the past few years interest in micelle-based passivation processes has been growing.[82,83] Metallic ions can be separated from the solution using reverse osmosis, but the permeate flux is limited and the capital and operating costs are quite high.[84] Metal ions and organic soluble molecules can be separated from the bulk solution by ultrafiltration if they are trapped by larger colloidal particles such as micelles.[85,86] *Micellar-enhanced ultrafiltration* (MEUF) is based on the addition to the wastewater of a surfactant at a concentration higher than its critical micelle concentration. The amphiphilic aggregates attract metal ions to the micelle surface and the organic molecules are solubilized in the micelle core. The main advantages of MEUF are the low energy requirements involved in ultrafiltration processes[87] and its high removal efficiency owing to the effective trapping of solutes by the micelles.[88] However, a problem which is far from being solved concerns the recycling of the micellar pseudophase after metal extraction. In most instances, the metal can be back-extracted in acid media.[89]

Acknowledgements

The authors gratefully acknowledge the financial support of the Generalitat Valenciana Project PROMETEO 2010-055.

References

1. A. Welch, *Am. Lab.* (Shelton, Conn.), February 2006.
2. J. W. Dolan, *LCGC Europe*, 2007, **20**, 254.
3. J. W. Dolan, *LCGC*, 1991, **10**, 426.
4. J. Srbek, P. Coufal, Z. Bosakova and E. Tesarova, *J. Sep. Sci.*, 2005, **28**, 1263.
5. O. Abreu and G. D. Lawrence, *Anal. Chem.*, 2000, **72**, 1749.
6. T. Yokoyama, A. Sakai and M. Zenki, *Chromatographia*, 2008, **67**, 535.
7. R. O. Sheth, *Closed Loop Recycling of Solvents Used in High Pressure Liquid Chromatograph (HPLC) Analysis*, Technical Note, Mason & Hanger-Silas Mason Co., Inc., Pantex Plant, Texas, USA.
8. T. Yokoyama, H. Maekubo, A. Sakai and M. Zenki, *J. Chromatogr. A*, 2005, **1089**, 82.
9. T. Yokoyama, K. Matsumoto, S. Hyohdoh and M. Zenki, in *Abstracts of 10th Int. Conf. Flow Analysis*, 2006, p. 41.
10. H. K. Teoha, E. Sorensen and N. Titchener-Hooker, *Chem. Eng. Sci.*, 2003, **58**, 4145.
11. R. M. Katusz, L. Bellew, J. A. Mangravite and R. F. Foery, *J. Chromatogr. A*, 1981, **213**, 331.
12. J. A. Mangravite, R. R. Roark, R. R. Roark Jr. and P. Van Triest, Solvent recycling by spinning band distillation: theory, equipment, and limitations, in *Pollution Prevention and Waste Minimization in Laboratories*, Lewis Publishers, New York, 1996, pp. 239–274.
13. P. Stepnowski, K. H. Blotevogel and B. Jastorff, *Environ. Sci. Pollut. Res.*, 2002, **9**, 34.
14. P. Stepnowski, K. H. Blotevogel, P. Ganczarek, U. Fischer and B. Jastorff, *Res. Conserv. Recycl.*, 2002, **35**, 163.
15. J. D. R. Thomas, *TrAC, Trends Anal. Chem.*, 1995, **14**, 186.
16. *ASTM International Method D7066-04e: Standard Test Method for Dimer/trimer of Chlorotrifluoroethylene (S-316) Recoverable Oil and Grease and Nonpolar Material by Infrared Determination*, http://www.astm.org/standards/D7066.htm.
17. V. J. Barwick, S. L. Ellison, M. J. Q. Rafferty and T. J. Farrant, *Int. J. Environ. Anal. Chem.*, 1998, **72**, 235.

18. M. J. Sanchez-Dasi, S. Garrigues, M. L. Cervera and M. de la Guardia, *Anal. Chim. Acta*, 1998, **361**, 253.
19. Z. Bouhsain, S. Garrigues and M. de la Guardia, *Analyst*, 1997, **122**, 441.
20. M. Zenki, *Bunseki Kagaku*, 2004, **53**, 245.
21. H. U. Bergmeyer, A. Hagen and Z. Fresenius, *Anal. Chem.*, 1972, **267**, 333.
22. V. V. S. Eswara Dun and H. A. Mottola, *Anal. Chem.*, 1975, **47**, 357.
23. M. Yamada and S. Suzuki, *Anal. Chim. Acta*, 1987, **193**, 337.
24. M. Ishii and M. Kawashima, *J. Flow Injection Anal.*, 1998, **15**, 25.
25. S. Hanaoka, J. M. Lin and M. Yamada, *Anal. Chim. Acta*, 2001, **426**, 57.
26. M. Zenki and Y. Iwadou, *Talanta*, 2002, **58**, 1055.
27. M. Zenki, A. Tanishita and T. Yokoyama, *Talanta*, 2004, **64**, 1273.
28. M. Zenki, K. Minamisawa and T. Yokoyama, *Talanta*, 2005, **68**, 281.
29. M. Zenki, T. Masutani and T. Yokoyama, *Anal. Sci.*, 2002, **18**, 1137.
30. N. Teshima, N. Fukui and T. Sakai, *Talanta*, 2005, **68**, 253.
31. R. A. S. Lapa, J. L. F. C. Lima, B. F. Reis and J. L. M. Santos, *Anal. Chim. Acta*, 1998, **377**, 103.
32. European Commission. *Integrated Pollution Prevention and Control. Draft Reference Document on the Best Available Techniques for Waste Incineration*. European Commission, Brussels, 2004.
33. G. Lunn, E. B. Sansone, M. De Meo, M. Laget and M. Castegnaro, *Am. Ind. Hyg. Assoc. J.*, 1994, **55**, 167.
34. T. Wei, Y. Wand and C. Wan, *J. Photochem. Photobiol., A*, 1990, **55**, 115.
35. J. Chen, W. H. Rulkens and H. Bruning, *Water Sci. Tech.*, 1997, **35**, 231.
36. A. Safarzadeh-Amiri, J. A. Bolton and S. R. Cater, *J. Adv. Oxid. Technol.*, 1996, **1**, 18.
37. P. R. Gogate and A. B. Pandit, *Adv. Environ. Res.*, 2004, **8**, 501.
38. J. A. Herrera Melián, E. Tello Rendón, J. Araña, J. M. Doña Rodríguez, O. González Díaz and J. Pérez Peña, *Toxicol. Environ. Chem.*, 2003, **85**, 61.
39. E. Chamarro, A. Marco and S. Esplugas, *Water Res.*, 2001, **35**, 1047.
40. D. W. Bahnemann, J. Cunningham, M. A. Fox, E. Pelizzetti, P. Pichat and N. Serpone, in: *Aquatic Surface Photochemistry*, ed. R. G. Zepp, G. R. Helz and D. G. Crosby, F. L. Lewis Publishers, Boca Raton, FL, 1994, p. 261.
41. J. M. Herrmann, *Helv. Chim. Acta*, 2001, **84**, 2731.
42. Z. Ding, X. J. Hu, G. Q. Lu, P. L. Yue and P. F. Greenfield, *Langmuir*, 2000, **16**, 6216.
43. N. Phonthammachai, E. Gulari, A. M. Jamieson and S. Wongkasemjit, *Appl. Organometal. Chem.*, 2006, **20**, 499.
44. Y. Peng and J. T. Richardson, *Appl. Catal., A*, 2004, **266**, 235.
45. P. Atienzar, A. Corma, H. Garcia and J. C. Scaiano, *Chem. Mater.*, 2004, **16**, 982.
46. F. L. Y. Lam and X. J. Hu, *Chem. Eng. Sci.*, 2003, **58**, 687.
47. K. Vinodgopal, U. Stafford, K. A. Gray and P. V. Kamat, *J. Phys. Chem.*, 1994, **98**, 6797.
48. D. M. Blake, *Bibliography of Work on the Heterogeneous Photocatalytic Removal of Hazardous Compounds from Water and Air*, National Renewable Energy Laboratory, Boulder, CO, 2001, pp. 1–158.

49. D. F. Ollis, E. Pelizzetti and N. Serpone, *Environ. Sci. Technol.*, 1991, **25**, 1522.
50. M. de la Guardia, K. D. Khalaf, V. Carbonell and A. Morales-Rubio, *Anal. Chim. Acta*, 1995, **308**, 462.
51. M. J. Escuriola, A. Morales-Rubio and M. de la Guardia, *Anal. Chim. Acta*, 1999, **390**, 147.
52. M. de la Guardia, K. D. Khalaf, B. A. Hasan, A. Morales-Rubio and V. Carbonell, *Analyst*, 1995, **120**, 231.
53. A. Zafra, J. Garcia, A. Milis and X. Domenech, *J. Mol. Catal.*, 1991, **70**, 343.
54. S. N. Frank and A. J. Bard, *J. Phys. Chem.*, 1977, **81**, 1484.
55. S. N. Frank and A. J. Bard, *J. Am. Chem. Soc.*, 1977, **99**, 303.
56. S. Horikoshi, H. Hidaka and N. Serpone, *J. Photochem. Photobiol., A*, 2004, **161**, 221.
57. S. Judd, *The MBR Book. Principles and Applications of Membrane Bioreactors in Water and Wastewater Treatment*, Elsevier, Oxford, 2006.
58. Z. F. Cui, S. Chang and A. G. Fane, *J. Memb. Sci.*, 2003, **2211**, 1.
59. M. A. Tarr, *Chemical Degradation Methods for Wastes and Pollutants*, Marcel Dekker, New York, 2003.
60. L. W. Parker and K. G. Doxtader, *J. Environ. Qual.*, 1983, **12**, 553.
61. L. D. Benefield, J. F. Judkins and B. L. Weand, *Process chemistry for Water and Wastewater Treatment*, Prentice-Hall, Englewood Cliff, NJ, 1982.
62. N. Kongsricharoern and C. Polprasert, *Water Sci. Technol.*, 1995, **31**, 109.
63. H. C. Lee, T. C. Chou and J. Chin, *Inst. Chem. Eng.*, 1994, **25**, 239.
64. J. Ge, J. Qu, P. Lei and H. Liu, *Sep. Purif. Technol.*, 2004, **36**, 33.
65. G. H. Chen, X. M. Chen and P. L. Yue, *J. Environ. Eng.*, 2000, **126**, 858.
66. M. Y. A. Mollah, R. Schennach, J. R. Parga and D. L. Cocke, *J. Hazard. Mater.*, 2001, **84**, 29.
67. W. T. Douglas, *US Patent 5013453—Method for removing heavy metals from aqueous solutions by coprecipitation*, No. 042565, filed on 04/16/1987.
68. W. Dyck and Can, *J. Chem.*, 1968, **46**, 1441.
69. T. Chohji, E. Hirai, Y. Hayashi and A. Touda, *Environ. Technol.*, 1990, **11**, 421.
70. P. Cava-Montesinos, E. Ródenas-Torralba, A. Morales-Rubio, M. L. Cervera and M. de la Guardia, *Anal. Chim. Acta*, 2004, **506**, 145.
71. N. T. Abdel-Ghani, M. M. Hefny and G. A. El-Chaghaby, *J. Chil. Chem. Soc.*, 2008, **53**, 1585.
72. L. Hu, A. A. Adeyiga, T. Greer, E. Miamee and A. Adeyiga, *Chem. Eng. Commun.*, 2002, **189**, 1587.
73. G. Sun and W. Shi, *Ind. Eng. Chem. Res.*, 1998, **37**, 1324.
74. W. S. Wan Ngah and M. A. K. M. Hanafiah, *Bioresour. Technol.*, 2008, **99**, 3935.
75. R. Say, E. Birlik, A. Denizli and A. Ersöz, *Appl. Clay Sci.*, 2006, **31**, 298.
76. A. Gamagea and F. Shahidi, *Food Chem.*, 2007, **104**, 989.

77. B. Salih, A. Denizli, B. Engin, A. Tuncel and E. Piskin, *J. Appl. Polym. Sci.*, 1996, **60**, 871.
78. S. P. Huang, K. S. Franz, R. L. Albright and R. H. Fish, *Ligands. Inorg. Chem.*, 1995, **34**, 2813.
79. S. S. Ullah, J. U. Ahmad, A. Kabir, T. A. Azam and R. J. Islam, *Bangladesh Acad. Sci.*, 1996, **20**, 167.
80. Y. Konishi, S. Asai, Y. Midoh and M. Oku, *Sep. Sci. Technol.*, 1993, **28**, 1691.
81. S. Chaudhari and V. Tare, *Water Sci. Technol.*, 1996, **34**, 161.
82. J. F. Scamehorn and J. H. Harwell, *Surfactant-Based Separation Processes*, Surfactant Science Series Vol. 33, Marcel Dekker, New York, 1989.
83. E. Pramauro and A. Bianco Prevot, *Pure Appl. Chem.*, 1995, **67**, 551.
84. C. C. Tung, Y. M. Yang, C. H. Chang and J. R. Maa, *Waste Manag.*, 2002, **22**, 695.
85. E. Pramauro, A. Bianco, E. Barni, G. Viscardi and W. L. Hinze, *Colloids Surf.*, 1992, **63**, 291.
86. P. Reiller, D. Lemordant, C. Moulin and C. Beaucaire, *J. Colloid Interface Sci.*, 1994, **163**, 81.
87. A. S. Jonsson and G. Tragardh, *Desalination*, 1990, **77**, 135.
88. R. S. Juang, Y. Y. Xu and C. L. Chen, *J. Membr. Sci.*, 2003, **218**, 257.
89. M. Ismael and C. Tondre, *Sep. Sci. Technol.*, 1994, **29**, 651.

Subject Index

Note: Figures are indicated by *italic page numbers*, Tables by **bold page numbers**

AAS, *see* atomic absorption spectrometry
accelerated solvent extraction (ASE), 49, 79–80, **97**, *226*, *see also* high pressure solvent extraction
accuracy, green analytical chemistry and, 4, *4*, 64, 286
advanced oxidation technologies (AOTs), 294
AES, *see* Auger electron spectroscopy
alternative solvents, 47
    for green electroanalysis, 199, 209–212
    green organic solvents, 55
analytical laboratory wastes, 287, 295–6
    degradation, 293–6
        biodegradation, 295–6
        chemical oxidation, 294
        photocatalytic, 294–5
        thermal, 294
    on-line decontamination, 286–298
    passivation, 296–8
    recycling, 287–293
analytical methods,
    clean, 14, 29,
    downsizing of, 83–90, 107–138, 186, 203
    environmentally friendly, 3, 191
    greening, 7, 10, *11*
    greenness of, 46, 58, 64, **77**
    miniaturization, 107–138
    strategies for greening, 8
APCI, *see* atmospheric pressure chemical ionization
APPI, *see* atmospheric pressure photoionization
arc optical emission, 30–1
ASE, *see* pressurized solvent extraction
atmospheric pressure chemical ionization (APCI), 272–3
atmospheric pressure glow discharge (AP–GD) source, 37
atmospheric pressure photoionization (APPI), 273
atomic absorption spectrometry (AAS), 72, 91, 233, 236
ATR, *see* attenuated total reflectance
attenuated total reflectance (ATR), 26–7
Auger electron spectroscopy (AES), 35
automation
    advantages of, 4, **5**, **7**
    future trends, 164
    solid phase microextraction, 128, 230, 273
    for waste passivation, 296

batch injection analysis (BIA), 205–6, *206*
BIA, *see* batch injection analysis

biodegradation, 291, 295–6, *295*
biopolymers, 218
bioreactor, 266, 295, *295*
biosensors, 251–270
　autonomous biosensor wireless networks, 268–270
　classes and fundamentals, 252–8
　　electrochemical transduction, 252–4
　　mass-sensitive sensors, 257–8
　　optical transducers, 254–7
　for environmental monitoring, 258–268
　　enzyme biosensors, 258–261
　　immunosensors, 261–2
　　nuclear receptors, 264
　　nucleic acid and biosensors, 262–4
　　whole-cell biosensors, 264–8
blister pack, measurements through, 19–22, *20, 22*
bottles, non-invasive measurements on, 19, 21–3

calibration, in laser ablation, 32
capillary electrophoresis, 179–185
　chiral analysis, 181
　column, 180
　flat sheet and, 91
　green alternative, 179–181, *180*
　microsystems, 137, *187*, 208–9
　portable instruments, 181–5, *184–5*
　solid-phase extraction, 123
　solid-phase microextraction, 128
　solvent replacement, 57
carbon dioxide
　photodegradation to, 294
　for SFE, 78, **91**, 243–4, 246
CCD, *see* charge coupled device
certified reference materials (CMRs), 30, 32
　fish muscle, 162
charge coupled device (CCD), 23, 31
chelating agents, 56, 234, 236, 297

chemical oxygen demand (COD), 72, 203–4
chemicals,
　replacement in flow methodologies, **153**
　reuse in flow-based systems, **156**
chemometrics, 9, 10
　direct methods, 39
　vibrational spectroscopy, 25, 28
cloud point extraction (CPE),
　flow analysis, 98, **153**, 162
　sample preparation, 94–5, **97**
　sequential injection analysis, 162
　steps, 95
COD, *see* chemical oxygen demand
computer screen photo-assisted technique (CSPT), 194
contactless conductivity detection, 183–4, 191
CPE, *see* cloud point extraction
CRMs, *see* certified reference materials
CSPT, *see* computer screen photo-assisted technique

DART, *see* direct analysis in real time
degradation,
　biodegradation, 295–6
　chemical oxidation, 294
　photocatalytic, 294–5
　thermal, 294
　wastes, 293–6
derivatization
　chemical, 7, *8*, 154
　greener reagents, 56–8
　in situ, 78, 118, 238, 273
DESI, *see* desorption electrospray ionization
desorption electrospray ionization (DESI), 29, *29*
　analysis of solids, 37–8
　detection of organic pollutants, 273
dialysis, 265
dielectric constant, 49, **50**, 66, 70, 80, 210, 245, 271

Subject Index

digital microfluidics (DMF), 186–9
diode array detector, 229, 247
direct analysis, *14*
  alkanes by DESI, 273
  chemometrics, 39
  elemental analysis, 31
  in real time (DART), 272–3
  molecular analysis, 24–8
    nuclear magnetic resonance, 24
    vibrational spectroscopy, 25–8
  non-invasive methods, 21
  solids, 23–4, 29, 31, 35, 64
  without sample damage, 14, 23–8, *38*
  without using reagents, 29–38
  arc optical emission spectrometry, 30–1
  desorption electrospray ionization, 37–8
    electrothermal atomic absorption spectrometry (ETAAS), 29–30
    glow discharge, 34–7
    laser ablation (LA), 31–3, **33**
    laser-induced breakdown spectroscopy, 33–4
    spark optical emission spectrometry, 29–30
  X-ray techniques, 23
direct sampling, 34, 85
  solid, 29–30
dispersive liquid-liquid micro-extraction (DLLME), 89–90
  coupled with GF-AAS, 234
  coupled with SPE, 234
  hyphenation with LSE, 235
  ionic liquids applied in, 235
  liquid phase microextraction techniques, 87
  literature about, 87, 234
  microextraction techniques in sample preparation, 89–90, **97, 112, 115**
  sample preparation, 89–90
  sequential injection, 234
  in solvent-based miniaturized extraction techniques, 118
  solvent-reduced techniques, 226, *226*, 233–5
DLLME, *see* dispersive liquid-liquid micro-extraction
DMF, *see* digital microfluidics
DNA
  amplification, 263
  analysis, 188
  biosensors, 216, 218, 252, **259**, 262–4
  immobilization, 218
  preservation methods, **59**
droplet miclofuidics, 186–9
dry-ash(ing), 28–9, **76**

ecological paradigm, 2
EHS, *see* environmental, health and safety
electrochemical biosensors, environmental applications, 215–6
  green analysis, 214–220
  microsystems-based biosensors, 218–220
  natural biopolymers, 218
  using liquid ionics, 216–8
electrochemical sensors, 202–9
  flow injection analysis, 203–7
  green analysis, 202–9
  microsystems, 207–9
electrode materials, 212–4
  hybrid nanocomposites, 213
  metal nanoparticles, 212–3
  oxide nanoparticles, 213–4
  polymers, 214
  solid amalgams, 214
electrospray ionization (ESI), 272–3
electrothermal atomic absorption spectrometry (ETAAS), 29–30, 68, 75
electrowetting-on-dielectric (EWOD), 188
emulsification, 94, 98–9
  ultrasonic, 98–9, 234

Subject Index

environmental, health and safety (EHS), 51
environmental protection agency (EPA), 3, 7, 45, 133, 170
   priority PAHs, 134, 238, 247
   triad approach, 13
enzyme(s),
   based immunoassays, 249, **250**
   biosensors, *253*, 258–261
   electrodes, 253, 260, 263
   immobilization, 155, 216, 260
   recycling, 292
   sample treatment, 73
   sensors, 212, 215–8, 260
   waste biodegradation, 295
EPA, *see* environmental protection agency
ESI, *see* electrospray ionization
ETAAS, *see* electrothermal atomic absorption spectrometry
EWOD, *see* electrowetting-on-dielectric

FAAS, *see* flame atomic absorption spectrometry
FAC, *see* field analytical chemistry
field analytical chemistry (FAC), 182
   instruments, 182
flame atomic absorption spectrometry (FAAS), 29, 162
flow analysis
   green analytical chemistry, 144–164
   minimization of reagent consumption, 155–163
   reduction of waste generation, 149–152, 155–163, **158–9**
   replacement of hazardous chemicals, 152–5
   reuse of chemicals, 155
   waste treatment, 164
flow systems
   description, 145–9
   flow injection analysis, 145, *146*
   minimization of waste, *158–9*
   monosegmented flow analysis, 147, *148*
   multicommutation approach, 147–9, *148*
   multipumping, 149, *149*
   multisyringe, 149
   segmented flow analysis, 145, *146*
   sequential injection analysis, 145–6, *147*

GD, *see* glow discharge
GF-AAS, *see* graphite furnace atomic absorption spectrometry
glow discharge (GD), 29, 34–7
   radiofrequency powered sources, 36–7
glow discharge optical emission spectrometry (GD-OES), 34–7,
graphite furnace atomic absorption spectrometry (GF-AAS), 29–30
   coupled with DLLME, 234
   solid sampling, 29–30
green analytical chemistry,
   concept of, 4, 64
   cost of, 10–1, *11*
   flow analysis, 144–164
   priorities of, 8
   publication on, 6
   strategies for, 5, 7, 9–10, *9*
green analytical separation methods, 168–195
green chemistry, 3
   concept of, 3, 286
   principles of, 7–9, *8*, 44, 56, 63, 87, 99, 170, 202, 213
green chromatography, 169–185
   gas-phase separations, 169–171
   liquid-phase separations, 171–185
      capillary electrophoresis, 179–185
      HPLC-methods, 171–5
      supercritical fluid chromatography, 175–7

green electroanalysis, 199–220
  alternative solvents, 209–212
    ionic liquids, 209–210
    supercritical fluids, 211–2
  electrochemical biosensors, 214–220
    environmental applications, 215–6
    microsystems-based biosensors, 218–220
    natural biopolymers, 218
    using liquid ionics, 216–8
  electrochemical sensors, 202–9
    flow injection analysis, 203–7
    microsystems, 207–9
  future trends, 220
  new electrode materials, 212–4
    hybrid nanocomposites, 213
    metal nanoparticles, 212–3
    oxide nanoparticles, 213–4
    polymers, 214
    solid amalgams, 214
  stripping voltammetric, 200–2
green pictograms, 6, *6*
green solvents, **8**, 45–8
  for electroanalysis, 220
  ionic liquids as, 69
  organic, 51–5
  for sample preparation, 225
green terminology, 5
greener,
  HPLC, 178
  methods, 4, 8, 46, 71, 94, 107, 209, 225
  reagents, 56–60
  solvents, 46–55, 231
greening separation and detection techniques, 247–273
greenness,
  criteria, 64, **76**
  measure of, 46
greenness criteria of Green Chemistry Institute, 64
greenness profiles, 64, *65*, 75,
greenness related issues, **76**, **96**, 99

hazardous
  chemicals, **8**, 45, 59, 82, 154, 160, 203, 247
  compounds, 294
  reagents, 9, 32, 46
  solvents and reagents
    replacement, 44–60, **153**
    waste, 45, 57, 64, 69, 93, 203, 287
hazards, 29, 51, 59, 213, 243, 268
health, 46, 51, 52, **52**, 58, 183
  hazards, 29
  human, 2, 6, 45, 58, 240
high-performance liquid chromatography (HPLC),
  amount of solvent, 45
  electrochemical detection, 210, 214
  green organic solvents, 51–5
  green separation, 168, 171–5
  preparative, 178
  reversed-phase, 169, 172, 177, 288
  solid phase microextraction, 238
  solvent recycling, **291**
  stir bar sorptive extraction, 238
high pressure solvent extraction (HPSE), 79, *see also* pressurized liquid extraction
HILIC, *see* hydrophilic interaction chromatography
hollow fibre, *91*, *see also* liquid phase microextraction
  membranes, 91, 232
  microextraction, **112**, 116–8, 232
HPLC, *see* high-performance liquid chromatography
HPSE, *see* high pressure solvent extraction
hydrophilic interaction chromatography (HILIC), 172–3

ICP-MS, *see* inductively coupled plasma-mass spectrometry
ICP-OES, *see* inductively coupled plasma-optical emission spectrometry

inductively coupled plasma-mass
  spectrometry (ICP-MS), *29*, 30–2,
  33, 36
inductively coupled plasma-optical
  emission spectrometry (ICP-OES),
  29, 30–1, 68, 72
  subcritical water with, 246
immobilized reagents, 156–7
immunochemical techniques,
  247–251
  chemiluminescent magnetic
    immunoassays, 251
  enzyme-based immunoassays, 249
  flow-injection immunoassays, 251
  fluorescence polarization
    immunoassays, 249–251
in-field, *14*, 93, 125
in situ derivatization, 78, 118, 238,
  273
internal reflection, *see also* attenuated
  total reflectance
ionic liquids, 69, 178, 199
  applied in DLLME, 235
  for green electroanalysis, 209–210
  greener solvents, 48–9
  room-temperature, 210, 216–8,
    235–6

killer application, 189–191

LA, *see* laser ablation
lab-on-a-chip, 161, 185, 191–2, 207,
  209, 219, *see* micro-total analysis
  systems
lab-on-valve (LOV), 136, 141, **159**,
  162, 204, *205*
laser ablation, 29, 31–3
  applications, **33**
  calibration, 32
  direct analysis, 31–3, **33**
  sampling, 32
laser induced breakdown
  spectrometry (LIBS), 29, 33
  analysis of solids, 33–4
  open-path, 34,
  Raman, 34, *35*
  stand-off, 34
leaching, 30, 136
LED, *see* light-emitting diode
LIBS, *see* laser induced breakdown
  spectrometry
light-emitting diode (LED), 184
  detector in CE, 191
  infrared, 155
  photometer based on, 161–4
liquid chromatography (LC), *see also*
  high-performance liquid
  chromatography
  liquid-phase microextraction, 233
  miniaturization, 109
  nano, 172
  recycling solvents, 288–291, *290*,
    291
  reversed phase, 288
  solid-phase extraction (SPE-LC),
    119–123, **121**
liquid phase microextraction (LPME)
  application, **114**, 234
  automation, 90
  hollow fibre (HF-LPME), *84*, 88–9,
    97, 232
  sample preparation, 87–90
  solvent-reduced techniques, 226
liquid-solid extraction (LSE),
  77, 235
LOV, *see* lab-on-valve
LPME, *see* liquid phase
  microextraction
LSE, *see* liquid-solid extraction

MAD, *see* microwave-assisted
  digestion
MAE, *see* microwave-assisted
  extraction
MALDI, *see* matrix-assisted
  laser-desorption ionization
MAME, *see* microwave-assisted
  micellar extraction
matrix-assisted laser-desorption
  ionization (MALDI), 49, 273
mechanization of analytical
  methods, 5

membrane-based extraction
methods, 90–4
method(s) of analysis
greening, 3,
non-invasive, 19–23
NIR spectroscopy, 19–22, *20*
Raman spectroscopy, 19, 21–3, *22*
side effects, 9
MEUF, *see* micellar enhanced ultrafiltration
micellar enhanced ultrafiltration (MEUF), 94, 297
microextraction,
dispersive liquid-liquid, 118, 233–5
coupled with GF-AAS, 234
coupled with SPE, 234
hyphenation with LSE, 235
ionic liquids applied in, 235
liquid phase techniques, 87
literature, 87, 234
in sample preparation, 89–90, **97, 112, 115**
sequential injection, 234
in solvent-based miniaturized extraction techniques, 118
solvent-reduced techniques, 226, *226*, 233–5
hollow fibre-protected, 116–8
liquid-phase, 231–3, 87–90
application, **114**, 234
automation, 90
hollow fibre (HF-LPME), *84*, 88–9, **97**, 232
sample preparation, 87–90
solvent-reduced techniques, 226
multicommutation, 151
in packed syringe, 238–240
single-drop, 112–6
applications, **114**, 234
characteristics, **97**, 113
direct, *84*, 88, 113
dynamic, 116, 234
headspace (HS-SDME), *84*, 88, 116

sample preparation, 87–8
solvent extraction techniques, 112–5
solvent reduced techniques, 231
solid-phase, 124–9, 227–230
applications, *228*
automation, 128, 230, 273
characteristics, **97**
derivatization, *125*, 128
device, *84, 124*
direct, *84*, 85
fibres, **127**, 228–230
headspace (HS-SPME), *84*, 85
in-tube, 124, 230–1
membrane protected, 85
passive samplers, 241
sample preparation methods, 83–5, *84*
selectivity, 125–6
solvent-reduced techniques, 226–230, *226*
working modes, 85, *125*
techniques, 83–90, 112–8
thin film, 231
microfluidic(s)
capillary electrophoresis, 189
continuous-flow, 186, 191
devices, 93
digital, 186–9
droplet, 186–9
electrochemical, 207–9, *208–9*
flow-systems, 162
non-instrumental, 190, 191–5
paper-based, 192–4
protein chip based sensors, 219
microspectroscopy, Raman, 27
micro-TAS, *see* micro-total analytical systems
micro-total analytical systems (μ-TAS), 10, 136–8, 190, 207, 273
capillary electrophoresis, 208
microwave-assisted
digestion (MAD), 67, **77**
compared with ultrasound, 72–3, 99
focused, 67, 69

Kjeldahl system, 69
  on-line, 68
  reagents for, 67
extraction (MAE), 44, 69–70, **96**, 133–4, 287
  advantages, 70
  dynamic, 135
  focused (FMAE), 70, **96**
  miniaturized, 134–5
  parameters, 69
  pressurized (PMAE), 70, **96**
hidrolysis (MAH), 69
micellar extraction (MAME), 94, **97**, 98
for sample preparation, 65–70
  applications, 66
  digestion, 66–9
  extraction, 69–70
microwave-induced plasma (MIP), for optical emission spectrometry (MIP-OES), 30
  techniques, 30
mid infrared, 25–6, 28, 288, 292
miniaturization
  alternative for GAC, 107–110
  analytical methods, 107–138
  analytical micro-systems, 136–8
  analytical techniques for treatment of liquid samples, 110–139
  analytical techniques for treatment of solid samples, 130–6
  enhanced fluid/solvent extraction techniques, 133–6
    microwave-assisted extraction, 134–5
    pressurized liquid extraction, 133–4
    ultrasonic-assisted extraction, 135–6
  lab-on-a-valve, 136–7
  matrix solid-phase dispersion, 130–3
  separation methods, 185–195
    challenges, 195
    continuous-flow microfluidics, 186
    digital microfluidics, 186–9
    droplet microfluidics, 186–9
    non-instrumental microfluidics devices, 191–5
  solvent-based extraction techniques, 110–8
    dispersive liquid-liquid microextraction, 118
    hollow fibre-protected microextraction, 116–8
    in-vial liquid-liquid extraction, 111–2
    microextraction techniques, 112–8
    single-drop microextraction, 112–6
  sorption-based extraction techniques, 118–129
    solid-phase extraction, 119–124
    solid-phase microextraction, 124–9
    stir bar sorptive extraction, 129
MIP, *see* microwave-induced plasma *or* molecularly imprinted polymer
MIR, *see* mid infrared
modified electrode, 17, *202*, 202, 204, 206, 208, 210, 260
molecularly imprinted polymer (MIP), 229, 237
monosegmented flow analysis (MSFA), 147, *see also* flow analysis
  automation, 151
  manifold, *148*
  microextraction, 151, **152**
  reagent consumption, *152*
  sequential injection (SIMSFA), 204
MSFA, *see* monosegmented flow analysis
multicommutation
  approach, 4, 147–8, *148*, 157–160
  microextraction, 151
  minimization of wastes, **158–9**
  reagent consumption, *152*
  sample dilution, 160

multipumping, 149, *149*
  flow systems, 160–1
multisyringe
  flow systems, 149, 151
  minimization reagents, 161–2

nanocomposite, 211, 213
  hybrid, 213
nanoparticles, 211, 257
  for biomolecule immobilization, 217
  functionalization, 217
  gold, 210, 213, 217, 249, 260, 263
  magnetic, 217
  metal, 212–3, 256, 260
  oxide, 213–4
  platinum, 213, 258
national environmental methods index (NEMI), *6*, 45–6, 64
near infrared spectroscopy (NIR),
  direct analysis without sample damage, 26–7.
  non-invasive measurements, 19–22, *20*
NEMI, see national environmental methods index
NIR, see near infrared spectroscopy
NMR, see nuclear magnetic resonance spectroscopy
non-biological techniques, 270–3
  detection techniques, 272–3
  separation techniques, 270–2
non-destructive
  direct determinations, 20
  elemental analysis, 24
  laser ablation techniques, 32
  measurements, 19
  total internal reflection fluorescence (TIRF), 254
non-invasive
  measurements, 14, 19
  methods of analysis, *8*, *9*, 10, 19–23
  NIR spectroscopy, 19–22, *20*
  Raman spectroscopy, 19, 21–3, *22*

nuclear magnetic resonance (NMR) spectroscopy,
  mobile analysers, 24
  molecular analysis, 24–5, *38*

on-line analysis, **14**, 45
on-line decontamination of wastes, 8, 286–298
  degradation, 293–6
    biodegradation, 295–6
    chemical oxidation, 294
    photocatalytic, 294–5
    thermal, 294
  passivation, 296–8
  recycling, 287–293
on-line detoxification of wastes, 5, *5*
on-line digestion, 68
on-line monitoring, 50, 202
on-line solvent recycling, 155, **156**
on-line treatment of wastes, 7, 10
operator(s) safety, 4–7, 9, *9*
organic pollutants,
  determination in the environment, 224–274
  future trends, 273–4
  green analytical methodologies, 224–5
  greening separation and detection techniques, 247–273
    autonomous biosensor wireless networks, 268–270
    biosensors, 251–270
    chemiluminescent magnetic immunoassays, 251
    electrochemical transduction, 252–4
    enzyme-based immunoassays, 249
    enzyme biosensors, 258–261
    flow-injection immunoassays, 251
    fluorescence polarization immunoassays, 249–251
    immunochemical techniques, 247–251
    immunosensors, 261–2

Subject Index

mass-sensitive sensors, 257–8
non-biological techniques, 270–3
nuclear receptors, 264
nucleic acid and biosensors, 262–4
optical transducers, 254–7
whole-cell biosensors, 264–8
sample preparation, 226–247
dispersive liquid-liquid microextraction, 233–5
immunoaffinity chromatography, 246–7
in-tube extraction, 230–1
ionic liquids for green extraction, 235–6
liquid-phase microextraction, 231–3
microextraction in packed syringe, 238–240
passive sampling, 240–3
solid-phase microextraction, 227–230
solvent-reduced techniques, 226–247
stir bar sorptive extraction, 236–8
subcritical water extraction, 244–6
supercritical fluid extraction, 243–4
thin-film extraction, 231

passivation,
electrode, 208
metal ion, 297
micelle-based, 297
on-line, *9*, 10,
wastes, *4*, *5*, *7*, 10, 287, 296–8,
PAT, *see* process analytical technology
PBT, *see* persistent, bioaccumulative and toxic
PDMS, *see* polydimethylsiloxane
persistent, bioaccumulative and toxic (PBT), *6*, 45–6, 64

photocatalytic degradation, 294–5
pictogram,
green, 6, *6*
NEMI, *6*
SWOT, *6*
PMAE, *see* pressurized microwave-assisted extraction
POC, *see* point of care
POCIS, *see* polar organic chemical integrative sampler
point of care (POC), 182, 185, 192, 194
polar organic chemical integrative sampler (POCIS), 241–2
polydimethylsiloxane (PDMS), 85, 93, 119, 126, **127**, 129, 186
electrochemical, 219
microfluidics, *209*
pumps, 193
solid phase microextraction, 86–7, 228
stir-bar sorptive extraction, 86–7, 236–7
thin film microextraction, 231
polymers,
analysis, 37
coating, 87
conducting, 217
electrode material, 214, 199
extraction membranes, 91
liquid, 47
molecular imprinted (MIPs), 119
non-instrumental microfluidics, 192
polyurethane (PU) foams, 236–7
stir-bar sorptive extraction, 236–7
portable
amperometric biosensor, 219
analysers, 31
capillary electrophoresis, 181–5, *184–5*
gas chromatograph, 93, 183
instruments, 37–8, 109, 191
multiarray optical biosensor, 267
NIR spectrometer, 20

portable (*continued*)
  Raman, 27
  real-time field, 34
  XRF instrument, 24
preservation methods, **58**, 60
preservatives, 58–9
pressurized hot solvent extraction (PHSE), 79
pressurized liquid extraction (PLE), 44, 79, **97**, 133, 226
  sample preparation, 79–81
pressurized microwave-assisted extraction (PMAE), 70, **96**
pressurized solvent extraction (PSE), 49, 287
process analytical technology (PAT), 20, 28
PSE, *see* pressurized solvent extraction
purge and trap, 226–7, *226*

radiofrequency powered glow discharge source (RF-GD), 36–7
Raman instruments, 27
Raman-LIBS, 34, *35*
Raman microspectroscopy, 27
Raman scattering, 17, 25
Raman sensor, 23
Raman spectroscopy, 19, 21–8, *22*
  direct analysis without sample damage, 26–7
Rayleigh scattering, 257
REACH, 242
refrigerated sorptive extraction (RSE), 238
remote sensing, 8, 9, *9*, 14–9, *14*, 38–9
  satellite, 15, *16*
  techniques, 17, *18*
reversed-phase HPLC, 169, 172, 177, 288
RF-GD, *see* radiofrequency powered glow discharge source
room temperature ionic liquids (RTILs), 210, 216–8

RP-HPLC, *see* reversed-phase
RSE, *see* refrigerated sorptive extraction
RTILs, *see* room temperature ionic liquids

sample preparation
  dry-ash(ing), 28–9, **76**
  green methods, 63–99
  greening, 63–5
  greenness indicator, *65*
  membrane-based extraction, 90–4
  microextraction techniques, 83–90
    dispersive liquid-liquid, 89–90
    hollow fibre liquid-phase, 88–9
    liquid-phase microextraction, 87–90
    single-drop, 87–8
    solid-phase microextraction, 83–5
    stir-bar, 86–7
  microwave-assisted, 65–6
    digestion, 66–9
    extraction, 69–70
  pressurized liquid extraction, 79–81
  solid-phase extraction, 81–3
  solvent-reduced techniques, 226–247
    dispersive liquid-liquid microextraction, 233–5
    immunoaffinity chromatography, 246–7
    in-tube extraction, 230–1
    ionic liquids for green extraction, 235–6
    liquid-phase microextraction, 231–3
    microextraction, 227–230
    microextraction in packed syringe, 238–240
    passive sampling, 240–3

Subject Index

stir-bar sorptive
   extraction, 236–8
subcritical water
   extraction, 244–6
supercritical fluid
   extraction, 243–4
thin-film extraction, 231
supercritical fluid extraction, 75–9
surfactant-based methods, 94–9
ultrasound-assisted, 70–2
   digestion, 72–3
   extraction, 73–5
sampling
   direct, 34, 68, 85
   frequency, 10, 39, 145, 147–150, 155
   headspace, 112
   in-field, *14*, 93, 125
   in laser ablation, 32
   passive, 231, 240–3
   problems, 9
   slurry, 30, 75
   solid, 29, **33**, 68
   in vivo, 230
SBME, *see* stir bar microextraction
SBSE, *see* stir bar sorptive extraction
screening methods, 28, 249, 256, 286
secondary ion mass spectrometry (SIMS), 35–7
secondary neutral mass spectrometry (SNMS), 35
separation methods
   green analytical, 168–195
sequential injection (SI), 136
sequential injection analysis (SIA), 4, 10
   dispersive liquid-liquid micro-extraction, 234
   electrochemical sensors, 204, *205*
   flow systems, 146–7
   manifold, *147*
   reagent consumption, *152*, 161–2
   waste generation, *152*, **158**, 161–2
SDME, *see* single-drop microextraction

segmented flow analysis (SFA), *see* flow analysis
sensors,
   autonomous biosensor wireless networks, 268–270
   biosensors, 251–270
      classes and fundamentals, 252–8
      electrochemical transduction, 252–4
      for environmental monitoring, 258–268
      enzyme biosensors, 258–261
      immunosensors, 261–2
      nuclear receptors, 264
      nucleic acid and biosensors, 262–4
      optical transducers, 254–7
      whole-cell biosensors, 264–8
   electrochemical, 202–9
      flow injection analysis, 203–7
      green analysis, 202–9
      microsystems, 207–9
   enzyme, 212, 215–8, 260
   green electroanalysis, 202–9
   immunosensors, 261–2
   mass-sensitive, 257–8
   protein chip based, 219
   Raman sensor, 23
SFA, segmented flow analysis, *see* flow analysis
SFC, *see* supercritical fluid chromatography
SFE, *see* supercritical fluid extraction
SIMS, *see* Secondary ion mass spectrometry
single-drop microextraction (SDME),
   applications, **114**, 234
   characteristics, **97**, 113
   direct, *84*, 88, 113
   dynamic, 116, 234
   headspace (HS-SDME), *84*, 88, 116
   sample preparation, 87–8
   solvent microextraction techniques, 112–5
   solvent reduced techniques, 231

SNMS, *see* Secondary neutral mass spectrometry
solid amalgam electrodes (SAE), 214
solid phase extraction (SPE),
  automation, 82
  characteristics, 97
  compared with LLE, 81, 95
  gas chromatography (SPE-GC), 119, 123
  liquid chromatography (SPE-LC), 119–123, **121**
  miniaturized, 119–124
  on-line, 119, *121*, **121**, 122, 135
  procedure, 82, 234
  sample preparation, 81–3
solid phase microextraction (SPME)
  applications, *228*
  automation, 128, 230, 273
  characteristics, **97**
  derivatization, *125*, 128
  device, *84*, *124*
  direct, *84*, 85
  fibres, **127**, 228–230
  headspace (HS-SPME), *84*, 85
  in-tube, 124, 230–1
  membrane protected, 85
  microextraction techniques, 83–5
  passive samplers, 241
  sample preparation methods, 83–5, *84*
  selectivity, 125–6
  solvent-reduced techniques, 226–230, *226*
  working modes, 85, *125*
solid sampling, 29, **33**
solvatochromic parameters, 47
solvent(s)
  alternative for electroanalysis, 209–212
  energy demand, 52, **53**
  environmental effects, 52, **53**
  green, **8**, 45–8, 99, 220, 225
  greener, 46–55, 231
  hazardous, replacement, 44–60
  recycling, 287–293, *290*
  replacement, **54**
  selection guide, 52, 54, **54**
sonication, 10, *see also* ultrasound-assisted
  compared with ASE, 79
  for leaching, 136
  potential for miniaturized, 135
  probe, 72–4
soxhlet extraction, 74
  compared with other techniques, 70, 74, 79, **96**
spark optical emission, 30–1, 35
SPE, *see* solid phase extraction
SPME, *see* solid phase microextraction
stir bar microextraction, 86–7, *see also* stir bar sorptive extraction
stir bar sorptive extraction (SBSE),
  characteristics, **97**
  coating, 237
  headspace (HS-SBSE), 86, 236
  in situ derivatization, 238
  miniaturized treatment of liquid samples, 129
  on-line, 238
  PDMS-coated, 86, 129
  polydimethylsiloxane, 86–7, 236–7
  sample preparation, 81, *84*, 86–7, 226, 236–8
strengths-weaknesses-opportunities-threats (SWOT), 6, *6*
stripping voltammetry,
  anodic, 68, 200–1, 204, 208
  cathodic, 200
  green, for trace analysis, 200–2
  lab-on-valve, 204
  miniaturized, 208
  SIA, 204
  square-wave anodic, 200–1, 205–6
subcritical water extraction (SWE)
  adsorptive (AdSV), 201,
  advantages, 80
  characteristics, **97**
  comparison with other techniques, **97**, *226*

Subject Index   315

concentration/extraction step for, 80
miniaturized treatment of samples, 133-4
organic pollutants determination, 226-7, 244-6
pressurized liquid extraction, 79-81
sample preparation, 79-81
supercritical fluid chromatography (SFC)
  chiral separations, 48, 176
  columns, 176
  preparative, 178
  solvating power, 176, *177*
supercritical fluid extraction (SFE)
  advantages, 78
  amount of solvent, 74
  automated, 78
  carbon dioxide for, 78, **91**, 243-4, 246
  comparison with other techniques, **96**, *226*
  green solvents for, 47-8
  organic pollutants analysis, 243-4
  for sample preparation, 75-9
supercritical fluids, 75, 175-6, 243
  alternative solvents for electroanalysis, 211-2
  greener solvents, 47-8
  references to, 75
surfactant-based sample preparation methods, 94-9
  emulsification, 98-9
  extraction, 94-8
SWE, *see* subcritical water extraction
SWOT, *see* strengths-weaknesses-opportunities-threats

TAS, *see* micro-total analytical systems
t-channel geometry, 187
teledetection systems, 14-9
thin film microextraction, 231
time of flight mass (TOF), 37, 195

TIRF, *see* total internal reflection fluorescence
TOF, *see* time of flight mass
total internal reflection fluorescence (TIRF), 254, **261**, 262
toxic release inventory (TRI), PBT in, 45
TRI, see toxic release inventory

ultrasonic bath, **96**, 135,
ultrasonic emulsification, 98-9
ultrasonic probe, 72-4, 135,
ultrasound-assisted,
  continuous extraction, 74
  digestion, 72-3
  for electrochemical biosensors, 216
  emulsification, 98-9
  emulsification-microextraction (USAEME), 234
  energy consumption, 99
  extraction (UAE), 73-5, **77**, **96**, 98, 135-6
  matrix solid-phase dispersion (UA-MSPD), 135
  micellar extraction (USME), 94, 98,
  sample preparation, 70-5
  miniaturized, 135-6
  for wastes degradation, 294

vibrational spectroscopy,
  chemometrics, 25, 28
  mid infrared, 25-6, 28, 288, 292
  near infrared, 19-22, 26-7
  Raman, 19, 21-8, *22*

waste minimization, 156, 160, 163
waste treatment in flow analysis, 163
wastes
  degradation of, 293-6
  generation of, 3, 39, 92, 94
  minimization of, 5, 29
  passivation of, *4*, *5*, *7*, *10*, 287, 296-8

water
  dielectric constant, 49, **50**, 271
  greener solvent, 49–51
world-to-chip, 189–191, 195

X-ray
  direct analysis, 23–4, *38*
  in situ analysis, 182
  microprobe, 34
  techniques for elemental analysis, 23–4
X-ray fluorescence (XRF), 23, 34
  energy dispersive (ED-XRF), 23
  total reflecting (TXRF), 68
  wavelength dispersive (WD-XRF), 23
XRF, *see also* X-ray fluorescence